Physiological and Molecular Mechanisms of Plant Resistance to Abiotic Stress

Physiological and Molecular Mechanisms of Plant Resistance to Abiotic Stress

Guest Editors

Lei Cao

Yao Zhang

Qiang Li

MDPI

Basel • Beijing • Wuhan • Barcelona • Belgrade • Novi Sad • Cluj • Manchester

Guest Editors
Lei Cao
College of Horticulture and Landscape Architecture
Northeast Agricultural University
Harbin
China

Yao Zhang
College of Life Sciences
Northeast Agricultural University
Harbin
China

Qiang Li
College of Life Sciences
Northeast Agricultural University
Harbin
China

Editorial Office
MDPI AG
Grosspeteranlage 5
4052 Basel, Switzerland

This is a reprint of the Special Issue, published open access by the journal *Plants* (ISSN 2223-7747), freely accessible at: https://www.mdpi.com/journal/plants/special_issues/UV74G2F1WX.

For citation purposes, cite each article independently as indicated on the article page online and as indicated below:

Lastname, A.A.; Lastname, B.B. Article Title. *Journal Name* **Year**, *Volume Number*, Page Range.

ISBN 978-3-7258-8021-8 (Hbk)
ISBN 978-3-7258-8022-5 (PDF)
https://doi.org/10.3390/books978-3-7258-8022-5

Contents

plants

MDPI

Article

Pan-Genome-Based Characterization of the PYL Transcription Factor Family in *Populus*

Xiaoli Han [1,2,3], Chen Qiu [1,2,3], Zhongshuai Gai [1,2,3], Juntuan Zhai [1,2,3], Jia Song [1,2,3], Jianhao Sun [1,2,3,*] and Zhijun Li [1,2,3,*]

[1] Xinjiang Production & Construction Corps Key Laboratory of Protection and Utilization of Biological Resources in Tarim Basin, Aral 843300, China; lilyan0509@163.com (X.H.); qiuchentea@163.com (C.Q.); gaizhongshuaitea@163.com (Z.G.); zhaijuntuan2022@163.com (J.Z.); nxy@163.com (J.S.)

[2] College of Life Science and Technology, Tarim University, Aral 843300, China

[3] Desert Poplar Research Center, Tarim University, Aral 843300, China

* Correspondence: sunjianhaotea@163.com (J.S.); lizhijun0202@126.com (Z.L.)

Abstract

Abscisic acid (ABA) is a key phytohormone involved in regulating plant growth and responses to environmental stress. As receptors of ABA, pyrabactin resistance 1 (PYR)/PYR1-like (PYL) proteins play a central role in initiating ABA signal transduction. In this study, a total of 30 *PopPYL* genes were identified and classified into three sub-families (PYL I–III) in the pan-genome of 17 *Populus* species, through phylogenetic analysis. Among these subfamilies, the PYL I subfamily was the largest, comprising 21 members, whereas PYL III was the smallest, with only four members. To elucidate the evolutionary dynamics of these genes, we conducted synteny and *Ka*/*Ks* analyses. Results indicated that most *PopPYL* genes had undergone purifying selection ($Ka/Ks < 1$), while a few were subject to positive selection ($Ka/Ks > 1$). Promoter analysis revealed 258 *cis*-regulatory elements in the *PYL* genes of *Populus euphratica* (EUP) and *Populus pruinosa* (PRU), including 127 elements responsive to abiotic stress and 33 ABA-related elements. Furthermore, six structural variations (SVs) were detected in *PYL_EUP* genes and significantly influenced gene expression levels ($p < 0.05$). To further explore the functional roles of *PYL* genes, we analyzed tissue-specific expression profiles of 17 *PYL_EUP* genes under drought stress conditions. *PYL6_EUP* was predominantly expressed in roots, *PYL17_EUP* exhibited leaf-specific expression, and *PYL1_EUP* showed elevated expression in stems. These findings suggest that the drought response of *PYL_EUP* genes is tissue-specific. Overall, this study highlights the utility of pan-genomics in elucidating gene family evolution and suggests that *PYL_EUP* genes contribute to the regulation of drought stress responses in EUP, offering valuable genetic resources for functional characterization of *PYL* genes.

Keywords: *Populus*; pan-genome; structural variation; *P. euphratica*; expression pattern; PYL

1. Introduction

Through evolutionary adaptation, plants have developed sophisticated regulatory networks to cope with abiotic stress [1–4], in which the phytohormone abscisic acid (ABA) plays a central role [5–8]. In plants, the ABA signaling cascade is initiated after ABA is perceived by pyrabactin resistance 1 (PYR1)/PYR1-like (PYL)/ABA receptors [9]. Under normal growth conditions, low ABA concentrations prevent PYR/PYL receptors from binding to PP2C phosphatases [10,11], thereby inhibiting *SnRK2* kinase activity and holding the ABA signaling pathway in an inactive state [12]. During drought stress, elevated ABA levels induce PYL expression and promote its interaction with PP2C, relieving the inhibition of *SnRK2* [13,14], activating downstream transcription factors and ion channels, and regulating stomatal closure and the expression of stress-responsive genes, ultimately enhancing plant drought resistance [15–17].

As members of the START protein superfamily, PYL proteins are characterized by a highly conserved START domain capable of binding hydrophobic ligands [18]. On the basis of this conserved gene sequence, 14 PYL genes containing the START domain have been identified in *Arabidopsis thaliana* and designated *AtPYR1–AtPYR13*. Phylogenetic analysis has classified these genes into three subfamilies [19,20]. Different subfamilies have been found to perform diverse functions in plant development and responses to abiotic stress. *AtPYL5* and *AtPYL9* have been identified as drought-resistant [21,22]. Under drought stress, *AtPYL5* has been shown to enhance the photosynthetic rate [23]; *AtPYR1*, *AtPYL2*, *AtPYL4*, and *AtPYL5* are involved in regulating stomatal and guard cell closure in response to CO_2 [24]; and *AtPYL8* and *AtPYL9* play key roles in root growth and leaf senescence [25]. In addition, expression analysis of 11 ABA-related rate-limiting enzyme genes in different tissues of grapevine revealed that *VvPYL1* exhibits tissue-specific expression, and the highest expression level has been observed in roots [26]. As a core regulator of ABA signaling, the function of PYL has been elucidated in many species, including rice (*Oryza sativa*) [27], tomato (*Solanum lycopersicum*) [7], soybean (*Glycine max*) [28], wheat (*Triticum aestivum*) [29], maize (*Zea mays*) [30], poplar (*Populus*) [31], rubber tree (*Hevea brasiliensis*) [32], strawberry (*Fragaria* × *ananassa*) [33], cotton (*Gossypium hirsutum*) [34], and castor bean (*Ricinus communis*) [35]. These studies demonstrate that the *PYL* gene family, as central components of the ABA signaling pathway, considerably enhances plant adaptation to abiotic stresses by regulating diverse physiological processes, making their functional characterization crucial for elucidating ABA signal transduction mechanisms [36].

With the advancement of plant pan-genomics, cross-species comparative genomic analysis has become feasible. Compared with conventional gene family identification approaches, pan-genomes overcome the limitation of single reference genomes by enabling the detection of gene family members that are absent from the reference genome but present in other genomes. This advancement provides new opportunities to investigate gene family conservation and divergence, as well as their roles in adaptive evolution [37]. Based on the pan-genome of 26 high-quality maize genomes, Sun et al. identified extensive presence–absence variations and structural variations (SVs), thereby laying the foundation for investigating the TPS gene family and its biological functions [38]. In addition, Wang et al. [39] constructed a genus-level super-pan-genome comprising 19 *Populus* genomes, identifying 142,202 cross-species SVs that intersected with numerous genes and significantly contributed to phenotypic and adaptive divergence. This study also provided comprehensive multi-omics resources associated with the *Populus* super-pan-genome. Although the structure and function of *PYL* genes have been extensively characterized in the model plant *Arabidopsis thaliana* and various crop species, a systematic study of the *PYL* gene family has not yet been conducted within the genus *Populus* of the family Salicaceae. *Populus euphratica* (EUP), known for its exceptional environmental adaptability and recognized as the most stress-resistant representative of the *Turanga* section of the genus *Populus*, is widely distributed across Central Asia, West Asia, and the arid regions of northwest China [40,41]. Its unique physiological and ecological adaptations enable remarkable survival under extreme drought and saline-alkali conditions, establishing it as an ideal model species for studying abiotic stress responses in woody plants [42]. In this study, we conducted a comprehensive identification and analysis of the *PYL* gene family across 17 species of the genus *Populus*, encompassing the sections *Leucoides*, *Tacamahaca*, *Aigeiros*, *Turanga*, and *Leucoides*. We conducted a comprehensive analysis of the gene structure, conserved motifs, phylogenetic relationships, promoter elements, SVs, and gene expression patterns of *PYL* proteins. In addition, tissue-specific expression profiles of *PYL* members in EUP were precisely quantified using qRT-PCR technology. These findings not only provide a timely foundation for functional studies of the *PopPYL* gene family but also significantly enhance our understanding of abiotic stress adaptation mechanisms in EUP, while offering valuable candidate genes for future genetic engineering and breeding programs in this species.

2. Results

2.1. Pan-Genome Distribution of the PopPYL Gene Family

Due to the limitations of a single reference genome in capturing the genetic diversity of a species, we collected 17 high-quality poplar genomes (Table 1) and conducted a pan-genome-level analysis of *PYL* genes to better understand their genetic variation. First, a comprehensive genome-wide comparative analysis of the *PYL* gene family was performed using BLASTP and HMM-based methods, resulting in the identification of 352 *PYL* genes in total. The number of *PYL* genes in each poplar genome ranged from 18 to 22 (Table S1). Orthologous genes across the 17 genomes were identified using OrthoFinder (v2.5.2; default parameters). The pan-gene IDs of the *PYL* gene family were extracted from the "Pan-genes.list" output file (Table S2), and genes sharing the same pan-gene ID across species were considered the same gene. Ultimately, 30 *PYL* pan-genes were obtained. Among these, 20 *PopPYL* genes were identified in *Populus yunnanensis*, *Populus wuana* (WUA), *Populus trichocarpa* (TRI), *Populus simonii* (SIM), and *Populus lasiocarpa*. Additionally, 18 *PopPYL* genes were detected in *Populus tremula* (TRE), *Populus szechuanica* (SZE), *Populus pseudoglauca* (PSE), *Populus koreana*, *Populus davidiana*, *Populus alba* var. *pyramidalis* (ALB), *Populus adenopoda* (ADE), *Populus rotundifolia*, and *Populus qiongdaoensis* (QIO). Notably, EUP (Table S3) and *Populus pruinosa* (PRU) exhibited the lowest number of *PYL* genes, with only 17 identified in each, whereas *Populus deltoides* (DEL) contained the highest number, with a total of 22 *PYL* genes.

Table 1. Statistics of data sources.

Species	Abbreviation	Assembly Size (Mb)	Genome Data	Reference
Populus euphratica	EUP	518.6	Unpublished	
Populus pruinosa	PRU	521.1	https://www.ncbi.nlm.nih.gov/search/all/?term=PRJNA863418Pprgenome_fa/20705107/2	[43]
Populus pseudoglauca	PSE	448.7	Database: https://bigd.big.ac.cn/gwh; Accession number: GWHBJUF00000000	[39]
Populus wuana	WUA	417.4	Database: https://bigd.big.ac.cn/gwh; Accession number: GWHBJUE00000000	[39]
Populus szechuanica	SZE	429.1	Database: https://bigd.big.ac.cn/gwh; Accession number: GWHBJUK00000000	[39]
Populus yunnanensis	YUN	433.7	Database: https://bigd.big.ac.cn/gwh; Accession number: GWHBJUD00000000	[39]
Populus koreana	KOR	401.4	Database: https://bigd.big.ac.cn/gwh; Accession number: GWHBHRS00000000	[44]
Populus trichocarpa v4.1	TRI	392.2	Database: https://phytozome-next.jgi.doe.gov/	Phytozome
Populus deltoids v2.1	DEL	446.8	Database: https://phytozome-next.jgi.doe.gov/	Phytozome

Table 1. *Cont.*

Species	Abbreviation	Assembly Size (Mb)	Genome Data	Reference
Populus simonii	SIM	408.0	Database: https://bigd.big.ac.cn/gwh; Accession number: GWHBJTX00000000	[39]
Populus lasiocarpa	LAS	419.5	Database: https://bigd.big.ac.cn/gwh; Accession number: GWHBJUB00000000	[45]
Populus davidiana	DAV	441.1	Database: https://bigd.big.ac.cn/gwh; Accession number: GWHBJTW00000000	[46]
Populus rotundifolia	ROT	414.3	Database: https://bigd.big.ac.cn/gwh; Accession number: GWHBJUA00000000	[39]
Populus tremula	TRE	408.8	Database: http://popgenie.org	[47]
Populus alba var. pyramidalis	ALB	408.1	Database: http://bigd.big.ac.cn/bioproject; Accession number: PRJCA002423	[48]
Populus qiongdaoensis	QIO	391.3	Database: http://bigd.big.ac.cn/bioproject; Accession number: PRJCA007862	[49]
Populus adenopoda	ADE	383.4	Database: https://bigd.big.ac.cn/gwh; Accession number: GWHBJTV00000000	[50]

Note: URL (Accessed on: 8 April 2025).

Notably, QIO contained the unique genes *Qdy09060.t1* (*PYL3*) and *Qdy37849.t2* (*PYL26*), whereas ALB harbored three unique genes: *pal_pou08344.t1* (*PYL8*), *pal_pou07031.t1* (*PYL21*), and *pal_pou33677.t1* (*PYL22*). Similarly, PRU possessed *PprTF01G0764.1* (*PYL23*), ADE contained *Poade06008.t2* (*PYL24*) and *Poade13605.t2* (*PYL25*), SIM carried *Posim17381.t2* (*PYL27*) and *Posim35007.t2* (*PYL28*), TRE carried *Potra2n8c17195.8* (*PYL29*), and WUA maintained *Powua15969.t2* (*PYL30*). These species-specific genes may be associated with adaptive evolution in their respective species. Conversely, multiple *PYL* genes demonstrated high conservation across all genomes. Core *PYL* genes, including *PYL1*, *PYL2*, *PYL4–7*, and *PYL9–17*, were universally present in all 17 examined genomes. These conserved genes likely play essential roles in the fundamental physiological processes of *Populus* species and have remained evolutionarily stable.

Phylogenetic analysis of the *Populus PYL* protein family revealed three distinct subfamilies (I, II, and III), which exhibited significant divergence in their physicochemical properties, including amino acid length, molecular weight, isoelectric point (pI), instability index, and aliphatic index (Table 2). Subfamily I members displayed the greatest variability, with the amino acid length ranging markedly from 54 residues to 706 residues. *PYL3* (706 amino acids) and *PYL8* (615 amino acids) represented the largest proteins within this subfamily, whereas *PYL18* (102–115 amino acids) was the smallest. This subfamily exhibited the broadest molecular weight distribution (9.7–77.3 kDa) and predominantly acidic isoelectric points (pI 4.4–10.2), and *PYL21* (pI 8.44) was the sole alkaline exception. In contrast, subfamily II proteins exhibited a more conserved amino acid length (186–306 amino acids) and molecular weight (20.6–34.1 kDa), but were distinguished by their markedly alkaline isoelectric points (pI 7.1–9.2), a feature potentially related to their subcellular localization or functional specialization. The protein length of subfamily III members ranged from 169 amino acids (*PYL20*) to 201–259 amino acids (*PYL7*), with the molecular weight spanning 13.5–28.7 kDa. Notably, *PYL7* exhibited the highest molecular weight within this subfamily (22.4–28.7 kDa). All members were characterized as acidic proteins (pI 4.93–7.01), and *PYL9* showed the strongest acidity (pI 4.93–5.53). Hydrophilicity analysis (GRAVY index 0.29–0.52) confirmed their hydrophilic nature, among which *PYL5* demonstrated the highest hydrophilicity.

Table 2. Summary of PYL protein properties.

	Sequence ID	Number of Amino Acid	Molecular Weight	Theoretical pI	Instability Index	Aliphatic Index
I	PYL1	54~189	9673.47~21,213.94	5.89~10.17	38.12~43.1	81.9~110
	PYL2	190~205	21,377.46~2,3149.56	5.6~7.73	34.04~39.96	88.53~91.63
	PYL3	706	77,272.84	5.22	30.48	85.44
	PYL4	169~248	18,686~27,574.39	4.44~5.15	26.15~44.62	66.73~77.65
	PYL24	202	22,653.53	4.7	47.65	74.75
	PYL14	168~381	18,895.38~43,039.84	4.46~4.8	25.8~31.73	72.66~78.99
	PYL21	188	21,413.6	8.44	34.03	81.38
	PYL19	108~143	12,205.97~16,086.06	4.39~4.59	40.25~43.88	76.36~92.96
	PYL15	169~329	18,665.8~36,424.99	4.71~5.55	29.9~42.91	72.12~86.8
	PYL18	102~115	11,253.82~12,760.6	5.57~5.86	26.7~36.85	77.91~85
	PYL12	103~163	11,430.91~18,163.62	4.46~4.9	23.22~45.19	77.04~93.74
	PYL29	163	18,053.43	4.64	29.94	86.01
	PYL26	186	20,986.81	6.44	39.15	84.25
	PYL8	615	67,740.22	7.23	38.38	93.07
	PYL30	190	21,527.63	6.06	38.55	91.63
	PYL25	190	21,487.52	6.06	40.01	89.05
	PYL27	190	21,463.46	6.06	40.31	88.53
	PYL23	220	24,606.14	5.62	43.17	92
	PYL13	191~229	21,355.14~25,681.5	5.78~6.11	40.64~48.96	94.29~98.25
	PYL28	191	21,369.16	5.78	48.96	94.29
	PYL16	134~282	15,264.5~32,181.38	5.55~8.03	38.94~44.91	95.9~100.9
II	PYL6	186~223	20,570.31~24,181.23	8.5~9.21	42.26~46.91	80.59~85.02
	PYL17	214~222	23,268.13~24,100.99	7.66~8.26	50.16~54.14	79.41~80.56
	PYL22	306	34,088.89	6.3	46.78	93.24
	PYL10	214~238	23,103.88~26,319.76	7.11~9.22	38.54~51.13	78.35~85.47
	PYL11	214	23,086.16~23,310.38	7.1~7.68	49.26~54.69	86.82~90.93
III	PYL5	124~191	13,517.37~21,341.06	5.2~7.01	2.05~51.39	82.43~89.69
	PYL9	203~210	22,319.97~23,190	4.93~5.53	32.74~42.8	82.1~86.85
	PYL7	201~259	22,392.17~28,658.25	5.08~5.62	29.23~34.58	79.57~84.33
	PYL20	169	18,961.56~18,961.58	5.94	38.88~40.02	85.27~85.86

2.2. Construction of the Phylogenetic Tree of PopPYL Gene Family Members

A phylogenetic tree was constructed based on the amino acid sequences of 14 *AtPYL* (*Arabidopsis thaliana*) and 30 *PopPYL* genes, revealing that the *PopPYL* proteins can be classified into three clades (PYL I, PYL II, and PYL III) following the established classification system for *AtPYL*, consistent with previous findings [51]. As shown in Figure 1, the PYL I subfamily was the most abundant, comprising 21 genes, followed by PYL II with 5 members, while PYL III was the smallest subfamily, containing only 4 members. This numerical disparity may reflect varying selective pressures and functional demands during the evolutionary diversification of these subfamilies. Within the PYL I subfamily, SZE uniquely possessed three copies of *PYL13*, whereas all other species retained only a single orthologous gene. Notably, *PYL3* and *PYL8* were detected exclusively in QIO and ALB, respectively, while *PYL21–PYL30* exhibited single-species-specific distribution patterns. Additionally, EUP and PRU exhibited the loss of *PYL15*, although PRU uniquely harbored a distinct *PYL23* gene. In the PYL II subfamily, *PYL22* was identified as a species-specific gene in ALB, whereas *PYL6*, *PYL10*, and *PYL11* were conserved as single-copy genes across all examined species. In contrast, *PYL17* was absent in ALB, ADE, and QIO. For the PYL III subfamily, *PYL7* and *PYL9* were conserved as single copies across all 17 species, while *PYL20* exhibited species specificity, being present only in WUA and PSE. The differential distribution of subfamily members across species suggests functional divergence within the *PYL* gene family, potentially linked to species-specific adaptations.

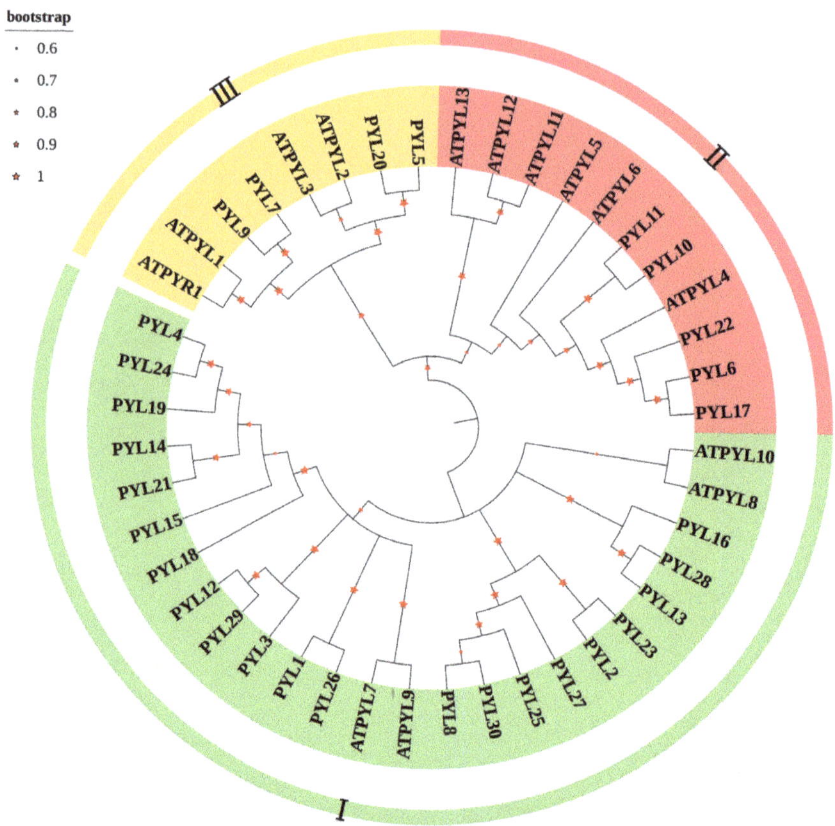

Figure 1. Phylogenetic analysis of *PYL* genes in *Populus*. An unrooted Neighbor-Joining phylogenetic tree was constructed from *Arabidopsis* and 30 *PopPYL*. The bootstrap test was set to 1000 replicates. Different colors represent different subfamilies of the *PYL* gene.

2.3. Chromosomal Location of PopPYL Genes in There Populus Species

Populus is a diploid species (2n = 18) with 18 chromosomes. To investigate the chromosomal distribution characteristics of *PYL* genes in poplar, we performed chromosomal localization analysis of 30 *PopPYL* genes (Figure 2). The results revealed both conserved features and species-specific variations in their distribution patterns. Overall, *PYL* genes were dispersed across multiple chromosomes without concentrated distribution on any single chromosome, reflecting genomic structural differences and evolutionary divergence among species. In the reference genome, we identified the following *PYL* genes: *PYL1*, *PYL2*, *PYL4*, *PYL5*, *PYL6*, *PYL7*, *PYL9*, *PYL10*, *PYL11*, *PYL12*, *PYL13*, *PYL14*, *PYL16*, and *PYL17*. Among these, *PYL1*, *PYL2*, *PYL5*, *PYL6*, *PYL7*, *PYL9*, *PYL10*, and *PYL11* were conserved across all 17 *Populus* genomes, while *PYL4*, *PYL12*, *PYL13*, *PYL14*, *PYL16*, and *PYL17* were present only in specific poplar species (Figure 2A). Additionally, 16 species-specific *PYL* genes were distributed across non-reference genomes (Figure 2B), with QIO, ALB, DEL, WUA, SIM, ADE, TRE, and PRU containing 2, 3, 3, 2, 2, 2, 1, and 1 gene, respectively. These genes compensate for the limitations of identifying gene family members using a single reference genome. In conclusion, the 30 homologous *PopPYL* genes collectively represent the *PYL* pan-genes across the 17 *Populus* species.

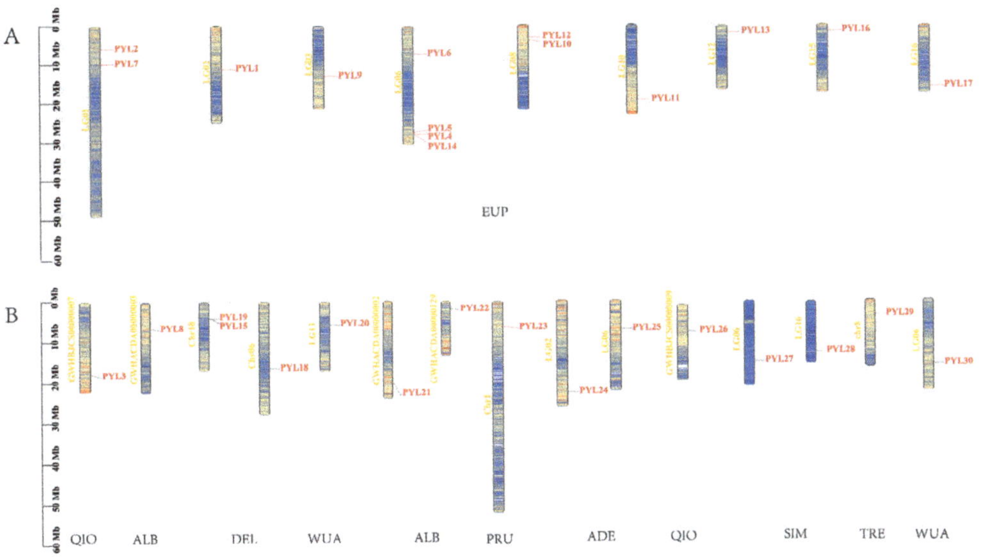

Figure 2. Distribution of the *PYL* genes in the reference genome (**A**) and non-reference genome (**B**) genome. PYL21-PYL30 were species-specific genes. The left axis shows the length of each chromosome and contig, as was estimated in megabases (Mb).

2.4. Gene Structure and Conserved Motif Analysis

Based on the evolutionary relationships between *PopPYL* and *AtPYL* genes, we analyzed the exon–intron structures of *PYL6*, *PYL7*, and *PYL13*, which are homologous to *AtPYL5*, *AtPYR1*, and *AtPYL8*, respectively, in 17 *Populus* species using TBtools software (v2.012). The results (Figure 3) revealed that *PYL6* genes lacked introns in all 17 *Populus* species, while *PYL13* genes consistently contained two introns. Members with identical or similar protein lengths shared conserved exon–intron structures, whereas genes with significant variation in protein length exhibited distinct gene architectures. For example, both *PYL7_TRE* and *PYL7_PRU*, which encode proteins of 255 and 259 amino acids, respectively, possessed two introns, whereas other *Populus PYL7* genes encoding shorter proteins (201–207 amino acids) exhibited no detectable introns. This structural diversity may reflect species-specific physiological functions and environmental adaptations. Notably, *PYL6*, *PYL7*, and *PYL13* genes all contained two exons across species, and homologous *PYL* genes maintained similar exon–intron patterns, further supporting their close evolutionary relationships and validating the current subfamily classification.

The conserved motifs of *PYL6*, *PYL7*, and *PYL13* genes across 17 *Populus* species were further analyzed. *PYL6* and *PYL13* members exhibited highly similar motif types and arrangements, and all the members contained five conserved motifs (*motif1–5*), except *PYL6_PSE*, which harbored an additional motif (motif6; Figure 3). In contrast, *PYL7* genes displayed considerable diversity in motif composition, and the number of motifs ranged from six to nine across species. Notably, *motif1–5* was universally present in all *PYL* proteins, indicating a high degree of evolutionary conservation. Moreover, *PopPYL* genes with varying protein lengths and gene structures showed distinct motif patterns. For example, within the *PYL7* subfamily, *PYL7_TRE* and *PYL7_PRU* exhibited longer protein lengths, divergent gene structures, and a greater number of conserved motifs compared to other *Populus* species. These findings suggest that *PopPYL* members within the same subfamily share conserved motif characteristics, supporting functional conservation, while the presence of unique motifs may confer specialized biological functions.

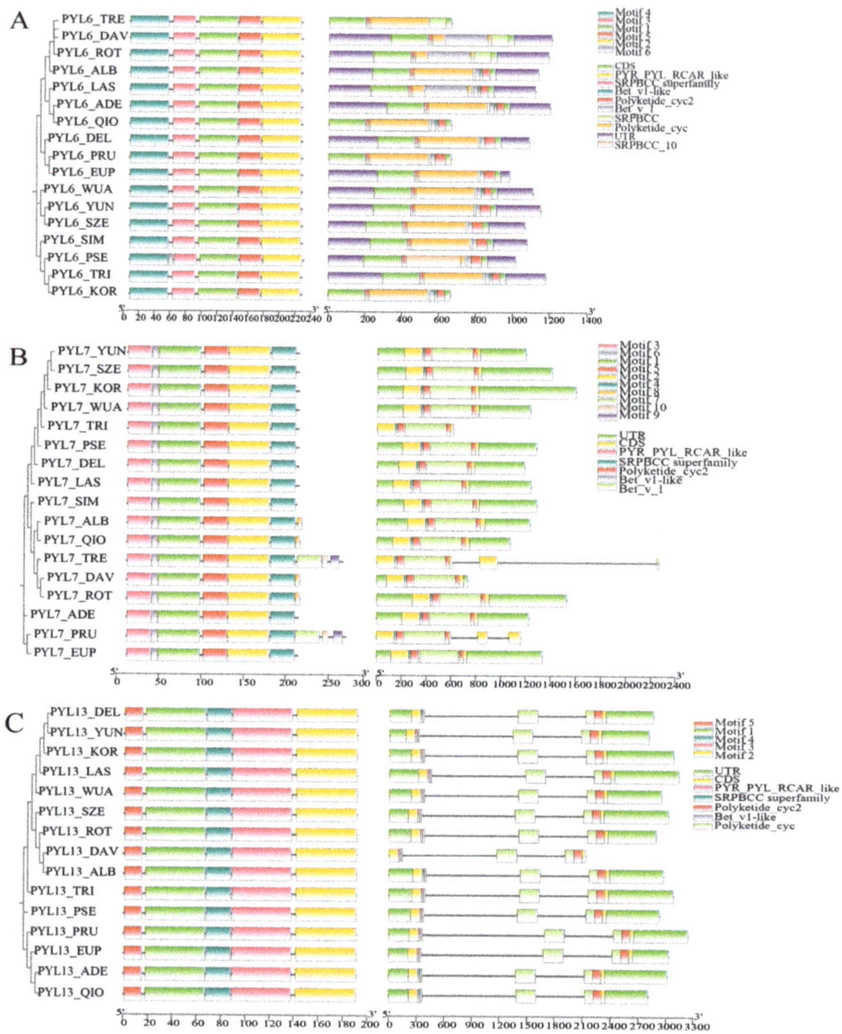

Figure 3. Comparison of the conserved motifs and gene structures of intermediate clade genes *PopPYL6* (**A**), *PopPYL7* (**B**), and *PopPYL13* (**C**) in 17 *Populus*. Left panel: the phylogenetic tree was constructed with the full length of PYL proteins using MEGA 7.0. Middle panel: the conserved motifs for PYL proteins. Motifs 1–10 are highlighted with different colors. Right panel: exon–intron structure of *PYL* genes.

2.5. Synteny Analysis of PopPYL Genes

Genome-wide synteny analysis of *PYL6*, *PYL7*, *PYL9*, and *PYL13* genes among four *Populus* species (EUP, TRI, PRU, and ADE) identified 61 syntenic gene pairs. Specifically, 14 homologous gene pairs were observed between TRI and ADE, while 11 pairs were detected between EUP and TRI, as well as between TRI and PRU. Additionally, 10 homologous pairs were identified between EUP and ADE, 9 between PRU and ADE, and 6 between EUP and PRU (Figure 4A). These findings suggest that the evolution and expansion of these genes primarily resulted from whole-genome duplication events.

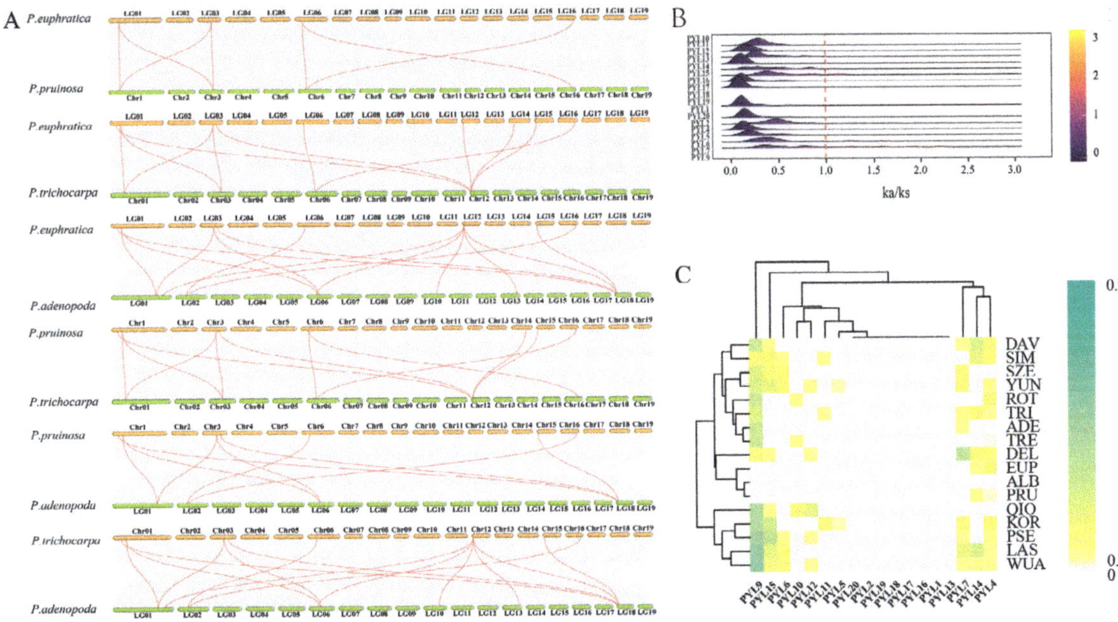

Figure 4. *Ka*/*Ks* values of *PopPYL*. (**A**) Synteny analysis of PYL among *P. euphratica* and other plant species. (**B**) Distribution of *Ka*/*Ks* values of *PopPYL* in 17 *Populus* varieties. (**C**) Heatmap of the frequency of occurrence of different *Populus* varieties at each *PYL* with a *Ka*/*Ks* ratio > 1.

Further *Ka*/*Ks* analysis of the *PYL* gene family across 17 *Populus* species revealed that most gene pairs exhibited *Ka*/*Ks* ratios significantly less than 1, indicating strong purifying selection during evolution (Figure 4B). This conservation of key functional domains is likely associated with their essential roles in plant stress adaptation. However, certain gene pairs, including *PYL4*, *PYL5*, *PYL6*, *PYL7*, *PYL9*, *PYL10*, *PYL11*, *PYL12*, *PYL14*, *PYL15*, and *PYL19*, displayed *Ka*/*Ks* values greater than 1, suggesting potential positive selection in some *Populus* lineages (Figure 4C). Notably, *PYL4*, *PYL6*, *PYL7*, *PYL9*, *PYL14*, and *PYL15* frequently exhibited *Ka*/*Ks* > 1 across multiple species, implying strong positive selection pressure during *Populus* evolution. In contrast, *PYL5*, *PYL10*, *PYL11*, and *PYL12* appeared to undergo positive selection only in specific species.

2.6. Promoter Analysis of PopPYL Genes

To elucidate the transcriptional regulatory mechanisms of *PYL_EUP* and *PYL_PRU*, a comparative analysis and PlantCARE-based screening of their promoter regions (2 kb upstream of the start codon) was performed (Figure 5A). A total of 258 *cis*-regulatory elements were identified and classified into five functional categories: plant growth and development (29 elements), abiotic stress response (127 elements), hormone signaling (39 elements), ABA-responsive elements (33 elements), and light response (30 elements). Notably, the AREB elements (ABA-responsive) were ubiquitously present, the highest abundance (five elements) was observed in *PYL17_PRU,* and only one was detected in *PYL7_EUP*/*PRU* (Figure 5B). Among abiotic stress-responsive elements, MYB elements associated with environmental adaptation exhibited significant variation in distribution. *PYL17_EUP*/*PRU* contained the fewest (2 elements), whereas *PYL9_EUP*/*PRU* exhibited the highest number (11 elements). Additionally, long terminal repeat (LTR) retrotransposon sequences were identified in *PYL7*/*9*/*13*/*17_EUP*. Hormone-responsive elements showed distinct patterns: auxin response elements (AREs) were widely distributed, with maximum enrichment in *PYL13_EUP* (ten elements) and a minimum of one element in *PYL7_EUP* (exclusively present in *P. euphratica*). JA/MeJA-responsive elements (TGACG motif and CGTCA motif) were restricted to *PYL17_EUP/PRU* and *PYL1_PRU* (≥1 element each), while SA-responsive (TCA element) and gibberellin-responsive elements (GARE motif) were sparsely distributed. Growth- and development-related CAT-box elements exhibited a broad distribution, peaking in *PYL17_EUP/PRU* (five elements) and with the lowest abundance in *PYL1_EUP* (one element, specific to EUP). Light-responsive G-box elements were found in all genes except *PYL13_PRU*, with maximum enrichment in *PYL1_EUP/PRU* (five elements). P-box elements were exclusively detected in *PYL7_EUP/PRU* and *PYL17_EUP/PRU*.

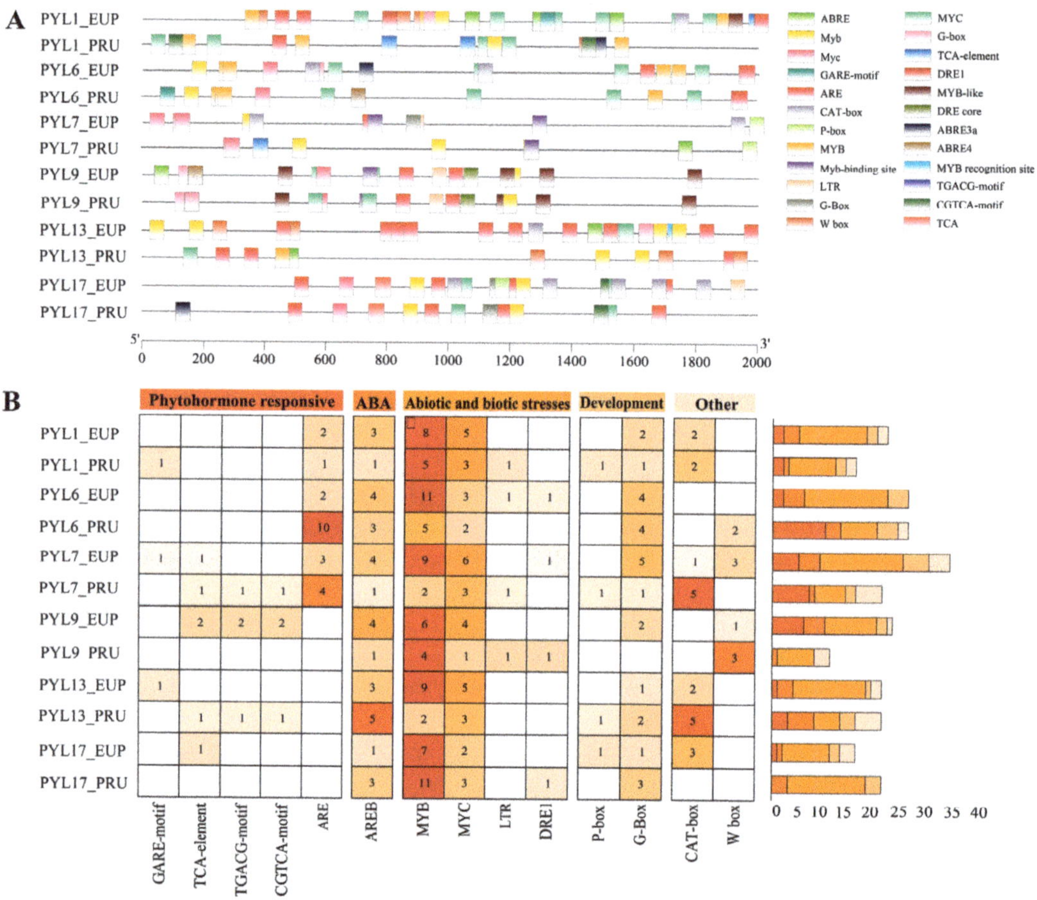

Figure 5. (**A**) Predicted cis-elements in promoter regions of *PYL_EUP* and *PYL_PRU* genes. (**B**) Cis-regulatory elements in PYL promoters of EUP and PRU. The depth of red square indicates the number of cis-acting elements for each gene, while white square indicates 0.

2.7. Structural Variation Analysis of PYL_EUP Genes

Structural variations (SVs), commonly defined as DNA sequence alterations exceeding 50 base pairs (bp) in length, encompass a range of mutation types including insertions (INS), deletions (DEL), inversions, translocations, and copy number variations. These genomic alterations can directly affect protein function or gene expression by modifying coding sequences, gene copy numbers, or regulatory regions, thereby contributing to species evolution and phenotypic diversity. In the present study, a pan-genome was constructed using *P. euphratica* (EUP) as the reference genome, and SV analysis was conducted across 17 *Populus* species. Within the *PYL* gene family of *P. euphratica*, six structural variation sites were identified, primarily located in upstream, downstream, and intronic regions, comprising two insertions and four deletions (Figure 6). Specifically, one insertion (562 bp) was located upstream of *Peu06G022430* on LG06, and another (32 bp) was found upstream of *Peu10G016340* on LG10, both potentially introducing novel *cis*-regulatory elements. Meanwhile, a 258 bp deletion occurred 459 bp downstream of *Peu01G012370* on LG01, and a 268 bp deletion occurred 271 bp downstream of *Peu02G014980* on LG02, which may attenuate transcriptional regulation of these genes. In addition, intronic regions of *Peu15G001800* and *Peu02G014980* harbored a 1 bp insertion and a 725 bp deletion, respectively. These variations can disrupt mRNA splicing efficiency and result in aberrant transcripts. To evaluate the functional impact of these SVs, we employed Mann–Whitney U tests to compare expression levels between PYL genes with and without SVs. The results revealed statistically significant differences ($p < 0.05$), suggesting that structural variations substantially influenced the expression patterns of *PYL_EUP* genes. Collectively, these findings provided novel molecular insights into the evolutionary dynamics and functional diversification of the *PYL* gene family in *Populus*.

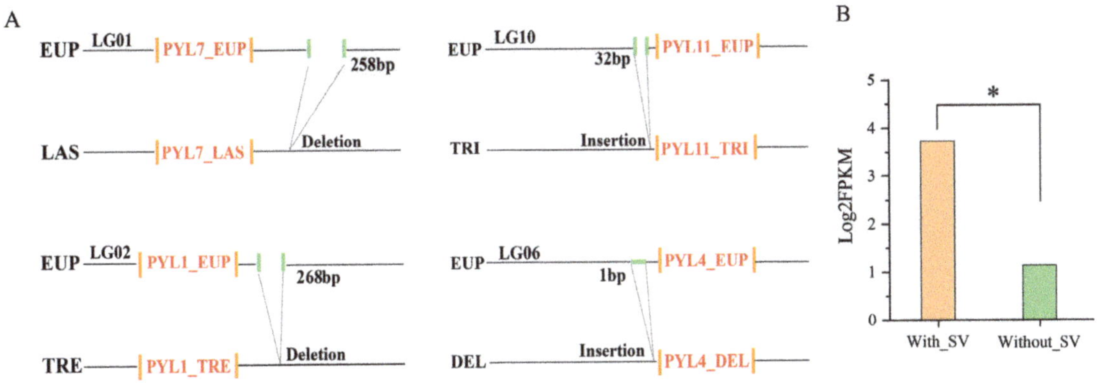

Figure 6. Effect of SVs on genes. (**A**) The effects of SV insertion and deletion on *PYL_EUP*. (**B**) The expression of *PYL_EUP* was significantly affected by SVs. (* $p < 0.05$).

2.8. Expression Analysis of the PYL_EUP Gene Family

Abscisic acid (ABA), a key phytohormone involved in regulating plant growth and stress responses, initiates its signaling pathway through recognition by the Pyrabactin Resistance 1-like (PYR/PYL) receptor family. Based on RNA-seq data analysis of *P. euphratica* under drought stress (Table S4), 16 *PYL_EUP* genes, excluding *PYL16_EUP*, exhibited differential expression patterns across root, stem, and leaf tissues (Figure 7A). Notably, *PYL1/2/6/9/11/13_EUP* were predominantly expressed in roots, suggesting their involvement in root-cap ABA signaling transduction. In contrast, *PYL5/7/12/17_EUP* showed leaf-specific upregulation, indicating potential roles in stomatal closure or leaf developmental processes. Meanwhile, the stem-preferential expression of *PYL1/4/10/14_EUP* implied a regulatory function in coordinating whole-plant drought responses, possibly by mediating ABA long-distance transport through phloem or xylem vascular tissues. This organ-specific divergence in expression clearly demonstrated the functional specialization of *PYL_EUP* family members across different tissues of *P. euphratica*, thereby optimizing ABA-mediated adaptation to drought stress.

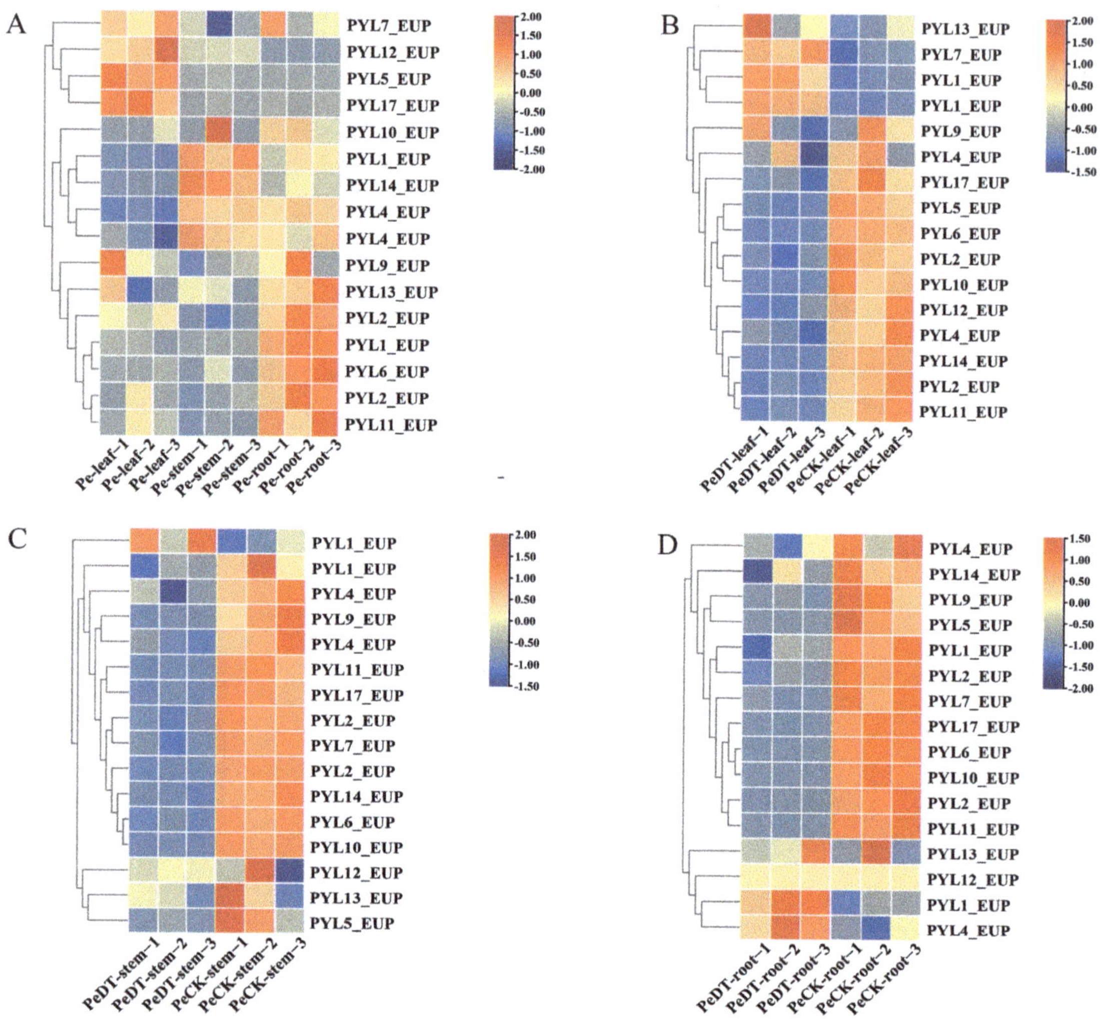

Figure 7. The expression pattern of *PYL* genes under drought stress was analyzed by heat mapping. (**A**) Pe-leaf vs. Pe-stem vs. Pe-root; (**B**) PeDT-leaf vs. PeCK-leaf; (**C**) PeDT-stem vs. PeCK-stem; (**D**) PeDT-root vs. PeCK-root.

To elucidate the potential roles of *PYL_EUP* genes in response to drought stress, their expression patterns were analyzed across different seedling tissues subjected to drought treatment. Heatmap analysis revealed distinct tissue-specific expression profiles among *PYL* family members under drought conditions (Figure 7B–D). These patterns suggest a functional differentiation of *PYL_EUP* genes in mediating ABA signaling and stress adaptation in a tissue-dependent manner. In leaves, *PYL1_EUP*, *PYL7_EUP*, and *PYL13_EUP* were upregulated under drought conditions, whereas *PYL2/4/5/6/10/11/12/14/17_EUP* exhibited downregulated expression (Figure 7B). In stems, only *PYL1_EUP* showed increased expression during drought stress, while all other *PYL_EUP* genes were downregulated (Figure 7C). In roots, *PYL1_EUP*, *PYL4_EUP*, *PYL12_EUP*, and *PYL13_EUP* were upregulated, and the remaining genes showed reduced expression levels (Figure 7D). Notably, *PYL1_EUP* demonstrated consistent upregulation in roots, stems, and leaves, highlighting its potential central role in the core regulatory network of the drought stress response in EUP. These results suggest that different *PYL_EUP* genes may contribute to drought adaptation through tissue-specific and functionally distinct regulatory mechanisms.

2.9. Verification of PYL Gene Expression Patterns in Different Tissues by RT-qPCR

To further elucidate the functional characteristics of the *PYL_EUP* gene family under drought stress, we performed a comparative expression analysis of six *PYL_EUP* genes across different plant tissues using quantitative real-time PCR (qRT-PCR). The results revealed significant tissue-specific expression differences among the selected *PYL* genes (Figure 8). Notably, *PYL7_EUP, PYL9_EUP,* and *PYL17_EUP* exhibited relatively higher expression levels in seedling leaves. In particular, *PYL7_EUP* showed peak expression in leaves and moderate expression in stems, whereas *PYL17_EUP*, although little expressed overall, displayed distinct preferential expression in leaf tissue. Conversely, *PYL13_EUP* was predominantly expressed in seedling stems. Additionally, *PYL1_EUP*, *PYL6_EUP,* and *PYL13_EUP* were primarily expressed in roots, and *PYL1_EUP* exhibited markedly higher expression in roots than in other tissues. *PYL6_EUP* demonstrated a graded expression pattern across organs, while *PYL13_EUP* showed its highest expression in root tissue. Validation of gene expression patterns demonstrated a strong correlation between qRT-PCR ($2^{-\Delta\Delta Ct}$) and RNA-seq (FPKM) data for most genes (*PYL1*, *PYL6*, *PYL7*, *PYL13*, and *PYL17*). These findings underscore the functional diversification of *PYL* family members during drought adaptation and provide valuable molecular insights for future mechanistic studies.

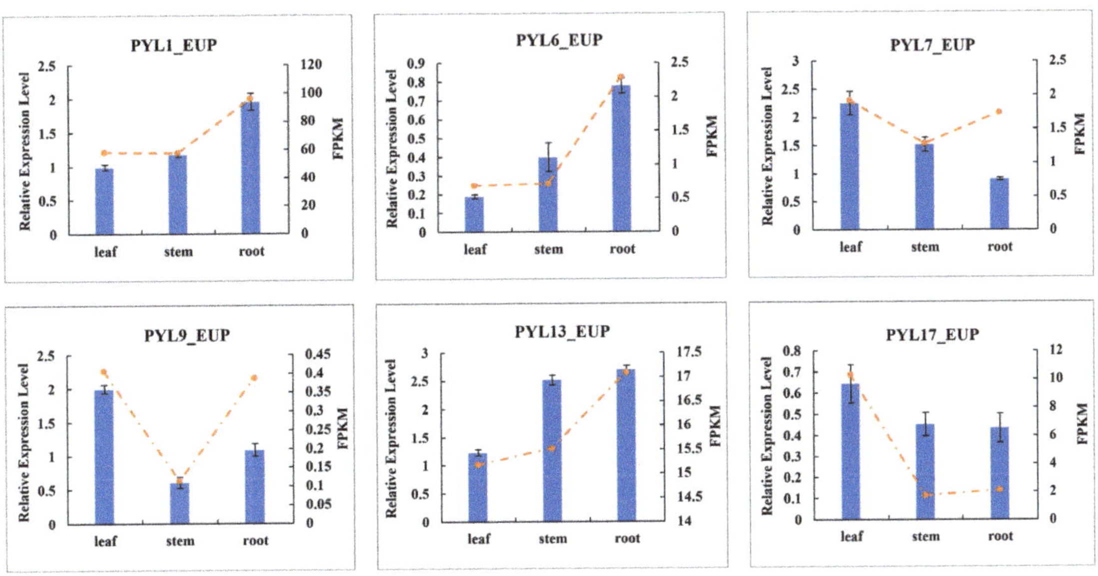

Figure 8. Expression of *PYL* genes analyzed by RT-qPCR in different tissues of *P. euphratica*. The bars are standard deviations (SDs) of three biological replicates.

3. Discussion

In this study, a high-quality pan-genome was constructed based on 17 *Populus* species, overcoming the limitations associated with relying on a single reference genome and enabling the comprehensive identification of full-spectrum genetic variations, including non-reference genes. By aligning 14 *Arabidopsis AtPYL* protein sequences, a total of 30 *Populus PYL* genes were identified in the pan-genome, comprising 17 reference-based genes and 13 non-reference genes. Among these, DEL possessed the highest number of *PYL* genes (22), while EUP and PRU exhibited the lowest gene counts (17). This variation in gene number may reflect divergent evolutionary trajectories among *Populus* species, and certain lineages underwent multiple gene duplication events [39] that contributed to the expansion of the *PYL* gene family. However, the duplicated genes have been subject to species-specific selective pressures, resulting in different evolutionary outcomes. Some *PYL* genes were conserved across most species due to essential functional roles, while others experienced functional divergence or were lost entirely in certain lineages. For instance, the eight *PopPYL* genes present in all species likely play central and conserved roles in fundamental biological processes, such as plant hormone signaling, cell growth, and development. In contrast, the 12 *PopPYL* genes unique to individual species might contribute to species-specific ecological adaptations or physiological functions, potentially aiding in responses to environmental stresses or specialized growth and developmental processes.

This study investigated the structural characteristics of *PYL6*, *PYL7*, and *PYL13* genes across 17 *Populus* species, revealing potential functional divergence within this gene family. The results showed that *PYL13* consistently contained two introns in all species, maintaining a highly conserved intron–exon structure across *Populus*. In contrast, *PYL7* exhibited substantial SV. Only *PYL7_TRE* and *PYL7_PRU* possessed two introns, whereas other members were intronless. This divergence may reflect functional differentiation following gene duplication, paralleling observations in *Arabidopsis*, where *AtPYL1–AtPYL8* typically contain 1–2 introns, while *AtPYL9–AtPYL13* are predominantly intronless, suggesting distinct evolutionary trajectories among subfamilies [52]. Motif analysis revealed that motifs 1–5 were universally present across all *PYL* proteins, likely corresponding to core functional domains of START-like ABA receptors, consistent with conserved ABA-binding sites reported in *Arabidopsis PYL* proteins [20]. However, *PYL7* members displayed notable diversity in motif composition (6–9 motifs), particularly due to the presence of unique additional motifs in *PYL7_TRE* and *PYL7_PRU*, which were associated with their extended protein length. Similar patterns were observed in the tomato *SlPYL* family, where C-terminal extensions conferred regulatory specificity to certain members [53]. These variations may reflect adaptive responses to drought or cold stress. Furthermore, significant differences in protein length and motif organization among *PYL7* members (e.g., *PYL7_TRE* with 259 amino acids versus 201–207 in others) suggest potential subfunctionalization or neofunctionalization. For instance, in maize, *ZmPYL10* acquired distinct ABA sensitivity through the insertion of a specific sequence [54].

Gene family expansion in plant evolution is primarily driven by whole-genome duplication (WGD) and tandem duplication [55]. Our multiple synteny analysis of four core *Populus PYL* genes revealed extensive syntenic relationships across species, consistent with the predominant mechanisms of plant gene family expansion. Previous studies have shown that the genus *Populus* underwent a unique WGD event, namely, the Salicoid duplication, approximately 60 million years ago, which contributed to the expansion of numerous gene families, including those involved in ABA signaling pathways [56]. Selection pressure analysis revealed that the majority of *PYL* gene pairs exhibited Ka/Ks ratios significantly less than 1 ($p < 0.01$), indicating strong purifying selection during evolution. This likely preserved their conserved functions in ABA perception and signal transduction. These findings are consistent with previous studies on the rice *OsPYL* family, in which the START-like domain of ABA receptor proteins demonstrated high conservation [57]. However, *PYL4*, *PYL6*, *PYL7*, *PYL9*, *PYL14*, and *PYL15* frequently showed $Ka/Ks > 1$ across multiple *Populus* species, suggesting that these genes may confer enhanced adaptive advantages under specific environmental stresses, such as drought or salinity. Similarly, the positive selection signals observed in *PYL6* and *PYL9* may reflect species-specific adaptation strategies to abiotic stresses, analogous to the drought tolerance conferred by positive selection in maize *ZmPYL10* [54].

Insertion and deletion (InDel) variations in genomes can directly alter gene coding sequences, resulting in amino acid changes or frameshift mutations that ultimately affect protein function [58]. For example, in peanut (*Arachis hypogaea*), an insertion variant in the 3′-UTR of *AhCKX6* reduced gene expression and increased active cytokinin levels, thereby promoting grain enlargement [59]. Similarly, a 358 bp insertion in the promoter region of maize *bZIP68* significantly upregulated its expression [60]. Structural variations (SVs) also play crucial roles in plant stress adaptation and evolutionary processes. In rice, for instance, long SVs are significantly enriched in promoter regions, and strong colocalization was observed between SVs and stress-responsive genes, suggesting that SVs contribute to the pleiotropic expression patterns commonly observed in stress-related genes [61]. In our investigation of the *PYL* gene family in EUP, we identified six SVs (two insertions and four deletions) predominantly located in regulatory regions. These included 1 bp and 32 bp insertions upstream, and 258 bp and 268 bp deletions downstream of *PYL* genes. These SVs may influence gene expression by modifying regulatory sequences or interfering with splicing, thus modulating ABA signaling pathways and environmental adaptation. Notably, SVs within intronic regions (a 1 bp insertion and a 725 bp deletion) can alter the spatial configuration of splice donor or acceptor sites, potentially increasing alternative splicing events. As demonstrated in rice *Ptr* gene studies, atypical gene structures induced by SVs (including variations in exon number or protein truncation) may still retain essential biological functions [62].

cis-acting elements in plant *PYL* genes play pivotal roles in environmental stress responses and signaling pathways that govern plant growth and development [36]. These regulatory elements can be classified into three major categories: hormone-responsive elements (e.g., ABRE) [63], growth-related elements (e.g., CAT-box) [64], and stress-responsive elements (e.g., MBS) [12]. Our systematic analysis of promoter regions (2 kb upstream) in *PYL* gene families from EUP and PRU identified 64 distinct *cis*-acting elements, encompassing light-responsive, abiotic stress-related, and hormone-responsive elements. As crucial components of ABA signaling, *PYL* gene promoters are particularly enriched with AREB elements, and *PYL17_PRU* exhibits the highest abundance. These findings suggest *PYL17_PRU*'s potential significance in ABA-mediated stress responses [65]. Notably, ARE elements involved in phytohormone responses exhibit substantial distribution variation, being exclusively present in EUP, with maximal enrichment in *PYL13_EUP*. Among abiotic stress-related elements, MYB transcription factors, which regulate plant cell morphogenesis [66], developmental processes [67], and stress adaptation [68], showed marked distribution variation. *PYL9_EUP/PRU* demonstrated the highest density, indicating its enhanced environmental stress sensitivity. Furthermore, JA/MeJA-responsive elements were uniquely distributed in *PYL17_EUP/PRU* and *PYL1_PRU*, implying their potential involvement in jasmonate signaling-mediated stress resistance. The distinct distribution patterns of these *cis*-elements reveal the sophisticated transcriptional regulatory mechanisms underlying the multifunctional roles of *PYL* genes in plant development, hormone signaling, and stress adaptation.

Gene expression patterns are closely related to biological functions. This study systematically analyzed the tissue-specific expression and drought stress response characteristics of the *PYL_EUP* gene family in EUP, providing insights into its physiological roles and regulatory mechanisms. Previous studies have demonstrated that plant *PYL* gene family members generally exhibit tissue-specific expression patterns. In soybean seeds, most *PYL* genes showed significantly higher expression levels than in other tissues [28]. During oilseed germination, *PYL* genes displayed high transcriptional abundance [69]. In rubber tree latex, *PYL* gene transcripts were particularly abundant [32]. Additionally, *PYL* genes exhibited predominant expression in Chinese cabbage callus [70]. Our research revealed distinct differential expression patterns of 17 *PYL_EUP* genes (excluding *PYL16_EUP*) in EUP across different tissues: *PYL1/2/6/9/11/13_EUP* was significantly enriched in roots, *PYL5/7/12/17_EUP* showed specifically high expression levels in leaves, and *PYL1/4/10/14_EUP* exhibiting predominant expression in stems.

Phylogenetic analysis of *PYL_EUP* genes and their *Arabidopsis thaliana* homologs revealed that *PYL1_EUP*, *PYL6_EUP*, *PYL7_EUP*, *PYL9_EUP*, *PYL13_EUP*, and *PYL17_EUP* were orthologous to *AtPYL9*, *AtPYL5*, *AtPYR1*, *AtPYL1*, *AtPYL8*, and *AtPYL4*, respectively [51]. Functional studies have demonstrated that *AtPYL9* overexpression accelerated senescence and death in older leaves while promoting summer dormancy-like responses in young tissues under severe water deficit [71]; *AtPYL4* overexpression significantly enhanced drought resistance [36]; *AtPYL8* regulated lateral root growth, with its mutation causing ABA insensitivity in roots during stress recovery [25]; and *AtPYL5* plays essential roles in stomatal responses to ABA and environmental signals by integrating various stimuli to balance water loss and CO_2 uptake [24]. Our transcriptomic and qPCR analyses revealed that *PYL_EUP* genes exhibited tissue-specific induction under drought stress. Specifically, *PYL1_EUP* was upregulated in roots, stems, and leaves, whereas *PYL7_EUP* and *PYL13_EUP* showed leaf-specific induction. In contrast, *PYL4/12/13_EUP* exhibited root-specific upregulation. These results suggested that *PYL_EUP* genes participated in the drought response of *P. euphratica* through tissue-specific regulatory mechanisms. Notably, *PYL1_EUP* appeared to function as a master regulator coordinating whole-plant responses, while other members executed tissue-specific functions, collectively establishing a robust drought response system.

4. Materials and Methods

4.1. Plant Materials and Growth Conditions

This experiment selected *P. euphratica* seeds from the Awati County *P. euphratica* Forest in Xinjiang (located at the northwestern edge of the Tarim Basin, characterized by a warm-temperate extreme continental arid desert climate). Mature capsules were collected in September 2022, followed by natural drying under indoor conditions, separation of pappus from seeds, and storage in sealed brown bottles at 4 °C for later use. The seeds were uniformly sown in nutrient soil trays and cultivated in a greenhouse at 30 °C, 50% humidity, and under a 16 h light/8 h dark photoperiod for two months. Subsequently, the seedlings were transplanted to larger trays, with one seedling retained per tray for continued growth. A completely randomized design was adopted, using one-year-old *P. euphratica* seedlings as experimental material for drought treatment. Sixty healthy seedlings with uniform growth and no pests or diseases were selected and evenly divided into two groups: one subjected to drought stress and the other as a well-watered control group, with three biological replicates per group. After treatment, both groups were maintained under identical growth conditions. When leaf wilting occurred and the soil moisture content reached 21.5%, root, stem, and leaf tissues from seedlings of both groups were collected on the 13th day, immediately frozen in liquid nitrogen, and stored at −80 °C for subsequent transcriptomic analysis.

4.2. Data Acquisition

The genomic sequences, genome annotation files, protein sequences, and bioinformatics data of PSE, WUA, SZE, YUN, KOR, SIM, LAS, DAV, ROT, ALB, QIO, and ADE were obtained from the China National Center for Bioinformation (CNCB) (https:

//bigd.big.ac.cn/gwh (accessed on 8 April 2025)). The genomic data of TRI and DEL were sourced from https://phytozome-next.jgi.doe.gov/ (accessed on 8 April 2025). The genomic data of TRE were obtained from http://popgenie.org (accessed on 8 April 2025). The genomic data of PRU were sourced from Figshare (https://figshare.com/articles/online_resource/Pprgenome_fa/20705107/2 (accessed on 8 April 2025)). The telomere-to-telomere (T2T) complete genome assembly of *Populus euphratica* used in this study was independently generated by our research group and has not been publicly released. The genomic and proteomic data of *Arabidopsis thaliana* PYL family members were retrieved from the TAIR database (https://www.arabidopsis.org/ (accessed on 10 April 2025)).

4.3. Identification of PopPYL Genes in Populus Species

In this study, two complementary approaches were employed to systematically identify *PYL* homologs. First, the BLASTP algorithm was utilized to search the protein databases of *Arabidopsis thaliana* and 17 *Populus* species, using 13 *Arabidopsis* PYL protein sequences as queries. The E-value threshold was set to $1e^{-5}$ to ensure stringent homology detection. Second, an HMM-based search was conducted by scanning the same protein datasets against the PYL-specific domain model (PF10604.hmm, retrieved from the Pfam database, https://www.ebi.ac.uk/interpro/ (accessed on 11 April 2025)) [72]. The union of candidate genes obtained from both methods was then subjected to further validation. SMART v8.0 (http://smart.embl-heidelberg.de/ (accessed on 11 April 2025)) and CD-Search v3.20 (https://www.ncbi.nlm.nih.gov/Structure/ (accessed on 11 April 2025)) were used to confirm the presence of complete PYL conserved domains, eliminating false-positive hits. The final high-confidence gene list was compiled, and TBtools was employed for batch calculation of molecular weights and isoelectric points (pI) of the identified PYL proteins.

4.4. Phylogenetic Analysis of PopPYL Family Protein Sequences

Multiple sequence alignment of *PopPYL* family protein amino acid sequences was conducted using the MUSCLE algorithm implemented in MEGA 7.0 software. Based on the alignment results, a phylogenetic tree was reconstructed through the Neighbor-Joining (NJ) method with 1000 bootstrap replicates to evaluate nodal support reliability. The final phylogenetic tree (Newick format: *.nwk) was subsequently visualized and annotated using the iTOL v6 online platform (http://itol.embl.de/ (accessed on 12 April 2025)).

4.5. Motif Analysis and Gene Structure

The conserved motifs in PopPYL proteins were predicted using the MEME Suite v 5.5.5 (https://memesuite.org/meme/tools/meme (accessed on 14 April 2025)) with parameters set to 10 motifs per sequence and an E-value cutoff $< 1 \times 10^{-5}$ [73]. Functional domain annotation was performed using NCBI's Batch CD-Search v3.20 tool (https://www.ncbi.nlm.nih.gov/Structure/bwrpsb/bwrpsb.cgi (accessed on 14 April 2025)) with default threshold settings. The predicted motifs and annotated domains were integrated and visualized using TBtools software (v2.309) to generate comprehensive schematic representations.

4.6. Gene Duplication Analysis and Ka/Ks Calculation

Gene Duplication Analysis: Whole-genome synteny analysis of *PYL* genes was conducted among EUP, PRU, TRI, and ADE lineages using TBtools software. Homologous gene pairs were identified through whole-genome alignment with the "MCScanX" module in TBtools, based on species-specific genome annotation files and genomic sequences.

Ka/Ks Analysis: Evolutionary selection pressure was assessed by calculating Ka and Ks values with *Ka/Ks* ratios using *Ka/Ks*_Calculator v2.0 [74]. Generally, *Ka/Ks* = 1 indicates neutral mutation, *Ka/Ks* > 1 suggests positive selection, and *Ka/Ks* < 1 represents purifying selection.

4.7. SVs and Gene Expression Analysis of PYL_EUP

This study constructed a pan-genome of 17 *Populus* species using the *P. euphratica* (EUP) reference genome and performed structural variant (SV) analysis, generating a novel SV dataset (17genome.sv.vcf). Functional annotation of SVs was conducted with ANNOVAR (2018Apr16) to extract key VCF features including chromosomal coordinates, variant boundaries, and reference/alternative alleles. Through custom scripts, we characterized SV distributions in genic regions (coding sequences, introns) and flanking regulatory regions (±2 kb). SVs associated with the *PYL_EUP* gene family were filtered from the annotated results. Gene expression profiles across all 17 *Populus* samples were obtained from the PSIR database (http://www.populus-superpangenome.com/ (accessed on 20 April 2025)). The Mann–Whitney U test was systematically employed to assess the association between SV presence/absence patterns and *PYL* gene expression levels. *PYL* genes with p-values < 0.05 were classified as showing SV-associated expression alterations.

4.8. Analysis of Promoter Regions of PopPYL Genes

The 2000 bp upstream sequences from the translation start sites of *PYL_EUP* and *PYL_PRU* genes were extracted using TBtools (v2.309). These promoter sequences were then analyzed in the PlantCARE database (http://bioinformatics.psb.ugent.be/webtools/plantcare/html/ (accessed on 23 April 2025)) to predict putative cis-regulatory elements. The identified cis-acting elements were visualized using TBtools (v2.309) [75].

4.9. Analysis of Gene Expression Profiles

Differential gene expression analysis between sample groups was conducted using DESeq2 to identify differentially expressed genes (DEGs) under two biological conditions [76]. The obtained P-values were adjusted for multiple testing using the Benjamini–Hochberg method to control the false discovery rate (FDR). DEGs were filtered with thresholds of $|\log_2\text{FoldChange}| \geq 1$ and FDR < 0.05. Expression patterns of *PYL_EUP* and *PYL_PRU* genes were visualized through heatmaps generated with TBtools software (v2.309).

4.10. qRT-PCR Validation

To further validate the accuracy of the sequencing data, we selected six significantly differentially expressed genes (DEGs) in different tissues of *P. euphratica* for qRT-PCR verification. Gene-specific primers for qRT-PCR were designed using the NCBI Primer-BLAST tool v6.25 (https://www.ncbi.nlm.nih.gov/tools/primer-blast/ (accessed on 25 April 2025)) (Table S5). Total RNA was extracted using the Total RNA Extractor (Trizol) Plant RNA Kit (B511311, Sangon Biotech (Shanghai, China)). Genomic DNA contamination was evaluated by 1.5% agarose gel electrophoresis, and RNA concentration and OD values were determined. First-strand cDNA was synthesized using Maxima Reverse Transcriptase (EP0743, Thermo Fisher Scientific). qPCR amplification was performed on an ABI 7500 Real-Time PCR System (Thermo Fisher Scientific, Waltham, MA, USA) in 20 μL reaction volumes containing 10 μL SGExcel FastSYBR Mixtures, 0.4 μL each of forward and reverse primers (10 μM), 7.2 μL RNase-Free ddH_2O, and 2 μL cDNA template. The thermal cycling conditions consisted of initial denaturation at 95 °C for 3 min, followed by 45 cycles of 95 °C for 15 sec and 60 °C for 45 sec (annealing/extension). Three biological and technical replicates were performed for each sample.

5. Conclusions

A pan-genomic analysis of 17 *Populus* species identified 30 *PopPYL* pan-genes exhibiting significant copy number variation, representing 325 distinct *PYL* gene homologues. These findings indicate frequent gene duplication and loss events during *Populus* evolution. Phylogenetic classification divided *PopPYL* into three clades (I–III), where clade III was the most expanded (21 members) and clade I the most conserved (four members). Species-specific distribution patterns were observed for *PYL3* and *PYL8*, while the absence of *PYL15* in EUP and PRU, and the unique acquisition of *PYL23*, reflected functional remodeling during adaptive evolution. Gene structure and conserved motif analyses revealed similar exon–intron architectures and motif compositions within subfamilies, although distinctive SVs in *PYL7_TRE/PRU* indicated potential neofunctionalization. Synteny and selection pressure analyses identified 61 homologous gene pairs resulting from whole-genome duplication events, and *PYL4/6/7/9* showed positive selection ($Ka/Ks > 1$) across multiple species, consistent with their roles in drought response. Promoter analysis revealed complex regulatory networks, featuring abundant MYB elements in *PYL9_EUP/PRU*, enriched ARE elements in *PYL13_EUP*, and the highest ABRE content in *PYL17_PRU*, providing a basis for tissue-specific expression. SV analysis detected six SV loci significantly affecting *PYL* gene expression patterns ($p < 0.05$). Expression profiling confirmed distinct functional specialization: root-enriched *PYL1/6/13_EUP* likely regulated root-cap ABA signaling; leaf-specific *PYL7/17_EUP* might participate in stomatal regulation; while stem-predominant *PYL1/4/10_EUP* potentially coordinated long-distance ABA transport. Crucially, *PYL1_EUP* showed drought-induced upregulation in roots, stems, and leaves, and *PYL1* and *PYL13* exhibited high expression levels across detection methods, demonstrating their roles as core regulatory nodes in poplar drought resistance. These findings advance our understanding of the biological functions of *PYL* genes and provide a foundation for breeding stress-resistant poplar varieties.

Supplementary Materials: The following supporting information can be downloaded at https://www.mdpi.com/article/10.3390/plants14162541/s1, Table S1. Identification of *PYL* genes in the 17 *Populors*; Table S2. Pan-gene list of the 17 *Populors*; Table S3. PYL gene family of *Populus euphratica* (EUP); Table S4. RNA-seq data of *PYL_EUP*; Table S5. Primer sequences.

Author Contributions: Conceptualization, Z.L. and J.S. (Jianhao Sun); methodology, Z.G. and J.S. (Jianhao Sun); formal analysis, X.H. and C.Q.; investigation, X.H. and J.Z.; data curation, X.H. and J.S. (Jia Song); writing—original draft preparation, X.H.; writing—review and editing, Z.L. and J.S. (Jianhao Sun); visualization, X.H.; project administration, Z.L.; funding acquisition, Z.L. All authors have read and agreed to the published version of the manuscript.

Funding: This work was financially supported by the General Program of the National Natural Science Foundation of China (32371838), the second batch of the 'Tianshan' Support Program—Technology Innovation Team (2023TSYCTD0019), the Guidance Program for Science & Technology Projects of XPCC (2023ZD091), and the Start-up Research Fund for Doctoral Scholars from the President's Fund of Tarim University (TDZKBS202532).

Data Availability Statement: Data are contained within the article and Supplementary Materials.

Acknowledgments: We sincerely acknowledge the State Key Laboratory of Tarim Basin Biological Resources Protection and Utilization of Xinjiang Production and Construction Corps for providing the advanced research platform and technical support for this study. All computations in this paper were performed on the bioinformatics computing platform of Tarim University.

Conflicts of Interest: The authors declare no conflicts of interest.

References

1. Shinozaki, K.; Yamaguchi-Shinozaki, K. Gene networks involved in drought stress response and tolerance. *J. Exp. Bot.* **2007**, *58*, 221–227. [CrossRef]
2. Cramer, G.R.; Urano, K.; Delrot, S.; Pezzotti, M.; Shinozaki, K. Effects of abiotic stress on plants: A systems biology perspective. *BMC Plant Biol.* **2011**, *11*, 163. [CrossRef]
3. Santos, A.P.; Serra, T.; Figueiredo, D.D.; Barros, P.; Lourenço, T.; Chander, S.; Oliveira, M.M.; Saibo, N.J. Transcription regulation of abiotic stress responses in rice: A combined action of transcription factors and epigenetic mechanisms. *Omics J. Integr. Biol.* **2011**, *15*, 839–857. [CrossRef]
4. Friedel, S.; Usadel, B.; von Wirén, N.; Sreenivasulu, N. Reverse engineering: A key component of systems biology to unravel global abiotic stress cross-talk. *Front. Plant Sci.* **2012**, *3*, 294. [CrossRef]
5. Cutler, S.R.; Rodriguez, P.L.; Finkelstein, R.R.; Abrams, S.R. Abscisic acid: Emergence of a core signaling network. *Annu. Rev. Plant Biol.* **2010**, *61*, 651–679. [CrossRef] [PubMed]
6. Fujii, H.; Zhu, J.K. Arabidopsis mutant deficient in 3 abscisic acid-activated protein kinases reveals critical roles in growth, reproduction, and stress. *Proc. Natl. Acad. Sci. USA* **2009**, *106*, 8380–8385. [CrossRef] [PubMed]
7. Gonzalez-Guzman, M.; Pizzio, G.A.; Antoni, R.; Vera-Sirera, F.; Merilo, E.; Bassel, G.W.; Fernández, M.A.; Holdsworth, M.J.; Perez-Amador, M.A.; Kollist, H.; et al. Arabidopsis PYR/PYL/RCAR receptors play a major role in quantitative regulation of stomatal aperture and transcriptional response to abscisic acid. *Plant Cell* **2012**, *24*, 2483–2496. [CrossRef]
8. Munemasa, S.; Hauser, F.; Park, J.; Waadt, R.; Brandt, B.; Schroeder, J.I. Mechanisms of abscisic acid-mediated control of stomatal aperture. *Curr. Opin. Plant Biol.* **2015**, *28*, 154–162. [CrossRef]
9. Belda-Palazon, B.; Gonzalez-Garcia, M.P.; Lozano-Juste, J.; Coego, A.; Antoni, R.; Julian, J.; Peirats-Llobet, M.; Rodriguez, L.; Berbel, A.; Dietrich, D.; et al. PYL8 mediates ABA perception in the root through non-cell-autonomous and ligand-stabilization-based mechanisms. *Proc. Natl. Acad. Sci. USA* **2018**, *115*, E11857–E11863. [CrossRef]
10. Schweighofer, A.; Hirt, H.; Meskiene, I. Plant PP2C phosphatases: Emerging functions in stress signaling. *Trends Plant Sci.* **2004**, *9*, 236–243. [CrossRef] [PubMed]
11. Saavedra, X.; Modrego, A.; Rodríguez, D.; González-García, M.P.; Sanz, L.; Nicolás, G.; Lorenzo, O. The nuclear interactor PYL8/RCAR3 of Fagus sylvatica FsPP2C1 is a positive regulator of abscisic acid signaling in seeds and stress. *Plant Physiol.* **2010**, *152*, 133–150. [CrossRef] [PubMed]
12. Zhou, J.; An, F.; Sun, Y.; Guo, R.; Pan, L.; Wan, T.; Hao, Y.; Cai, Y. Genome-wide identification of the ABA receptor PYL gene family and expression analysis in *Prunus avium* L. *Sci. Hortic.* **2023**, *313*, 111919. [CrossRef]
13. Yuan, X.; Yin, P.; Hao, Q.; Yan, C.; Wang, J.; Yan, N. Single Amino Acid Alteration between Valine and Isoleucine Determines the Distinct Pyrabactin Selectivity by PYL1 and PYL2. *J. Biol. Chem.* **2010**, *285*, 28953–28958. [CrossRef]
14. Fujita, Y.; Nakashima, K.; Yoshida, T.; Katagiri, T.; Kidokoro, S.; Kanamori, N.; Umezawa, T.; Fujita, M.; Maruyama, K.; Ishiyama, K.; et al. Three SnRK2 protein kinases are the main positive regulators of abscisic acid signaling in response to water stress in Arabidopsis. *Plant Cell Physiol.* **2009**, *50*, 2123–2132. [CrossRef]
15. Choi, H.-i.; Hong, J.-h.; Ha, J.-o.; Kang, J.-y.; Kim, S.Y. ABFs, a Family of ABA-responsive Element Binding Factors. *J. Biol. Chem.* **2000**, *275*, 1723–1730. [CrossRef]
16. Yang, W.; Zhang, W.; Wang, X. Post-translational control of ABA signalling: The roles of protein phosphorylation and ubiquitination. *Plant Biotechnology Journal.* **2017**, *15*, 4–14. [CrossRef]
17. Wang, P.; Zhao, Y.; Li, Z.; Hsu, C.-C.; Liu, X.; Fu, L.; Hou, Y.-J.; Du, Y.; Xie, S.; Zhang, C.; et al. Reciprocal Regulation of the TOR Kinase and ABA Receptor Balances Plant Growth and Stress Response. *Mol. Cell* **2018**, *69*, 100–112.e106. [CrossRef] [PubMed]
18. Iyer, L.M.; Koonin, E.V.; Aravind, L. Adaptations of the helix-grip fold for ligand binding and catalysis in the START domain superfamily. *Proteins* **2001**, *43*, 134–144. [CrossRef]
19. Ma, Y.; Szostkiewicz, I.; Korte, A.; Moes, D.; Yang, Y.; Christmann, A.; Grill, E. Regulators of PP2C phosphatase activity function as abscisic acid sensors. *Science* **2009**, *324*, 1064–1068. [CrossRef]
20. Park, S.Y.; Fung, P.; Nishimura, N.; Jensen, D.R.; Fujii, H.; Zhao, Y.; Lumba, S.; Santiago, J.; Rodrigues, A.; Chow, T.F.; et al. Abscisic acid inhibits type 2C protein phosphatases via the PYR/PYL family of START proteins. *Science* **2009**, *324*, 1068–1071. [CrossRef]
21. Zhao, Y.; Chan, Z.; Gao, J.; Xing, L.; Cao, M.; Yu, C.; Hu, Y.; You, J.; Shi, H.; Zhu, Y.; et al. ABA receptor PYL9 promotes drought resistance and leaf senescence. *Proc. Natl. Acad. Sci. USA* **2016**, *113*, 1949–1954. [CrossRef]
22. Santiago, J.; Rodrigues, A.; Saez, A.; Rubio, S.; Antoni, R.; Dupeux, F.; Park, S.Y.; Márquez, J.A.; Cutler, S.R.; Rodriguez, P.L. Modulation of drought resistance by the abscisic acid receptor PYL5 through inhibition of clade A PP2Cs. *Plant J. Cell Mol. Biol.* **2009**, *60*, 575–588. [CrossRef]
23. Quan, W.; Hu, Y.; Mu, Z.; Shi, H.; Chan, Z. Overexpression of AtPYL5 under the control of guard cell specific promoter improves drought stress tolerance in Arabidopsis. *Plant Physiol. Biochem. PPB* **2018**, *129*, 150–157. [CrossRef]

24. Dittrich, M.; Mueller, H.M.; Bauer, H.; Peirats-Llobet, M.; Rodriguez, P.L.; Geilfus, C.M.; Carpentier, S.C.; Al Rasheid, K.A.S.; Kollist, H.; Merilo, E.; et al. The role of Arabidopsis ABA receptors from the PYR/PYL/RCAR family in stomatal acclimation and closure signal integration. *Nat. Plants* **2019**, *5*, 1002–1011. [CrossRef]
25. Zhao, Y.; Xing, L.; Wang, X.; Hou, Y.J.; Gao, J.; Wang, P.; Duan, C.G.; Zhu, X.; Zhu, J.K. The ABA receptor PYL8 promotes lateral root growth by enhancing MYB77-dependent transcription of auxin-responsive genes. *Sci. Signal.* **2014**, *7*, ra53. [CrossRef]
26. Li, H.; Gao, Z.; Chen, Q.; Li, Q.; Luo, M.; Wang, J.; Hu, L.; Zahid, M.S.; Wang, L.; Zhao, L.; et al. Grapevine ABA receptor VvPYL1 regulates root hair development in Transgenic Arabidopsis. *Plant Physiol. Biochem. PPB* **2020**, *149*, 190–200. [CrossRef] [PubMed]
27. Yadav, S.K.; Santosh Kumar, V.V.; Verma, R.K.; Yadav, P.; Saroha, A.; Wankhede, D.P.; Chaudhary, B.; Chinnusamy, V. Genome-wide identification and characterization of ABA receptor PYL gene family in rice. *BMC Genom.* **2020**, *21*, 676. [CrossRef]
28. Bai, G.; Yang, D.H.; Zhao, Y.; Ha, S.; Yang, F.; Ma, J.; Gao, X.S.; Wang, Z.M.; Zhu, J.K. Interactions between soybean ABA receptors and type 2C protein phosphatases. *Plant. Mol. Biol.* **2013**, *83*, 651–664. [CrossRef] [PubMed]
29. Gordon, C.S.; Rajagopalan, N.; Risseeuw, E.P.; Surpin, M.; Ball, F.J.; Barber, C.J.; Buhrow, L.M.; Clark, S.M.; Page, J.E.; Todd, C.D.; et al. Characterization of Triticum aestivum Abscisic Acid Receptors and a Possible Role for These in Mediating Fusairum Head Blight Susceptibility in Wheat. *PLoS ONE* **2016**, *11*, e0164996. [CrossRef]
30. Fan, W.; Zhao, M.; Li, S.; Bai, X.; Li, J.; Meng, H.; Mu, Z. Contrasting transcriptional responses of PYR1/PYL/RCAR ABA receptors to ABA or dehydration stress between maize seedling leaves and roots. *BMC Plant Biol.* **2016**, *16*, 99. [CrossRef]
31. Yu, J.; Yang, L.; Liu, X.; Tang, R.; Wang, Y.; Ge, H.; Wu, M.; Zhang, J.; Zhao, F.; Luan, S.; et al. Overexpression of Poplar Pyrabactin Resistance-Like Abscisic Acid Receptors Promotes Abscisic Acid Sensitivity and Drought Resistance in Transgenic Arabidopsis. *PLoS ONE* **2016**, *11*, e0168040. [CrossRef]
32. Guo, D.; Zhou, Y.; Li, H.-L.; Zhu, J.-H.; Wang, Y.; Chen, X.-T.; Peng, S.-Q. Identification and characterization of the abscisic acid (ABA) receptor gene family and its expression in response to hormones in the rubber tree. *Sci. Rep.* **2017**, *7*, 45157. [CrossRef] [PubMed]
33. Chai, Y.-m.; Jia, H.-f.; Li, C.-l.; Dong, Q.-h.; Shen, Y.-y. FaPYR1 is involved in strawberry fruit ripening. *J. Exp. Bot.* **2011**, *62*, 5079–5089. [CrossRef]
34. Zhang, G.; Lu, T.; Miao, W.; Sun, L.; Tian, M.; Wang, J.; Hao, F. Genome-wide identification of ABA receptor PYL family and expression analysis of PYLs in response to ABA and osmotic stress in Gossypium. *PeerJ* **2017**, *5*, e4126. [CrossRef]
35. Hou, H.; Lv, L.; Huo, H.; Dai, H.; Zhang, Y. Genome-Wide Identification of the ABA Receptors Genes and Their Response to Abiotic Stress in Apple. *Plants* **2020**, *9*, 1028. [CrossRef]
36. Liu, X.; Zhao, X.; Yan, Y.; Shen, M.; Feng, R.; Wei, Q.; Zhang, L.; Zhang, M. Genome-wide analysis of the PYL gene family and identification of PYL genes that respond to cold stress in *Triticum monococcum* L. Subsp. Aegilopoides. *Sci. Rep.* **2024**, *14*, 26627. [CrossRef]
37. Sun, Y.; Xiao, W.; Wang, Q.-n.; Wang, J.; Kong, X.-d.; Ma, W.-h.; Liu, S.-x.; Ren, P.; Xu, L.-n.; Zhang, Y.-J. Multiple variation patterns of terpene synthases in 26 maize genomes. *BMC Genom.* **2023**, *24*, 46. [CrossRef]
38. Hufford, M.B.; Seetharam, A.S.; Woodhouse, M.R.; Chougule, K.M.; Ou, S.; Liu, J.; Ricci, W.A.; Guo, T.; Olson, A.; Qiu, Y.; et al. De novo assembly, annotation, and comparative analysis of 26 diverse maize genomes. *Science* **2021**, *373*, 655–662. [CrossRef]
39. Shi, T.; Zhang, X.; Hou, Y.; Jia, C.; Dan, X.; Zhang, Y.; Jiang, Y.; Lai, Q.; Feng, J.; Feng, J.; et al. The super-pangenome of Populus unveils genomic facets for its adaptation and diversification in widespread forest trees. *Mol. Plant* **2024**, *17*, 725–746. [CrossRef] [PubMed]
40. Jiao, P.; Wu, Z.; Wang, X.; Jiang, Z.; Wang, Y.; Liu, H.; Qin, R.; Li, Z. Short-term transcriptomic responses of Populus euphratica roots and leaves to drought stress. *J. For. Res.* **2021**, *32*, 841–853. [CrossRef]
41. Ottow, E.A.; Polle, A.; Brosché, M.; Kangasjärvi, J.; Dibrov, P.; Zörb, C.; Teichmann, T. Molecular characterization of PeNhaD1: The first member of the NhaD Na^+/H^+ antiporter family of plant origin. *Plant Mol. Biol.* **2005**, *58*, 75–88. [CrossRef]
42. Ma, T.; Wang, J.; Zhou, G.; Yue, Z.; Hu, Q.; Chen, Y.; Liu, B.; Qiu, Q.; Wang, Z.; Zhang, J.; et al. Genomic insights into salt adaptation in a desert poplar. *Nat. Commun.* **2013**, *4*, 2797. [CrossRef]
43. Sun, J.; Xu, J.; Qiu, C.; Zhai, J.; Zhang, S.; Zhang, X.; Wu, Z.; Li, Z. The chromosome-scale genome and population genomics reveal the adaptative evolution of Populus pruinosa to desertification environment. *Hortic. Res.* **2024**, *11*, uhae034. [CrossRef] [PubMed]
44. Sang, Y.; Long, Z.; Dan, X.; Feng, J.; Shi, T.; Jia, C.; Zhang, X.; Lai, Q.; Yang, G.; Zhang, H.; et al. Genomic insights into local adaptation and future climate-induced vulnerability of a keystone forest tree in East Asia. *Nat. Commun.* **2022**, *13*, 6541. [CrossRef] [PubMed]
45. Long, Z.; Sang, Y.; Feng, J.; Zhang, X.; Shi, T.; Zhang, L.; Mao, K.; Rieseberg, L.H.; Liu, J.; Wang, J. Evolutionary genomics predicts adaptive genetic and plastic gene expression responses to climate change in a key alpine forest tree species. *bioRxiv* **2024**. [CrossRef]
46. Chen, L.; Jiang, Y.; Shi, T.; Jia, C.; Long, Z.; Zhang, X.; Sang, Y.; Liu, J.; Wang, J. Chromosomal-level genome assembly of *Populus davidiana*. *bioRxiv* **2023**. [CrossRef]

47. Schiffthaler, B.; Delhomme, N.; Bernhardsson, C.; Jenkins, J.; Jansson, S.; Ingvarsson, P.; Schmutz, J.; Street, N. An Improved Genome Assembly of the European Aspen *Populus tremula*. *bioRxiv* **2019**. [CrossRef]
48. Zhang, L.; Zhao, J.; Bi, H.; Yang, X.; Zhang, Z.; Su, Y.; Li, Z.; Zhang, L.; Sanderson, B.J.; Liu, J.; et al. Bioinformatic analysis of chromatin organization and biased expression of duplicated genes between two poplars with a common whole-genome duplication. *Hortic. Res.* **2021**, *8*, 62. [CrossRef]
49. Li, Y.; Wang, D.; Wang, W.; Yang, W.; Gao, J.; Zhang, W.; Shan, L.; Kang, M.; Chen, Y.; Ma, T. A chromosome-level Populus qiongdaoensis genome assembly provides insights into tropical adaptation and a cryptic turnover of sex determination. *Mol. Ecol.* **2023**, *32*, 1366–1380. [CrossRef]
50. Liu, S.; Wang, Z.; Shi, T.; Dan, X.; Zhang, Y.; Liu, J.; Wang, J. Chromosomal-level genome assembly of *Populus adenopoda*. *bioRxiv* **2023**. [CrossRef]
51. Li, J.; Shi, C.; Sun, D.; He, Y.; Lai, C.; Lv, P.; Xiong, Y.; Zhang, L.; Wu, F.; Tian, C. The HAB1 PP2C is inhibited by ABA-dependent PYL10 interaction. *Sci. Rep.* **2015**, *5*, 10890. [CrossRef]
52. Hao, Q.; Yin, P.; Li, W.; Wang, L.; Yan, C.; Lin, Z.; Wu, J.Z.; Wang, J.; Yan, S.F.; Yan, N. The Molecular Basis of ABA-Independent Inhibition of PP2Cs by a Subclass of PYL Proteins. *Mol. Cell* **2011**, *42*, 662–672. [CrossRef]
53. Sun, L.; Wang, Y.-P.; Chen, P.; Ren, J.; Ji, K.; Li, Q.; Li, P.; Dai, S.-J.; Leng, P. Transcriptional regulation of SlPYL, SlPP2C, and SlSnRK2 gene families encoding ABA signal core components during tomato fruit development and drought stress. *J. Exp. Bot.* **2011**, *62*, 5659–5669. [CrossRef] [PubMed]
54. Zhan, X.; Lu, Y.; Zhu, J.K.; Botella, J.R. Genome editing for plant research and crop improvement. *J. Integr. Plant Biol.* **2021**, *63*, 3–33. [CrossRef] [PubMed]
55. Cannon, S.B.; Mitra, A.; Baumgarten, A.; Young, N.D.; May, G. The roles of segmental and tandem gene duplication in the evolution of large gene families in Arabidopsis thaliana. *BMC Plant Biol.* **2004**, *4*, 10. [CrossRef]
56. Tuskan, G.A.; Difazio, S.; Jansson, S.; Bohlmann, J.; Grigoriev, I.; Hellsten, U.; Putnam, N.; Ralph, S.; Rombauts, S.; Salamov, A.; et al. The genome of black cottonwood, *Populus trichocarpa* (Torr. & Gray). *Science* **2006**, *313*, 1596–1604. [CrossRef]
57. Tian, X.; Wang, Z.; Li, X.; Lv, T.; Liu, H.; Wang, L.; Niu, H.; Bu, Q. Characterization and Functional Analysis of Pyrabactin Resistance-Like Abscisic Acid Receptor Family in Rice. *Rice* **2015**, *8*, 28. [CrossRef]
58. Zhang, Y.; Yang, Z.; He, Y.; Liu, D.; Liu, Y.; Liang, C.; Xie, M.; Jia, Y.; Ke, Q.; Zhou, Y.; et al. Structural variation reshapes population gene expression and trait variation in 2,105 Brassica napus accessions. *Nat. Genet.* **2024**, *56*, 2538–2550. [CrossRef]
59. Zhao, K.; Xue, H.; Li, G.; Chitikineni, A.; Fan, Y.; Cao, Z.; Dong, X.; Lu, H.; Zhao, K.; Zhang, L.; et al. Pangenome analysis reveals structural variation associated with seed size and weight traits in peanut. *Nat. Genet.* **2025**, *57*, 1250–1261. [CrossRef]
60. Li, Z.; Fu, D.; Wang, X.; Zeng, R.; Zhang, X.; Tian, J.; Zhang, S.; Yang, X.; Tian, F.; Lai, J.; et al. The transcription factor bZIP68 negatively regulates cold tolerance in maize. *Plant Cell* **2022**, *34*, 2833–2851. [CrossRef] [PubMed]
61. Fuentes, R.R.; Chebotarov, D.; Duitama, J.; Smith, S.; De la Hoz, J.F.; Mohiyuddin, M.; Wing, R.A.; McNally, K.L.; Tatarinova, T.; Grigoriev, A.; et al. Structural variants in 3000 rice genomes. *Genome Res.* **2019**, *29*, 870–880. [CrossRef] [PubMed]
62. Zhao, H.; Wang, X.; Jia, Y.; Minkenberg, B.; Wheatley, M.; Fan, J.; Jia, M.H.; Famoso, A.; Edwards, J.D.; Wamishe, Y.; et al. The rice blast resistance gene Ptr encodes an atypical protein required for broad-spectrum disease resistance. *Nat. Commun.* **2018**, *9*, 2039. [CrossRef]
63. Hao: Structural Biological Studies of the Abscisic Acid Receptor PYL Protein Family. Tsinghua University. Issue 9. 2015. Available online: https://kns.cnki.net/ (accessed on 8 April 2025).
64. Fernández-Cancelo, P.; Muñoz, P.; Echeverría, G.; Larrigaudière, C.; Teixidó, N.; Munné-Bosch, S.; Giné-Bordonaba, J. Ethylene and abscisic acid play a key role in modulating apple ripening after harvest and after cold-storage. *Postharvest Biol. Technol.* **2022**, *188*, 111902. [CrossRef]
65. Zhao, H.; Nie, K.; Zhou, H.; Yan, X.; Zhan, Q.; Zheng, Y.; Song, C.-P. ABI5 modulates seed germination via feedback regulation of the expression of the PYR/PYL/RCAR ABA receptor genes. *New Phytologist.* **2020**, *228*, 596–608. [CrossRef]
66. Ramya, M.; Kwon, O.K.; An, H.R.; Park, P.M.; Baek, Y.S.; Park, P.H. Floral scent: Regulation and role of MYB transcription factors. *Phytochem. Lett.* **2017**, *19*, 114–120. [CrossRef]
67. Lau, S.E.; Schwarzacher, T.; Othman, R.Y.; Harikrishna, J.A. dsRNA silencing of an R2R3-MYB transcription factor affects flower cell shape in a Dendrobium hybrid. *BMC Plant Biol.* **2015**, *15*, 194. [CrossRef]
68. Wang, H.X.; Liu, T.Y.; Zhuang, W.B.; Wang, Z.Z.L.; Qu, S.C.; Zhai, H.H. Research advances in the function of anthocyanin in plant stress response. *J. Agric. Biotechnol.* **2020**, *28*, 174–183.
69. Di, F.; Jian, H.; Wang, T.; Chen, X.; Ding, Y.; Du, H.; Lu, K.; Li, J.; Liu, L. Genome-Wide Analysis of the PYL Gene Family and Identification of PYL Genes That Respond to Abiotic Stress in *Brassica napus*. *Genes* **2018**, *9*, 156. [CrossRef]
70. Li, Y.; Wang, D.; Sun, C.; Hu, X.; Mu, X.; Hu, J.; Yang, Y.; Zhang, Y.; Xie, C.G.; Zhou, X. Molecular characterization of an AtPYL1-like protein, BrPYL1, as a putative ABA receptor in Brassica rapa. *Biochem. Biophys. Res. Commun.* **2017**, *487*, 684–689. [CrossRef] [PubMed]

71. Asad, M.A.U.; Zakari, S.A.; Zhao, Q.; Zhou, L.; Ye, Y.; Cheng, F. Abiotic Stresses Intervene with ABA Signaling to Induce Destructive Metabolic Pathways Leading to Death: Premature Leaf Senescence in Plants. *Int. J. Mol. Sci.* **2019**, *20*, 256. [CrossRef] [PubMed]
72. Mistry, J.; Finn, R.D.; Eddy, S.R.; Bateman, A.; Punta, M. Challenges in homology search: HMMER3 and convergent evolution of coiled-coil regions. *Nucleic Acids Res.* **2013**, *41*, e121. [CrossRef]
73. Bailey, T.L.; Boden, M.; Buske, F.A.; Frith, M.; Grant, C.E.; Clementi, L.; Ren, J.; Li, W.W.; Noble, W.S. MEME SUITE: Tools for motif discovery and searching. *Nucleic Acids Res.* **2009**, *37*, W202–W208. [CrossRef] [PubMed]
74. Wang, D.; Zhang, Y.; Zhang, Z.; Zhu, J.; Yu, J. KaKs_Calculator 2.0: A toolkit incorporating gamma-series methods and sliding window strategies. *Genom. Proteom. Bioinform.* **2010**, *8*, 77–80. [CrossRef] [PubMed]
75. Chen, C.; Xia, R.; Chen, H.; He, Y. TBtools, a Toolkit for Biologists integrating various HTS-data handling tools with a user-friendly interface. *bioRxiv* **2018**, 289660. [CrossRef]
76. Love, M.I.; Huber, W.; Anders, S. Moderated estimation of fold change and dispersion for RNA-seq data with DESeq2. *Genome Biol.* **2014**, *15*, 550. [CrossRef]

plants

MDPI

Article

GsCYP93D1, a Cytochrome P450 Gene from Wild Soybean, Mediates the Regulation of Plant Alkaline Tolerance and ABA Sensitivity

Chao Chen [1,2], Jianyue Dai [2], Nuo Xu [2], Wanying Zhou [2], Liankun Xu [2], Qiuying Pang [3], Huizi Duanmu [1,*] and Haiying Li [1,*]

1 Engineering Research Center of Agricultural Microbiology Technology, Ministry of Education, Heilongjiang University, Harbin 150500, China; chchao@hrbnu.edu.cn
2 Department of Chemistry and Molecular Biology, School of Life Science and Technology, Harbin Normal University, Harbin 150025, China; daijianyueya@163.com (J.D.); xunuonx@163.com (N.X.); zhouwanyingz@163.com (W.Z.); liankunxu@163.com (L.X.)
3 Key Laboratory of Saline-Alkali Vegetation Ecology Restoration, Ministry of Education, College of Life Sciences, Northeast Forestry University, Harbin 150040, China; qiuying@nefu.edu.cn
* Correspondence: 2016065@hlju.edu.cn (H.D.); lihaiying@hlju.edu.cn (H.L.); Tel.: +86-0451-86609753 (H.L.)

Abstract

Cytochrome P450 enzymes (CYPs) are crucial catalysts responsible for the oxidative modification of diverse substrates, including plant hormones, antioxidants, and compounds involved in abiotic stress responses. While CYP functions in drought and salt stress adaptation have been extensively studied, their contribution to alkaline stress tolerance, particularly concerning specific cytochrome P450 genes in wild soybean (*Glycine soja*), remains less explored. In this study, a cytochrome P450 gene, *GsCYP93D1*, was identified and isolated, and its regulatory role under alkaline stress was elucidated. Transgenic *GsCYP93D1* increased *Arabidopsis* and soybean hairy root resistance to alkaline stress, but the *Arabidopsis atcyp93d1* mutant showed a reduced capacity for alkaline tolerance. Subsequent investigation showed the enhanced antioxidant defense capabilities of *GsCYP93D1* transgenic plants, as evidenced by reduced superoxide radical (O_2^-) production under exposure to alkaline stress. Furthermore, compared to the *atcyp93d1* mutant, transgenic lines of *GsCYP93D1* showed sensitivity to ABA. Moreover, transcript levels of genes associated with alkaline stress response and ABA signaling pathways were elevated in both *GsCYP93D1* transgenic and mutant lines. Collectively, our findings demonstrate that *GsCYP93D1* positively modulates plant tolerance to alkaline stress and enhances ABA sensitivity.

Keywords: *Glycine soja*; *GsCYP93D1*; alkaline; ABA

1. Introduction

Plants are frequently subjected to various biotic and abiotic environmental stresses. Alkaline stress pose serious threat to plant growth and agricultural productivity [1]. Alkaline stress imposes greater damage on plants than salt stress [2,3]. In addition to ionic toxicity and osmotic stress, high pH levels severely disrupt cellular pH homeostasis, damage cell membrane structure, weaken root activity and photosynthetic efficiency under alkaline stress [1,4–6].

Under alkaline stress, reactive oxygen species (ROS) levels increase within plants, which can cause oxidative damage, disrupt membrane systems, and even lead to plant death in severe cases [1,7]. To counteract the toxicity of excessive ROS, plants have evolved a complex antioxidant system. This system includes antioxidant enzymes such as superoxide dismutase (SOD), peroxidase (POD), catalase (CAT), and ascorbate peroxidase (APX), as well as non-enzymatic antioxidants like ascorbic acid (ASA) and glutathione [8,9]. Recent years have seen significant progress in understanding genetic regulators involved in ion homeostasis and ROS scavenging under saline–alkaline conditions [10,11]. Plants can also enhance stress tolerance by increasing the synthesis of endogenous hormones, such as abscisic acid and jasmonic acid (JA) [12]. ABA plays a critical role in plant responses to abiotic stress. Under alkaline conditions, high levels of endogenous or exogenous ABA can activate the antioxidant defense system, reducing ROS accumulation and alleviating oxidative damage [13]. For example, silencing the gene *OsABA8ox1* elevates endogenous ABA levels in rice, thereby enhancing its alkaline tolerance [14].

Cytochrome P450s (CYPs) are a class of oxidoreductases utilizing heme as a cofactor. CYPs mediate NADPH-dependent oxidation reactions and play a fundamental role in both primary and secondary metabolic pathways across diverse plant species [15]. CYPs exert pivotal functions within metabolic networks by catalyzing the oxidative modification of structurally diverse substrates, including fatty acid derivatives, plant hormones, specialized defense compounds, the biopolymer lignin and protective phytoalexins [16,17]. CYPs have also been found to play a role in hormone signaling, ROS homeostasis, and controlling how plants react to different types of stress [18,19]. For example, overexpression of the *CYP709B3* gene enhanced salt tolerance in transgenic *Arabidopsis* as indicated by *cyp709b3* low resistance levels with a higher damaged seedling percentage [20]. The expression of genes from the *CYP709* family was significantly upregulated in black locust (*Robinia pseudoacacia*) under salt stress [21]. *CYP71* gene expression was activated in ginseng following exposure to nickel and cadmium [22]. During metal stress adaptation, maize plants use CYP88A-mediated gibberellin production to synchronize connections between phytohormone signaling pathways [23]. In conclusion, how CYP-mediated hormone synthesis integrates with stress signaling (such as ABA under alkaline condition) is not fully understood yet. Therefore, there is a need to investigate how CYP-regulated hormones fine-tune antioxidant defense under alkaline stress.

Wild soybean, the wild progenitor of cultivated soybean (*Glycine max*), exhibits strong adaptability to saline–alkali stress [24]. In our previous studies, a highly adaptable alkaline-tolerant wild soybean line was identified [25]. By using transcriptome data, we identified a candidate cytochrome P450 homolog *GsCYP93D1* in response to alkaline stress. In this study, the *GsCYP93D1* gene was isolated in wild soybeans, and we detected the expression profiles under alkaline stress. We further confirmed the positive roles of *GsCYP93D1* in response to alkaline stress in *Arabidopsis* and soybean hairy roots by promoting reactive oxygen species scavenging. Moreover, *GsCYP93D1* displayed a sensitive role in response to exogenous ABA treatment. In total, our results suggested that *GsCYP93D1* played significant regulatory roles in plant responses to alkaline stress and ABA signal transduction.

2. Results

2.1. Spatial and Temporal Expression Patterns of GsCYP93D1 in Wild Soybean

To investigate the spatial expression dynamics of the *GsCYP93D1* gene in wild soybeans, their transcript levels were analyzed across different tissues (young roots, young stems, young leaves, mature roots, mature leaves, mature stems, pods, and flowers) using quantitative real-time PCR (qRT-PCR). The results revealed that the *GsCYP93D1* gene showed the highest expression in mature roots, followed by young roots. Expression levels were also comparatively higher in pods, exhibiting moderate abundance, whereas other tissues displayed only minimal expression (Figure 1A). These results suggest that

GsCYP93D1 may play a primary role in root-associated functions and a secondary role in pod development in wild soybeans.

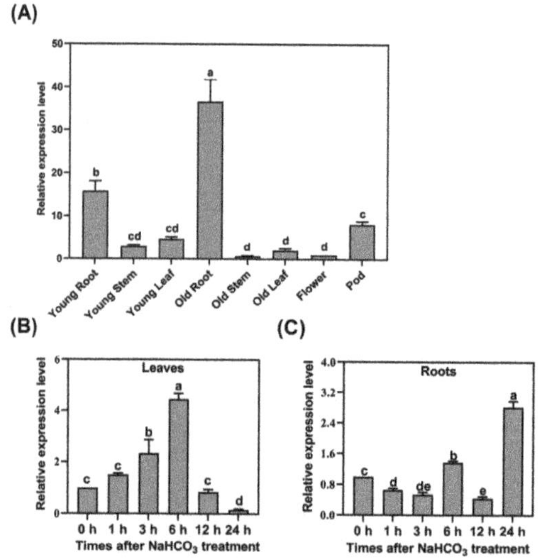

Figure 1. Expression analysis of *GsCYP93D1* in wild soybean. (**A**) Spatial expressions of *GsCYP93D1* in different tissues of wild soybean. Different lettering on bars indicates significant differences. (**B**,**C**) Transcript expressions of *GsCYP93D1* in wild soybean leaves and roots at different time points in response to 50 mM $NaHCO_3$ treatment. Statistical significance of differences was analyzed using one-way analysis of variance (ANOVA) followed by Duncan's multiple range test. Data are expressed as means ± SD. Different letters (a–e) indicate significant differences with $p < 0.05$.

To further characterize the response of *GsCYP93D1* to alkaline stress, its temporal expression dynamics were analyzed in wild soybean roots and leaves at different time points. The results showed that alkaline stress induced higher transcriptional levels of *GsCYP93D1* in leaves compared to roots, particularly during the early stress phase (Figure 1B,C). In leaves, expression peaked at 6 h with a 4-fold increase, followed by a 2-fold elevation at 3 h. In roots, maximal induction (3-fold) occurred later at 24 h, with a secondary peak of 1.5-fold at 6 h. Collectively, the expression patterns demonstrated that *GsCYP93D1* expression peaks earlier in leaves than in roots under alkaline stress, indicating distinct temporal regulation between the two tissues.

2.2. GsCYP93D1 Enhanced Alkaline Tolerance in Arabidopsis

To investigate the role of the *GsCYP93D1* gene in *Arabidopsis*, the wild-type (WT), the *GsCYP93D1* overexpression lines, OE4 and OE5, and the *atcyp93d1* mutant were analyzed. The results showed that all lines exhibited similar growth phenotypes under control conditions (Figure 2A). However, under treatment with 0.6 or 0.8 mmol/L $NaHCO_3$, the transgenic lines *GsCYP93D1* OE4 and OE5 exhibited significantly greater stress resistance compared to that in the WT, as evidenced by longer root lengths and increased fresh weights. In contrast, the *atcyp93d1* line displayed increased sensitivity (Figure 2B,C). Under alkali stress, chlorophyll content of the transgenic lines OE4 and OE5 was higher than that of WT and *atcyp93d1*, while malondialdehyde (MDA) content was significantly lower than that of WT and *atcyp93d1* (Figure 2D,E). MDA content was related to the degree of cell membrane damage, indicating that plants adapt to the external environment by adjusting MDA content, reducing the adverse effects of alkali stress on plants. Overall, these findings support that *GsCYP93D1* is a positive regulator of alkaline stress tolerance, functioning through enhancement of root growth, oxidative stress mitigation, and maintenance of photosynthetic capacity.

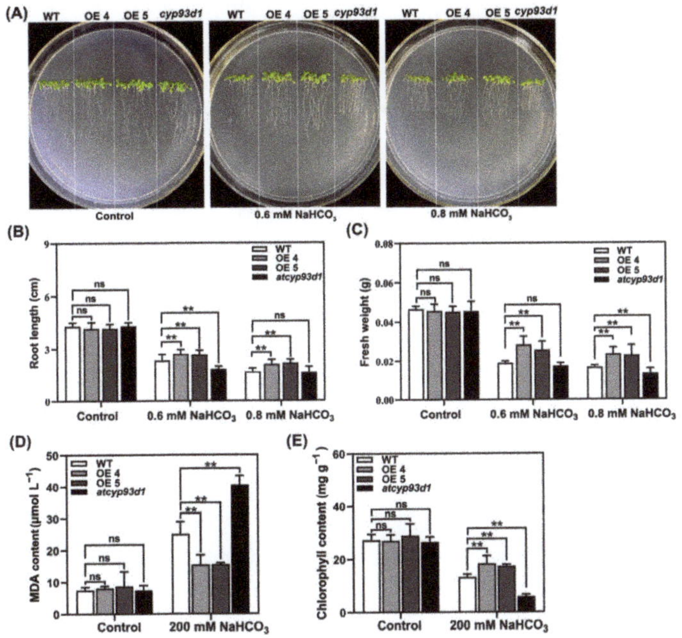

Figure 2. Effect of alkaline stress on growth and germination indices in WT, *atcyp93d1*, and *GsCYP93D1* overexpression lines. (**A**) Phenotype of WT, *atcyp93d1*, and *GsCYP93D1*-overexpression plants grown in $\frac{1}{2}$ strength MS medium with addition of 0.6 mM or 0.8 mM $NaHCO_3$. (**B**) Root lengths, (**C**) fresh weight, (**D**) MDA contents, and (**E**) chlorophyll contents were determined in these lines under control or alkaline stress conditions. Data are expressed as means ± SD. ns indicates no significant difference, ** $p < 0.01$.

2.3. Modulation of Redox Homeostasis by GsCYP93D1 Confers Stress Resilience

Under external abiotic stress, plants experience a significant accumulation of ROS, including superoxide anion radicals (O_2^-), which induce structural and functional damage to cellular components. To monitor O_2^- dynamics, we used a nitroblue tetrazolium (NBT) staining histochemical assay that relies on O_2^--mediated reduction of NBT to insoluble blue formazan precipitates (Figure 3A). Under untreated control conditions, all genotypes displayed a similar baseline staining intensity. However, under 0.6 mmol/L $NaHCO_3$, WT plants displayed visible light blue pigmentation indicating O_2^- accumulation. The overexpression line of *GsCYP93D1* demonstrated significantly reduced staining intensity and limited lesion expansion relative to WT, whereas *atcyp93d1* loss-of-function mutants displayed intensified dark blue staining. Under 0.8 mmol/L $NaHCO_3$, all leaf samples displayed darkened to deep blue staining, but transgenic plants still exhibited a lighter staining intensity with unstained regions remaining visible, whereas mutant plants displayed the darkest staining, with nearly all tissues turning blue. These results collectively demonstrate that *GsCYP93D1* overexpression enhances the superoxide scavenging capacity in *Arabidopsis* under alkaline stress, while the hypersensitive phenotype of *atcyp93d1* mutants confirms the enzyme's endogenous role in ROS homeostasis.

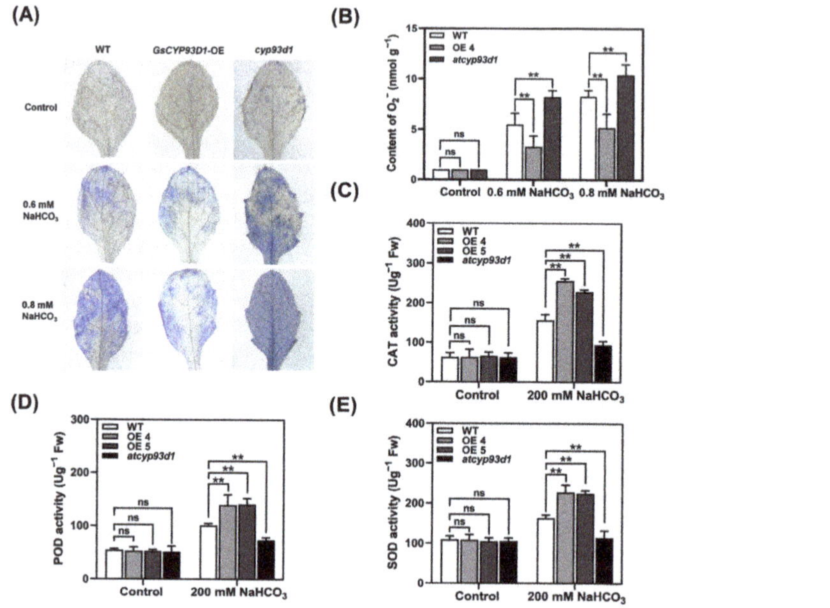

Figure 3. Effect of alkaline stress on superoxide radicals and antioxidant enzymes in WT, *atcyp93d1*, and *GsCYP93D1*-overexpression lines. (**A**,**B**) Detection of superoxide radicals in *Arabidopsis* leaves using NBT staining under 0.6 mM or 0.8 mM $NaHCO_3$. (**C**–**E**) CAT, POD, and SOD enzyme activity analysis in *Arabidopsis* leaves in response to alkaline stress. Data are expressed as means ± SD. ns indicates no significant difference, ** $p < 0.01$.

To further validate the NBT staining results, the content of superoxide anion in alkali-stressed *Arabidopsis* was measured (Figure 3B). The results indicated that the content of O_2^- in all lines was almost the same as that in control plants. Under 0.6 mmol/L $NaHCO_3$ treatment, the content of O_2^- in each line was significantly increased. Compared with the WT, the O_2^- accumulation in *GsCYP93D1* transgenic lines was significantly lower than that in WT, while the content of O_2^- in *atcyp93d1* was even higher than that in the WT. O_2^- was significantly increased under 0.8 mmol/L $NaHCO_3$ treatment, while similar to the situation at 0.6 mmol/L $NaHCO_3$ treatment, the content of O_2^- in transgenic *Arabidopsis* was lower, and the content of O_2^- in the mutant was higher.

To further investigate the role of *GsCYP93D1* in oxidative stress regulation under alkaline conditions, the enzymatic activities of CAT, SOD, and POD were quantified. As shown in Figure 3C–E, no significant differences in antioxidant enzyme activity were observed among the transgenic lines, WT, or the *atcyp93d1* mutant under non-stress conditions. However, following exposure to alkaline stress, the transgenic lines exhibited a marked and statistically significant increase in CAT, SOD, and POD activities. In comparison, the WT plants showed a moderate induction of enzyme activity, while the *atcyp93d1* mutant displayed the lowest levels of induction. These findings suggest that *GsCYP93D1* enhances the antioxidant defense system in plants, thereby mitigating cellular damage and reducing the detrimental effects of alkaline stress. These results indicate that *GsCYP93D1* contributes to the maintenance of ROS homeostasis under alkaline stress, likely by modulating antioxidant pathways that limit O_2^- accumulation and tissue damage. The contrasting phenotypes between overexpression and mutant lines further support the notion that *GsCYP93D1* functions as a positive regulator of oxidative stress tolerance in plants facing alkaline conditions.

2.4. *GsCYP93D1* Regulated the Expression Levels of Stress Responsive Genes

To further elucidate the role of *GsCYP93D1* in alkaline stress response, six stress-responsive marker genes, *COR15A*, *COR47*, *RD29A*, *KIN1*, *H⁺-ATPase*, and *NADP-ME*, were selected for expression analysis. Transcript levels were quantified using qRT-PCR in *Arabidopsis* plants treated with 50 mmol/L $NaHCO_3$ for 0, 3, and 6 h (Figure 4A–F). In *GsCYP93D1* transgenic lines, all six marker genes exhibited elevated expression levels following alkaline stress. Specifically, *COR15A*, *RD29A*, and *KIN1* reached peak expression at 3 h post-treatment, while *COR47*, *H^+-ATPase*, and *NADP-ME* displayed a progressive increase, with maximal expression observed at 6 h.

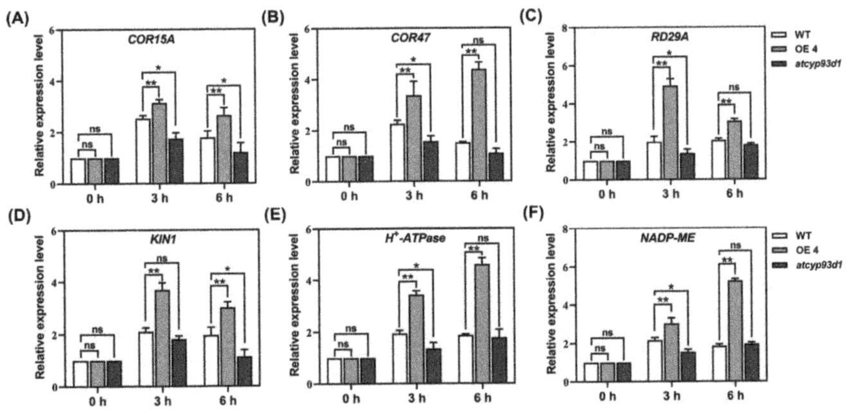

Figure 4. Expression analysis of stress responsive marker genes in WT, *atcyp93d1*, and *GsCYP93D1*-overexpression lines under alkaline stress. Two-week-old *Arabidopsis* seedlings were supplemented with 50 mM $NaHCO_3$ for 0 h, 3 h, and 6 h. (**A–F**) The expression patterns of stress responsive genes *COR15A*, *COR47*, *RD29A*, *KINI*, *H^+-ATPase*, and *NADP-ME* were determined by qRT-PCR analysis. *Actin2* was used as an internal control. Data are expressed as means ± SD. ns indicates no significant difference, * $p < 0.05$, ** $p < 0.01$.

In contrast, the *atcyp93d1* mutant showed overall lower transcript levels of the marker genes compared to the WT. Additionally, *COR15A*, *COR47*, and *KIN1* in the mutant exhibited a transient induction at 3 h, followed by a decline at 6 h, suggesting a compromised and less sustained transcriptional response to alkaline stress. These results collectively suggest that *GsCYP93D1* positively regulates the expression of key stress-related genes, contributing to an enhanced and sustained defense response under alkaline conditions.

2.5. *GsCYP93D1* Enhanced Alkaline Tolerance in Hairy Roots of Soybean

To further assess the role of *GsCYP93D1* in conferring alkaline stress tolerance in soybean, we obtained transgenic soybean hairy roots of *GsCYP93D1*. The transgenic and control hairy roots (K599 Agrobacterium rhizogenes) were subjected to either 0 mmol/L (control) or 60 mmol/L $NaHCO_3$ treatment. Under non-stress conditions, no significant differences in root growth were observed between the transgenic lines and the K599 control. However, following alkaline stress treatment, all hairy roots exhibited growth inhibition to varying degrees. Notably, *GsCYP93D1* transgenic hairy roots displayed significantly longer root lengths and greater biomass compared to the control, indicating a reduced sensitivity to alkali stress (Figure 5B,C). These findings are consistent with the stress-resilient phenotype observed in *GsCYP93D1* transgenic *Arabidopsis* lines, collectively supporting the conclusion that *GsCYP93D1* enhances tolerance to alkaline stress by promoting root growth and reducing growth inhibition under adverse conditions.

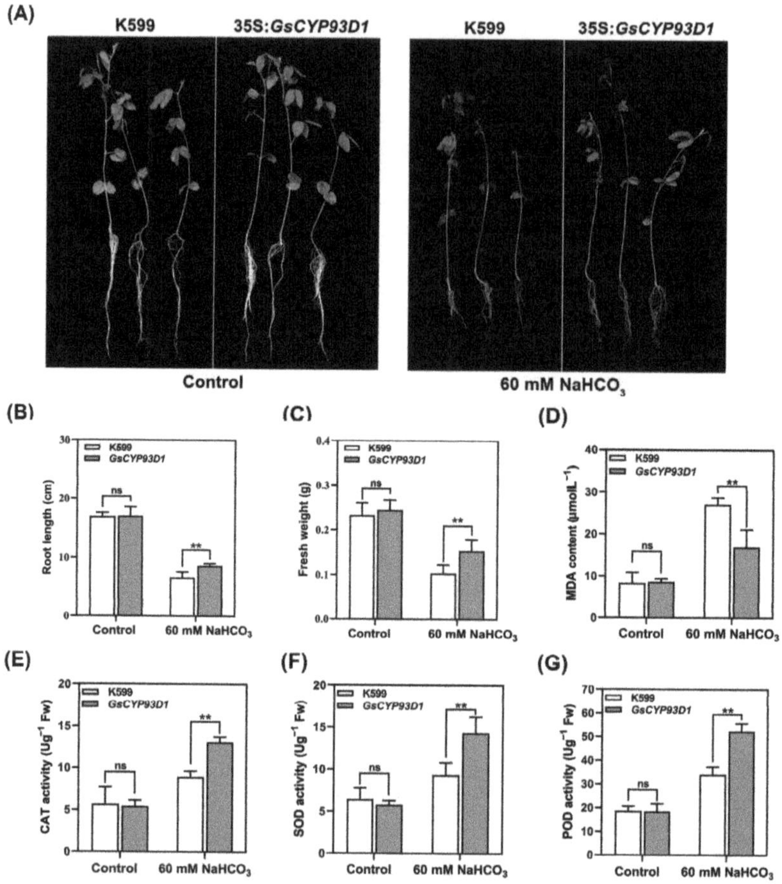

Figure 5. Effect of alkaline stress on growth parameters of K599 and transgenic soybean hairy roots. (**A**) Phenotype of K599 and transgenic *GsCYP93D1* soybean hairy roots under normal conditions or 60 mM $NaHCO_3$ stress. (**B**,**C**) Root lengths and fresh weight were determined in these lines under control or alkaline stress conditions. (**D**) Determination of MDA contents. (**E**) CAT, (**F**) SOD, and (**G**) POD enzyme activity analysis in soybean in response to 60 mM $NaHCO_3$. Data are expressed as means ± SD. ns indicates no significant difference, ** $p < 0.01$.

To further validate the alkali stress tolerance of *GsCYP93D1* transgenic soybean hairy roots, we measured MDA content and activities of antioxidant enzymes (Figure 5D). Under non-stress conditions, no significant differences were observed between transgenic hairy roots and K599. Under alkali stress treatment, MDA accumulation in *GsCYP93D1* transgenic hairy roots was significantly lower than in K599. The activities of the three antioxidant enzymes (CAT, SOD, and POD) in transgenic hairy roots increased significantly with stress intensity, whereas K599 hairy roots exhibited much smaller increases in enzyme activity (Figure 5E–G). These findings align with previous physiological index data from *GsCYP93D1* transgenic *Arabidopsis*, confirming that *GsCYP93D1* enhances antioxidant enzyme activity, alleviates oxidative stress impacts, reduces cellular damage, and thereby improves plant tolerance to alkali stress.

2.6. *GsCYP93D1* Enhanced ABA Sensitivity in *Arabidopsis*

To investigate whether *GsCYP93D1* is involved in abscisic-acid-related signaling pathways, we detected the roles of *GsCYP93D1* exposed to exogenous ABA (Figure 6A). Under control conditions, all lines exhibited similar growth phenotypes. However, at 0.3 μmol/L ABA, *GsCYP93D1* transgenic plants displayed more severe growth attenuation and stronger inhibition compared to WT, whereas *atcyp93d1* mutants exhibited enhanced ABA resistance. When the ABA concentration increased to 0.5 μmol/L, transgenic plants showed greater sensitivity and intolerance to ABA than did WT, with consistent trends in phenotype, root length, and fresh weight (Figure 6B,C). In contrast, *atcyp93d1* mutants displayed stronger resistance to ABA under these conditions. These results indicate that *GsCYP93D1* positively regulates plant sensitivity to ABA, potentially functioning as a downstream component or modulator of ABA signaling. The hypersensitivity of transgenic lines to ABA, coupled with the ABA resistance of mutants, supports a role for *GsCYP93D1* in amplifying ABA-mediated growth inhibition, possibly contributing to fine-tuning plant stress responses under adverse conditions such as alkalinity.

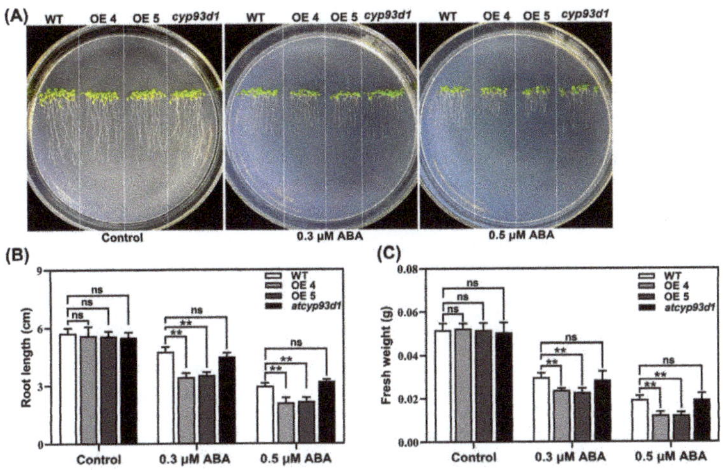

Figure 6. Effect of ABA stress on growth and germination indices in WT, *atcyp93d1*, and *GsCYP93D1*-overexpression lines. (**A**) Phenotype of WT, *atcyp93d1*, and *GsCYP93D1*-overexpression plants grown in $\frac{1}{2}$ strength MS medium with addition of 0.3 μM or 0.5 μM of ABA. (**B**) Root lengths and (**C**) fresh weight were determined in these lines under control or ABA treatment conditions. Data are expressed as means ± SD. ns indicates no significant difference, ** $p < 0.01$.

2.7. *GsCYP93D1* Altered Expression Patterns of ABA Signal-Related Genes

To further analyze the role of *GsCYP93D1* in the ABA signal transduction pathway, the ABA signal transduction pathway genes *ABI1*, *ABI2*, *ABI4*, *ABI5*, *ABF2*, and *RAS1* were selected, and the expression levels of these genes were detected by qRT-PCR (Figure 7A–F). The results showed that the relative expression levels of *ABI1*, *ABI2*, *ABI5*, and *RAS1* in transgenic plants were much lower than those in WT, except for *ABI1* at 3 h. The levels of *ABI4* and *ABF2* in transgenic *Arabidopsis* were higher than those in WT, and the expression levels of *ABI4* and *ABF2* were extremely significant. The relative expression levels of *ABI1*, *ABI2*, *ABI5*, and *RAS1* in *atcyp93d1* were significantly higher than those in WT, and the differences of *ABI1* at 3 h were not significant. In total, we speculated that *GsCYP93D1* might suppress ABA sensitivity by regulating the expression of the ABA-signaling-related genes.

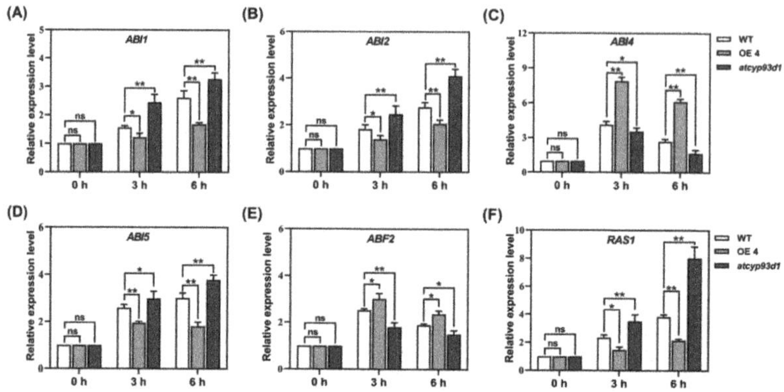

Figure 7. Expression analysis of stress-responsive marker genes in WT, *atcyp93d1*, and *GsCYP93D1*-overexpression lines under ABA stress. Two-week-old *Arabidopsis* seedlings were supplemented with 50 μM ABA for 0 h, 3 h, and 6 h. (**A–F**) The expression patterns of stress-responsive genes *ABI1, ABI2, ABI4, ABI5, ABF2,* and *RAS1* were determined by qRT-PCR analysis. *Actin2* was used as an internal control. Data are expressed as means ± SD. ns indicates no significant difference, * $p < 0.05$, ** $p < 0.01$.

3. Discussion

Soil alkalinization is one of the major environmental factors affecting plant growth, development, and reducing global crop yield. Cytochrome P450 is one of the largest enzyme protein families. In plants, cytochrome P450 families are involved in the biosynthesis of plant hormones, the formation of secondary metabolites, and the response to external environmental stresses [26,27]. Studies have demonstrated that the cytochrome P450 gene can confer salt tolerance on plants [28]. However, the role of the cytochrome P450 genes in wild soybean under alkaline stress remains unclear. In this study, the crucial role of the *GsCYP93D1* gene was detected in plant alkaline stress response and ABA signal transduction. *GsCYP93D1* was found to be upregulated under alkaline stress; especially, the *GsCYP93D1* gene showed predominant expression both in mature roots and seedling roots (Figure 1), aligning with prior studies showing the significant induction and sustained high expression of *GsCYP82C4* under alkaline stress [29].

In the present study, *GsCYP93D1 Arabidopsis* lines demonstrated significantly enhanced root growth and biomass accumulation compared to WT plants under alkaline stress, particularly during the seedling stage. In contrast, *atcyp93d1* mutant lines exhibited a hypersensitive phenotype, characterized by markedly shorter root lengths and reduced fresh biomass, indicating a compromised ability to cope with alkaline conditions (Figure 2A–C). Moreover, transgenic plants maintained higher chlorophyll content and exhibited lower malondialdehyde (MDA) levels than both WT and mutant plants (Figure 2D,E), suggesting that *GsCYP93D1* contributes to the maintenance of photosynthetic efficiency and protection against lipid peroxidation. Similarly, overexpression of *LrCYP78A5* was found to induce intensified leaf pigmentation and higher chlorophyll concentrations in Lycium, suggesting a potential role of this gene in promoting photosynthetic efficiency [27].

Alkaline-stress-induced root damage in rice is associated with excessive reactive oxygen species generation [30]. The above findings support the role of *GsCYP93D1* as a positive regulator of alkaline stress tolerance, likely through enhancing root system architecture, stabilizing photosynthetic machinery, and reducing oxidative-stress-induced cellular damage. It is becoming more acknowledged that cytochrome P450 genes have a role in abiotic stress responses, specifically via influencing secondary metabolism, hormone crosstalk, and antioxidant defenses [26]. Thus, *GsCYP93D1* may act as a key component in coordinating stress-responsive pathways, especially those related to ROS homeostasis and photosynthetic stability, providing new insights into the molecular mechanisms of plant adaptation to alkaline soil.

In upland cotton (*Gossypium hirsutum*), the cytochrome P450 gene enhances the plant's antioxidant defense mechanisms by modulating the activity or synthesis of antioxidant enzymes [31]. Our findings are in line with this notion, demonstrating that overexpression of *GsCYP93D1* elevates the antioxidant defense in wild soybean by enhancing the activities of the POD, SOD, and CAT antioxidant enzymes (Figure 3C–E). Further research has revealed that the *GsCYP93D1* gene plays a role in plant cell membrane damage under alkaline stress. The degree of injury in the leaves of transgenic and mutant *Arabidopsis* seedlings was observed using the NBT staining method. Compared to the WT, the mutant had higher O_2^- levels, while transgenic *Arabidopsis* had relatively lower O_2^- levels. This is attributed to the capabilities of *GsCYP93D1* to scavenge the excessive reactive oxygen species produced under alkaline stress, thereby alleviating cellular damage (Figure 3A,B). The *CBRLK* gene mediates ROS homeostasis in wild soybean under alkaline stress by dynamically regulating the expression of oxidative-stress-responsive genes [32]. In addition, studies also have confirmed that the wheat gene *TaCYP81D5* can enhance the activity of antioxidant enzymes by accelerating the clearance of ROS, thereby increasing the salt tolerance of wheat [19].

Previous studies have proved that *H^+-ATPase* and *NADP-ME* confer salt and alkaline tolerance [33,34]. A number of studies have also showed that *COR15A*, *COR47*, *RD29A*, and *KIN1* are involved in various stresses, such as salt or drought stress [35,36]. To elucidate the molecular functional basis of *GsCYP93D1* in the alkaline stress response, we analyzed the expression of these stress-responsive marker genes in transgenic *Arabidopsis* plants (Figure 4). qRT-PCR revealed that *GsCYP93D1* overexpression significantly upregulated the expression of stress-related genes, including *COR15A*, *COR47*, *RD29A*, *KIN1*, *H^+-ATPase*, and *NADP-ME* under alkaline stress. Conversely, the *atcyp93d1* mutant exhibited downregulation of these genes, further supporting the role of *GsCYP93D1* in enhancing alkaline stress tolerance. Consistently, studies have demonstrated that overexpression of the wild soybean gene *SKP21* in *Arabidopsis* leads to substantial upregulation of these marker genes, contributing to enhanced alkali tolerance [37].

Agrobacterium-rhizogene-mediated genetic transformation is an effective way to obtain transgenic plants. Hairy roots of soybean transformed with *GmEF8* show tolerance to drought stress [38]; hairy roots transformed with the *GmPKS4* gene exhibit strong resistance to salt and alkaline stress [39]. In this study, the function of transgenic *GsCYP93D1* soybean hairy roots was verified under alkaline stress. Compared with the non-transgenic K599 hairy roots, over-expressing the *GsCYP93D1* gene under alkaline stress could alleviate the growth inhibition of soybeans by increasing root length, fresh biomass, and antioxidant enzyme activity (Figure 5).

ABA is a crucial phytohormone that plays a pivotal role in regulating plant growth, development, and adaptive responses to environmental stresses [12,40]. Emerging evidence indicates that cytochrome P450 (CYP) family genes participate in the ABA signaling pathway, thereby enhancing plant tolerance to abiotic stresses, including alkali and drought [13,41]. Our findings indicate that *GsCYP93D1* displays enhanced sensitivity to ABA, as evidenced by pronounced growth inhibition, diminished root elongation, and reduced fresh biomass relative to the *atcyp93d1* mutant (Figure 6). The ABA signaling pathway in plants may be affected by alkaline stress [42]. Previous studies have demonstrated that *GsSKP21* negatively modulates the ABA signaling pathway through upregulation of *ABI1* and *ABI2*, consequently decreasing plant sensitivity to ABA while improving alkali stress tolerance [37]. Conversely, *AtMYB44* enhances ABA sensitivity and abiotic stress resistance by suppressing *ABI1/ABI2* expression, thereby alleviating their inhibitory effect on ABA signaling [43]. Our findings indicate that *GsCYP93D1* displays enhanced sensitivity to ABA, as evidenced by pronounced growth inhibition, diminished root elongation, and reduced fresh biomass relative to the *atcyp93d1* mutant (Figure 7), resulting in reduced ABA stress tolerance but enhanced alkaline stress resistance in plants.

Taken together, the positive roles of *GsCYP93D1* in response to alkaline stress were confirmed by promoting reactive oxygen species scavenging, and *GsCYP93D1* displayed a sensitive role in response to exogenous ABA treatment. Further investigations should specify the roles of the GsCYP93D1 protein under alkaline stress, as well as the role of *GsCYP93D1* in the crosstalk of ABA and alkaline-mediated signaling systems. Secondly, metabolomic profiling of overexpression and mutant lines under stress conditions could help identify downstream metabolic products regulated by *GsCYP93D1*, thereby clarifying its biochemical function. Further, to assess its translational potential, *GsCYP93D1* overexpression lines should be tested under field conditions with naturally occurring alkaline soils. This will validate the gene's function beyond controlled environments and inform its application in agricultural stress management strategies.

4. Materials and Methods

4.1. Materials and Growing Conditions

Wild soybean (G07256) seeds were sterilized in a 6% NaClO solution for 10 min, followed by rinsing 4–5 times with sterile water. Seeds were then placed on moist filter paper and germinated in the dark at 26 °C for 3 days. When the radicles of the seedlings reached approximately 1–2 cm in length, seedlings were transplanted into hydroponic containers filled with a 1/4-strength Hoagland nutrient solution. Seedlings were secured using aerated cotton, ensuring roots were submerged in the solution. The containers were placed in an artificial climate greenhouse set to the following conditions: 28 °C/23 °C day/night temperature, a 16 h light/8 h dark photoperiod, and relative humidity maintained at 65–75%. The young roots, young stems, young leaves, mature roots, mature stems, mature leaves, pods, and flowers were collected for tissue-specific localization analyses of *GsCYP93D1* gene expression. Additionally, three-week-old seedlings were subjected to stress treatment with 50 mM $NaHCO_3$ [44]. Leaf and root samples were collected at 0, 1, 3, 6, 12, and 24 h after the initiation of stress for dynamic expression analysis of *GsCYP93D1*. Specific primers were designed based on the complete CDS sequence of the *GsCYP93D1* gene obtained from the Phytozome database. The full-length cDNA sequence was amplified and obtained using RT-PCR.

4.2. Analysis of Tissue Localization and Expression Characteristics Under Alkaline Stress

Total RNA was isolated from various tissues of wild soybean, as well as from leaf and root samples collected at different time points after treatment with 50 mM $NaHCO_3$ stress, using a Plant Total RNA Isolation Kit (Foregene). The total RNA was reverse-transcribed into cDNA. The expression level of the *GsCYP93D1* gene was detected using qRT-PCR, with *GsGAPDH* serving as the reference gene. The specific primer sequences used in this study are listed in Supplementary Table S1. Three biological replicates are prepared for each sample, with each biological replicate subjected to three technical replicates during qRT-PCR analysis. The $2^{-\Delta\Delta Ct}$ method was used to analyze the results of the gene transcripts. Statistical analysis of the data was performed using SPSS 20.0 software.

4.3. Generation of GsCYP93D1 Transgenic Plants

The full-length cDNA of *GsCYP93D1* was amplified from wild soybean leaves. The cDNA fragment was cloned into the pCAMBIA1300 vector using restriction enzymes to construct the recombinant vector. The recombinant vector was transformed into *Agrobacterium tumefaciens* strain GV3101 for *Arabidopsis* genetic transformation (primer sequences *GsCYP93D1*-1300-F and *GsCYP93D1*-1300-R are listed in Table S1). *Arabidopsis* transformation was performed using the floral dip method, following the protocol as described previously [45]. Two overexpression lines, OE4 and OE5, were selected for phenotypic experiments (Figure S1A). Homozygous T-DNA insertion mutant seeds of *cyp93d1* (*WiscDsLoxHs035_01E*) were purchased from Fuzhou Biersen Technology Co., Ltd (Fuzhou, China). (Figure S1A).

4.4. Functional Analysis of Transgenic Arabidopsis Under Alkaline Stress

To evaluate alkali-stress-induced root phenotypic changes, seeds of vernalized WT *Arabidopsis*, homozygous *GsCYP93D1* overexpressing lines (OE4, OE5), and the *cyp93d1* mutant were sown in 1/2 MS medium. After germination and growth for 5 days, the seedlings were transferred into 1/2 MS solid medium containing either 0.6 mM $NaHCO_3$ or 0.8 mM $NaHCO_3$ for stress treatment [29]. After 7 days of vertical growth under light, primary root length and seedling fresh weight were measured and calculated. Seeds were vernalized at 4 °C for 3 days and then sown in a soil/vermiculite mixture (1:1, *w*/*w*) (approximately 10–15 seeds per pot). Plants were grown for 30 days in a greenhouse under the following conditions: 22/20 °C day/night temperature, 16 h light/8 h dark photoperiod, and irrigated with standard Hoagland nutrient solution. Uniform 30-day-old plants were selected and subjected to alkaline stress treatment with 200 mM $NaHCO_3$ solution. Leaves (0.1 g) from WT, transgenic, and mutant *Arabidopsis* plants were collected, placed into pre-labeled 1.5 mL centrifuge tubes, rapidly frozen in liquid nitrogen, and stored at −80 °C for subsequent physiological and biochemical analyses. Chlorophyll content was determined as described previously [46]. CAT and POD activities were assayed as described previously [47,48]. SOD activity was measured according to the method as described previously [49]. MDA content was assayed using the method described previously [50].

4.5. NBT Staining and Superoxide Anion Determination of Transgenic Arabidopsis Under Alkali Stress

Leaves of WT, *GsCYP93D1* transgenic (OE4 and OE5), and *cyp93d1* mutant *Arabidopsis* plants were selected and immersed in either 0.6 mM or 0.8 mM $NaHCO_3$ solution for 12 h. Leaves immersed in distilled water served as the control. Following treatment, NBT staining was performed according to the method of [51] to detect the superoxide anion (O_2^-) content in the leaves.

4.6. Alkaline Stress Responses in Hairy Roots of Transgenic Soybean Lines

Soybean seeds (DN50) were sown in sterilized vermiculite with a Hoagland nutrient solution. The pots were covered with plastic wrap to maintain humidity and placed in a light incubator under 26 °C with a 16 h light/8 h dark photoperiod. Seeds were grown for 5 days until cotyledon-stage seedlings suitable for infection were obtained. pCAMBIA1300-*GsCYP82C4* recombinant vector was transformed into the *Agrobacterium rhizogenes*. Seedling hypocotyls were infected using *Agrobacterium rhizogenes* [52]. The infected seedlings were then transplanted into moist, sterilized vermiculite, covered with plastic wrap, and cultured under identical conditions in the light incubator. Following the establishment of the plants and the stable growth of the hairy roots (Figure S1B), an alkaline stress treatment using 60 mM $NaHCO_3$ was administered, whereas plants with no supplementation of $NaHCO_3$ were placed in the control group. Regular phenotypic monitoring was maintained throughout the cultivation process until notable distinctions between the control group and the stress treatment group were noted. Phenotypic changes in soybean hairy roots were observed and recorded, and root length and fresh weight were measured. Hairy root samples (0.3 g) from each treatment group were collected, placed into pre-labeled 1.5 mL centrifuge tubes, rapidly frozen in liquid nitrogen, and stored at −80 °C for subsequent physiological and biochemical analysis. MDA content was determined as described previously [53]. Antioxidant enzyme activities were assayed as described previously [47,50].

4.7. Analysis of *GsCYP93D1*-Regulated Gene Expression in Response to Alkaline Stress

To further elucidate the role of *GsCYP93D1* in plant alkaline stress response, the alkaline-stress-responsive marker genes (*COR15A*, *COR47*, *RD29A*, *KIN1*, *H^+-APase*, and *NADP-ME*) were selected. The expression levels at different time points under 50 mM $NaHCO_3$ stress were analyzed using qRT-PCR. The transgenic lines OE4 and OE5 exhibited comparable expression levels of *GsCYP93D1*. Based on this equivalence, OE4 was randomly chosen as the representative specimen for qRT-PCR analyses [29]. WT, *GsCYP93D1*-overexpression lines, and the *cyp93d1* mutant seedlings were subjected to 50 mM $NaHCO_3$ stress treatment. The samples were collected at 0, 3, and 6 h under stress. The relative expression levels of marker genes in each sample were quantified using qRT-PCR. *Actin2* was used as an internal control [54].

4.8. Functional Characterization of Transgenic *Arabidopsis* Under ABA Treatment

Seeds of WT, *GsCYP93D1* transgenic lines (OE4 and OE5), and the *cyp93d1* mutant were sown on 1/2 MS solid medium. Fifteen consistent and sturdy seedlings per line were chosen after five days of growth and moved to plates with exogenous ABA at concentrations of 0.3 μM or 0.5 μM. Seedlings grown in 1/2 MS medium without ABA served as the control. The plates were positioned vertically and cultured under light for 7 days. Phenotypes were then photographed, and the root length and fresh weight of ten seedlings per treatment were measured. To investigate the role of the *GsCYP93D1* gene in the ABA signaling pathway, the expression levels of six ABA-stress-related marker genes (*ABI1*, *ABI2*, *ABI4*, *ABI5*, *ABF2*, and *RAS1*) were detected using qRT-PCR. WT, *GsCYP93D1* transgenic plants, and the *cyp93d1* mutant were treated with 50 μM ABA. Samples were collected at 0, 3, and 6 h after ABA application to detect the expression changes of these genes.

4.9. Statistical Analysis

All experiments included at least three independent biological replicates. Statistical significance of differences was analyzed using a one-way analysis of variance (ANOVA) followed by Duncan's multiple range test ($p < 0.05$). Statistical analysis and graph generation were performed using GraphPad Prism software 9.5.1.

Supplementary Materials: The following supporting information can be downloaded at: https://www.mdpi.com/article/10.3390/plants14172623/s1, Table S1: Gene-specific primers used in this study. Figure S1: qRT-PCR analysis of *GsCYP93D1* expression in transgenic *Arabidopsis*, mutant *Arabidopsis*, and soybean hairy roots by qRT-PCR. (A) qRT-PCR analysis of *GsCYP93D1* expression in transgenic *Arabidopsis* and mutant *Arabidopsis*. (B) qRT-PCR analysis of *GsCYP93D1* expression in soybean hairy roots. Data are represented as means ± SD. * $p < 0.05$, ** $p < 0.01$.

Author Contributions: Conceptualization, C.C., Q.P., H.D. and H.L.; methodology, C.C. and L.X.; software, J.D.; validation, L.X. and W.Z.; formal analysis, C.C.; investigation, L.X.; resources, C.C., H.D. and H.L.; data curation, J.D.; writing—original draft preparation, J.D. and N.X.; writing—review and editing, J.D., H.D. and C.C.; visualization, W.Z.; supervision, C.C. and H.L.; project administration, C.C., H.D. and H.L.; funding acquisition, C.C. All authors have read and agreed to the published version of the manuscript.

Funding: This research was funded by the Natural Science Foundation of Heilongjiang Province, grant number YQ2023C034, and the Science and Technology Innovation Ascent Program, grant number XPPY202305. The APC was funded by the China Postdoctoral Science Foundation, grant number 2022M711431.

Data Availability Statement: The datasets generated and analyzed during the current study are available from the corresponding authors upon reasonable request.

Acknowledgments: The authors wish to thank Zaib-un Nisa for language editing assistance.

Conflicts of Interest: The authors declare no conflicts of interest.

Abbreviations

The following abbreviations are used in this manuscript:

CYPs	Cytochrome P450 enzymes
qRT-PCR	Quantitative real-time PCR
ROS	Reactive oxygen species
ABA	Abscisic acid
SOD	Superoxide dismutase
POD	Peroxidase
CAT	Catalase
MDA	Malondialdehyde
NBT	Nitroblue tetrazolium
O_2^-	Superoxide anion
WT	Wild type

References

1. Yang, S.; Xu, Y.; Tang, Z.; Jin, S.; Yang, S. The Impact of Alkaline Stress on Plant Growth and Its Alkaline Resistance Mechanisms. *Int. J. Mol. Sci.* **2024**, *25*, 13719. [CrossRef]
2. Fang, S.; Hou, X.; Liang, X. Response mechanisms of plants under saline-alkali stress. *Front. Plant Sci.* **2021**, *12*, 667458. [CrossRef]
3. Shaban, A.S.; Safhi, F.A.; Fakhr, M.A.; Pruthi, R.; Abozahra, M.S.; El-Tahan, A.M.; Subudhi, P.K. Comparison of the morpho-physiological and molecular responses to salinity and alkalinity stresses in rice. *Plants* **2023**, *13*, 60. [CrossRef]
4. Li, J.; Yang, Y. How do plants maintain pH and ion homeostasis under saline-alkali stress? *Front. Plant Sci.* **2023**, *14*, 1217193. [CrossRef] [PubMed]
5. An, D.; Chen, J.-G.; Gao, Y.-Q.; Li, X.; Chao, Z.-F.; Chen, Z.-R.; Li, Q.-Q.; Han, M.-L.; Wang, Y.-L.; Wang, Y.-F. *AtHKT1* drives adaptation of *Arabidopsis thaliana* to salinity by reducing floral sodium content. *PLoS Genet.* **2017**, *13*, e1007086. [CrossRef]
6. Wang, X.; Ajab, Z.; Liu, C.; Hu, S.; Liu, J.; Guan, Q. Overexpression of transcription factor *SlWRKY28* improved the tolerance of *Populus davidiana*× *P. bolleana* to alkaline salt stress. *BMC Genet.* **2020**, *21*, 103. [CrossRef]
7. Nie, W.; Gong, B.; Wen, D.; Qiao, P.; Guo, H.; Shi, Q. Brassinosteroid Enhances Cucumber Stress Tolerance to $NaHCO_3$ by Modulating Nitrogen Metabolism, Ionic Balance and Phytohormonal Response. *Plants* **2024**, *14*, 80. [CrossRef]
8. Singh, P.; Choudhary, K.K.; Chaudhary, N.; Gupta, S.; Sahu, M.; Tejaswini, B.; Sarkar, S. Salt stress resilience in plants mediated through osmolyte accumulation and its crosstalk mechanism with phytohormones. *Front. Plant Sci.* **2022**, *13*, 1006617. [CrossRef] [PubMed]
9. Wang, J.; Ding, C.; Cui, C.; Song, J.; Ji, G.; Sun, N.; Qi, S.; Li, J.; Xu, Z.; Zhang, H. Physiological and Molecular Responses of Poplar to Salt Stress and Functional Analysis of *PagGRXC9* to Salt Tolerance. *Tree Physiol.* **2025**, *45*, tpaf039. [CrossRef] [PubMed]
10. Zhao, S.; Zhang, Q.; Liu, M.; Zhou, H.; Ma, C.; Wang, P. Regulation of plant responses to salt stress. *Int. J. Mol. Sci.* **2021**, *22*, 4609. [CrossRef]
11. Feng, Z.; Xu, Y.; Xie, Z.; Yang, Y.; Lu, G.; Jin, Y.; Wang, M.; Liu, M.; Yang, H.; Li, W. Overexpression of abscisic acid biosynthesis gene *OsNCED3* enhances survival rate and tolerance to alkaline stress in rice seedlings. *Plants* **2024**, *13*, 1713. [CrossRef]
12. Kushiro, T.; Okamoto, M.; Nakabayashi, K.; Yamagishi, K.; Kitamura, S.; Asami, T.; Hirai, N.; Koshiba, T.; Kamiya, Y.; Nambara, E. The *Arabidopsis* cytochrome P450 CYP707A encodes ABA 8′-hydroxylases: Key enzymes in ABA catabolism. *EMBO J.* **2004**, *23*, 1647–1656. [CrossRef]

13. Liu, X.-L.; Zhang, H.; Jin, Y.-Y.; Wang, M.-M.; Yang, H.-Y.; Ma, H.-Y.; Jiang, C.-J.; Liang, Z.-W. Abscisic acid primes rice seedlings for enhanced tolerance to alkaline stress by upregulating antioxidant defense and stress tolerance-related genes. *Plant Soil* **2019**, *438*, 39–55. [CrossRef]
14. Liu, X.; Xie, X.; Zheng, C.; Wei, L.; Li, X.; Jin, Y.; Zhang, G.; Jiang, C.-J.; Liang, Z. RNAi-mediated suppression of the abscisic acid catabolism gene *OsABA8ox1* increases abscisic acid content and tolerance to saline–alkaline stress in rice (*Oryza sativa* L.). *Crop J.* **2022**, *10*, 354–367. [CrossRef]
15. Mizutani, M.; Sato, F. Unusual P450 reactions in plant secondary metabolism. *Arch. Biochem. Biophys.* **2011**, *507*, 194–203. [CrossRef]
16. Schuler, M.A.; Werck-Reichhart, D. Functional genomics of P450s. *Annu. Rev. Plant Biol.* **2003**, *54*, 629–667. [CrossRef]
17. Lv, X.; Zhao, X.; Wang, F.; Wang, H.; Zhang, Y.; Ruan, B.; Dong, G.; Yu, Y.; Wu, L.; Chen, F. Rice Cytochrome P450 Protein CYP71P1 Is Required for Heat Stress Tolerance by Regulating Serotonin Biosynthesis and ROS Homeostasis. *Plants* **2025**, *14*, 1072. [CrossRef]
18. Heitz, T.; Widemann, E.; Lugan, R.; Miesch, L.; Ullmann, P.; Désaubry, L.; Holder, E.; Grausem, B.; Kandel, S.; Miesch, M. Cytochromes P450 CYP94C1 and CYP94B3 catalyze two successive oxidation steps of plant hormone jasmonoyl-isoleucine for catabolic turnover. *J. Biol. Chem.* **2012**, *287*, 6296–6306. [CrossRef] [PubMed]
19. Wang, M.; Yuan, J.; Qin, L.; Shi, W.; Xia, G.; Liu, S. *TaCYP81D5*, one member in a wheat cytochrome P450 gene cluster, confers salinity tolerance via reactive oxygen species scavenging. *Plant Biotechnol. J.* **2020**, *18*, 791–804. [CrossRef] [PubMed]
20. Mao, G.; Seebeck, T.; Schrenker, D.; Yu, O. CYP709B3, a cytochrome P450 monooxygenase gene involved in salt tolerance in *Arabidopsis thaliana*. *BMC Plant Biol.* **2013**, *13*, 169. [CrossRef] [PubMed]
21. Xu, F.; Peng, M.; Luo, Q.; Jiang, M.; Zhang, X.; Zong, X.; Meng, F.; Li, Y. Isolation and detection of transcript-derived fragments (TDFs) in NaCl-stressed black locust (*Robinia pseudoacacia* L.) using cDNA-AFLP analysis. *Acta Physiol. Plant.* **2015**, *37*, 168. [CrossRef]
22. Seitz, C.; Eder, C.; Deiml, B.; Kellner, S.; Martens, S.; Forkmann, G. Cloning, functional identification and sequence analysis of flavonoid 3′-hydroxylase and flavonoid 3′, 5′-hydroxylase cDNAs reveals independent evolution of flavonoid 3′, 5′-hydroxylase in the Asteraceae family. *Plant Mol. Biol.* **2006**, *61*, 365–381. [CrossRef]
23. Guo, P.; Bai, G.; Carver, B.; Li, R.; Bernardo, A.; Baum, M. Transcriptional analysis between two wheat near-isogenic lines contrasting in aluminum tolerance under aluminum stress. *Mol. Genet. Genom.* **2007**, *277*, 1–12. [CrossRef] [PubMed]
24. Cao, L.; Yu, Y.; DuanMu, H.; Chen, C.; Duan, X.; Zhu, P.; Chen, R.; Li, Q.; Zhu, Y.; Ding, X. A novel *Glycine soja* homeodomain-leucine zipper (HD-Zip) I gene, *Gshdz4*, positively regulates bicarbonate tolerance and responds to osmotic stress in *Arabidopsis*. *BMC Plant Biol.* **2016**, *16*, 184. [CrossRef] [PubMed]
25. Ge, Y.; Li, Y.; Lv, D.-K.; Bai, X.; Ji, W.; Cai, H.; Wang, A.-X.; Zhu, Y.-M. Alkaline-stress response in *Glycine soja* leaf identifies specific transcription factors and ABA-mediated signaling factors. *Funct. Integr. Genom.* **2011**, *11*, 369–379. [CrossRef]
26. Pandian, B.A.; Sathishraj, R.; Djanaguiraman, M.; Prasad, P.V.; Jugulam, M. Role of cytochrome P450 enzymes in plant stress response. *Antioxidants* **2020**, *9*, 454. [CrossRef] [PubMed]
27. Zhao, Y.; Rao, S.; Dai, G.; Chen, J. Genome-Wide Identification of the *CYP78A* Gene Family in *Lycium* and Functional Characterization of *LrCYP78A5*. *Plants* **2025**, *14*, 1152. [CrossRef]
28. Zhang, X.; Xue, L.; Chen, R.; Ma, Q.; Ma, D.; Liu, X. Genome-wide identification of the cytochrome P450 family and analysis of CYP regarding salt tolerance in *Medicago sativa* L. *Grass Res.* **2023**, *3*, 21. [CrossRef]
29. Wu, J.; Fang, Y.; Xu, L.; Jin, X.; iqbal, A.; Nisa, Z.-u.; Ali, N.; Chen, C.; Shah, A.A.; Gatasheh, M.K. The Glycine soja cytochrome P450 gene GsCYP82C4 confers alkaline tolerance by promoting reactive oxygen species scavenging. *Physiol. Plant.* **2024**, *176*, e14252. [CrossRef]
30. Zhang, H.; Liu, X.-L.; Zhang, R.-X.; Yuan, H.-Y.; Wang, M.-M.; Yang, H.-Y.; Ma, H.-Y.; Liu, D.; Jiang, C.-J.; Liang, Z.-W. Root Damage under Alkaline Stress Is Associated with Reactive Oxygen Species Accumulation in Rice (*Oryza sativa* L.). *Front. Plant Sci.* **2017**, *8*, 1580. [CrossRef]
31. Sun, K.; Fang, H.; Chen, Y.; Zhuang, Z.; Chen, Q.; Shan, T.; Khan, M.K.R.; Zhang, J.; Wang, B. Genome-wide analysis of the cytochrome P450 gene family involved in salt tolerance in Gossypium hirsutum. *Front. Plant Sci.* **2021**, *12*, 685054. [CrossRef]
32. Sun, X.L.; Sun, M.Z.; Jia, B.W.; Qin, Z.W.; Yang, K.J.; Chen, C.; Yu, Q.Y.; Zhu, Y.M. A *Glycine soja* methionine sulfoxide reductase B5a interacts with the Ca^{2+}/CAM-binding kinase GsCBRLK and activates ROS signaling under carbonate alkaline stress. *Plant J.* **2016**, *86*, 514–529. [CrossRef]
33. Li, M.; Guo, P.; Nan, N.; Ma, A.; Liu, W.; Wang, T.-J.; Yun, D.-J.; Xu, Z.-Y. Plasma membrane-localized H+-ATPase OsAHA3 functions in saline–alkaline stress tolerance in rice. *Plant Cell Rep.* **2024**, *43*, 9. [CrossRef] [PubMed]
34. Liu, S.; Cheng, Y.; Zhang, X.; Guan, Q.; Nishiuchi, S.; Hase, K.; Takano, T. Expression of an NADP-malic enzyme gene in rice (*Oryza sativa.* L) is induced by environmental stresses; over-expression of the gene in Arabidopsis confers salt and osmotic stress tolerance. *Plant Mol. Biol.* **2007**, *64*, 49–58. [CrossRef] [PubMed]

35. Zhang, L.; Cheng, J.; Sun, X.; Zhao, T.; Li, M.; Wang, Q.; Li, S.; Xin, H. Overexpression of *VaWRKY14* increases drought tolerance in *Arabidopsis* by modulating the expression of stress-related genes. *Plant Cell Rep.* **2018**, *37*, 1159–1172. [CrossRef]
36. Cai, J.; Liu, T.; Li, Y.; Ow, D.W. A C-terminal fragment of Arabidopsis OXIDATIVE STRESS 2 can play a positive role in salt tolerance. *Biochem. Biophys. Res. Commun.* **2021**, *556*, 23–30. [CrossRef]
37. Liu, A.; Yu, Y.; Duan, X.; Sun, X.; Duanmu, H.; Zhu, Y. *GsSKP21*, a *Glycine soja* S-phase kinase-associated protein, mediates the regulation of plant alkaline tolerance and ABA sensitivity. *Plant Mol. Biol.* **2015**, *87*, 111–124. [CrossRef]
38. Zhang, H.-Y.; Hou, Z.-H.; Zhang, Y.; Li, Z.-Y.; Chen, J.; Zhou, Y.-B.; Chen, M.; Fu, J.-D.; Ma, Y.-Z.; Zhang, H. A soybean EF-Tu family protein GmEF8, an interactor of GmCBL1, enhances drought and heat tolerance in transgenic *Arabidopsis* and soybean. *Int. J. Biol. Macromol.* **2022**, *205*, 462–472. [CrossRef] [PubMed]
39. Ketehouli, T.; Zhou, Y.-G.; Dai, S.-Y.; Carther, K.F.I.; Sun, D.-Q.; Li, Y.; Nguyen, Q.V.H.; Xu, H.; Wang, F.-W.; Liu, W.-C. A soybean calcineurin B-like protein-interacting protein kinase, GmPKS4, regulates plant responses to salt and alkali stresses. *J. Plant Physiol.* **2021**, *256*, 153331. [CrossRef]
40. Pirredda, M.; Fañanás-Pueyo, I.; Oñate-Sánchez, L.; Mira, S. Seed longevity and ageing: A review on physiological and genetic factors with an emphasis on hormonal regulation. *Plants* **2023**, *13*, 41. [CrossRef]
41. Li, S.; Yuan, J.; Zhou, F.; Liu, Y.; Xie, H.; Jia, W.; Chao, Y.; Han, L. Modulating ABA-dependent growth and development by overexpressing cytochrome P450 ABA 8′-hydroxylase in *Medicago truncatula*. *Environ. Exp. Bot.* **2025**, *229*, 106060. [CrossRef]
42. Degenhardt, B.; Gimmler, H.; Hose, E.; Hartung, W. Effect of alkaline and saline substrates on ABA contents, distribution and transport in plant roots. *Plant Soil* **2000**, *225*, 83–94. [CrossRef]
43. Jung, C.; Seo, J.S.; Han, S.W.; Koo, Y.J.; Kim, C.H.; Song, S.I.; Nahm, B.H.; Do Choi, Y.; Cheong, J.-J. Overexpression of *AtMYB44* enhances stomatal closure to confer abiotic stress tolerance in transgenic Arabidopsis. *Plant Physiol.* **2008**, *146*, 323–324. [CrossRef]
44. Nisa, Z.-u.; Wang, Y.; Ali, N.; Chen, C.; Zhang, X.; Jin, X.; Yu, L.; Jing, L.; Chen, C.; Elansary, H.O. Strigolactone signaling gene from soybean *GmMAX2a* enhances the drought and salt-alkaline resistance in *Arabidopsis* via regulating transcriptional profiles of stress-related genes. *Funct. Integr. Genom.* **2023**, *23*, 216. [CrossRef]
45. Clough, S.J.; Bent, A.F. Floral dip: A simplified method for Agrobacterium-mediated transformation of *Arabidopsis thaliana*. *Plant J.* **1998**, *16*, 735–743. [CrossRef]
46. Arnon, D.I. Copper enzymes in isolated chloroplasts. Polyphenoloxidase in Beta vulgaris. *Plant Physiol.* **1949**, *24*, 1–15. [CrossRef]
47. Chance, B.; Maehly, A. *Assay of Catalases and Peroxidases*; Wiley: Hoboken, NJ, USA, 1955; Volume 2, pp. 764–775. [CrossRef]
48. Asada, K. Production and scavenging of active oxygen in photosynthesis. In *Photo-Inhibition*; Elsevier: Amsterdam, The Netherlands, 1987.
49. Beauchamp, C.; Fridovich, I. Superoxide dismutase: Improved assays and an assay applicable to acrylamide gels. *Anal. Biochem.* **1971**, *44*, 276–287. [CrossRef]
50. Dhindsa, R.S.; Plumb-Dhindsa, P.; Thorpe, T.A. Leaf senescence: Correlated with increased levels of membrane permeability and lipid peroxidation, and decreased levels of superoxide dismutase and catalase. *J. Exp. Bot.* **1981**, *32*, 93–101. [CrossRef]
51. Chen, J.; Li, H.; Yang, K.; Wang, Y.; Yang, L.; Hu, L.; Liu, R.; Shi, Z. Melatonin facilitates lateral root development by coordinating *PAO*-derived hydrogen peroxide and *Rboh*-derived superoxide radical. *Free Radic. Biol. Med.* **2019**, *143*, 534–544. [CrossRef] [PubMed]
52. Kereszt, A.; Li, D.; Indrasumunar, A.; Nguyen, C.D.; Nontachaiyapoom, S.; Kinkema, M.; Gresshoff, P.M. *Agrobacterium rhizogenes*-mediated transformation of soybean to study root biology. *Nat. Protoc.* **2007**, *2*, 948–952. [CrossRef] [PubMed]
53. Landi, M. Commentary to: "Improving the thiobarbituric acid-reactive-substances assay for estimating lipid peroxidation in plant tissues containing anthocyanin and other interfering compounds" by Hodges et al., Planta (1999) 207: 604–611. *Planta* **2017**, *245*, 1067. [CrossRef] [PubMed]
54. Wang, W.; Wang, X.; Liu, X.; Wang, Y.; Li, Y.; Hussain, S.; Jing, X.; Chen, S.; Wang, S. *AtAUEs*, a Small Family of ABA Up-Regulated EAR Motif-Containing Proteins Regulate ABA Responses in Arabidopsis. *Plants* **2024**, *13*, 3282. [CrossRef] [PubMed]

plants

MDPI

Review

Multi-Scale Remote-Sensing Phenomics Integrated with Multi-Omics: Advances in Crop Drought–Heat Stress Tolerance Mechanisms and Perspectives for Climate-Smart Agriculture

Xiongwei Liang [1,2], Shaopeng Yu [1], Yongfu Ju [1], Yingning Wang [1,2] and Dawei Yin [3,*]

1 Cold Region Wetland Ecology and Environment Research Key Laboratory of Heilongjiang Province, Harbin University, Harbin 150086, China; liangxiongwei007@163.com (X.L.); shaopengyu1972@163.com (S.Y.); juyongfu@163.com (Y.J.); 14b927013@hit.edu.cn (Y.W.)
2 State Key Laboratory of Urban Water Resource and Environment, Harbin Institute of Technology, Harbin 150086, China
3 College of Agricultural Science, Heilongjiang Bayi Agricultural University, Daqing 163319, China
* Correspondence: yindazhiyindawei@126.com

Abstract

Climate change is intensifying the co-occurrence of drought and heat stresses, which substantially constrain global crop yields and threaten food security. Developing climate-resilient crop varieties requires a comprehensive understanding of the physiological and molecular mechanisms underlying combined drought–heat stress tolerance. This review systematically summarizes recent advances in integrating multi-scale remote-sensing phenomics with multi-omics approaches—genomics, transcriptomics, proteomics, and metabolomics—to elucidate stress response pathways and identify adaptive traits. High-throughput phenotyping platforms, including satellites, UAVs, and ground-based sensors, enable non-invasive assessment of key stress indicators such as canopy temperature, vegetation indices, and chlorophyll fluorescence. Concurrently, omics studies have revealed central regulatory networks, including the ABA–SnRK2 signaling cascade, HSF–HSP chaperone systems, and ROS-scavenging pathways. Emerging frameworks integrating genotype × environment × phenotype (G × E × P) interactions, powered by machine learning and deep learning algorithms, are facilitating the discovery of functional genes and predictive phenotypes. This "pixels-to-proteins" paradigm bridges field-scale phenotypes with molecular responses, offering actionable insights for breeding, precision management, and the development of digital twin systems for climate-smart agriculture. We highlight current challenges, including data standardization and cross-platform integration, and propose future research directions to accelerate the deployment of resilient crop varieties.

Keywords: drought–heat stress; remote sensing phenomics; multi-omics integration; ABA signaling pathway; climate-smart agriculture

1. Introduction

Climate change is exacerbating abiotic stresses such as drought and heat, which increasingly occur in combination and severely threaten global food security [1]. Drought and heat stress lead to extensive yield losses in major crops by disrupting plant physiological processes—drought limits water availability for photosynthesis, while heat waves damage enzymes and cellular structures [2]. For instance, elevated temperatures above optimal levels can suppress photosynthesis, induce excessive evapotranspiration, and cause oxidative damage, together reducing crop productivity [3,4]. Severe drought remains a major challenge for cereal and oilseed production. In critical crops like cereals and oilseeds, drought events have been shown to lead to significant yield reductions, as highlighted in assessments by the FAO and IPCC [1,5], it is consistently identified as one of the most damaging climate extremes for agriculture, with substantial yield reductions reported across diverse crops and regions, depending on crop type, geography, and stress intensity [6]. Each additional degree of warming is projected to increase the likelihood and severity of such compound drought–heat events, heightening the risk that photosynthetic gains cannot compensate for hydraulic and thermal limitations. Developing climate-resilient varieties capable of withstanding drought and heat stress is therefore a critical priority for safeguarding future global food and oilseed production.

Global economic costs of stress-related yield losses. Multi-country datasets show drought disasters reduce national cereal output by ~9–10% on average, with heat waves adding further losses and compounding impacts in low-latitude breadbaskets [7]. Meta-analyses estimate per-degree-warming yield penalties of ~−6% (wheat), −7% (maize), and ~−3% (rice/soybean) absent adaptation [7,8]. Accounting for price responses and supply chains, FAO assessments attribute tens of billions of USD in agricultural losses annually to weather extremes, with compound drought–heat events increasingly dominant [1,5].

Abiotic stress encompasses diverse non-living constraints, with drought and heat being the most critical for global agriculture. Drought reduces stomatal conductance and leaf water potential, triggering osmotic adjustment (e.g., proline, soluble sugars) and hydraulic remodeling (root allocation, aquaporins) [9,10]. Heat adds thermal load that can overwhelm evaporative cooling, elevating canopy and leaf temperatures and promoting reactive oxygen species (ROS) accumulation [11,12]. Under combined drought–heat conditions, plants must balance carbon gain against hydraulic safety and thermal homeostasis. Tolerance emerges when adaptive responses stabilize key functional traits—such as canopy temperature depression (CTD), chlorophyll fluorescence (Fv/Fm), stem water potential, and membrane integrity—that are quantifiable by remote sensing (thermal, red-edge/PRI, solar-induced fluorescence, SIF) and linkable to omics signatures (e.g., ABA–SnRK2 signaling, HSF–HSP chaperones, antioxidant pathways) [13,14]. Notably, co-stress responses are often emergent and non-additive—due to hormone crosstalk, amplified ROS bursts when evaporative cooling is constrained, elevated proteostasis demands (HSF–HSP), and hydraulic–thermal feedbacks—producing greater injury than either stress alone and justifying explicit phenotyping under synchronized water and thermal deficits [2,13–15]. This integrated view yields measurable indicators for selection and management under field variability [16].

At the molecular level, several core signaling pathways mediate plant responses to drought and heat. The abscisic acid (ABA) signaling pathway is central in drought tolerance: under water deficit, ABA accumulates and initiates a signaling cascade via *pyrabactin resistance1/PYR-like* (PYR/PYL) receptors, protein phosphatase 2C (*PP2C*) inactivation, and SnRK2 kinase activation, which leads to expression of numerous drought-responsive genes and stomatal closure [17,18]. Small-molecule studies have confirmed the pivotal role of ABA pathways; for example, synthetic ABA receptor agonists that mimic ABA can significantly improve drought resilience in crops by activating downstream stress responses. In parallel, the heat shock factor–heat shock protein (HSF–HSP) network is crucial for heat tolerance. Upon high temperatures, heat shock transcription factors (e.g., *HsfA1, HsfA2*) are released from chaperones and upregulate heat shock proteins (HSP70, HSP90, HSP101, etc.), which function as molecular chaperones to prevent protein unfolding and aggregation [19,20]. This HSF–HSP system equips cells with thermotolerance by stabilizing proteins and refolding denatured enzymes during heat stress. Notably, drought and heat signaling pathways cross-talk: ABA-responsive element binding factors can also regulate genes involved in heat resistance [21], and heat stress can elevate ABA levels that in turn modulate stress-responsive genes [22]. Both stresses also converge on ROS signaling—drought and heat induce accumulation of ROS like hydrogen peroxide, which at moderate levels serve as secondary messengers to trigger defense genes, but at high levels cause oxidative damage if not scavenged by antioxidant enzymes [23,24]. Therefore, an interplay between hormonal signals (ABA and others), chaperone systems (HSPs), and redox homeostasis underpins the physiological and molecular mechanisms of drought–heat tolerance.

To build climate-resilient crop systems, plant scientists are exploiting new technologies to better understand and harness these complex tolerance mechanisms. Traditional breeding for drought or heat tolerance has been challenging due to the multigenic nature of these traits and difficulty of phenotyping under field stress conditions. However, recent advances in remote sensing-based high-throughput phenotyping (phenomics) and multi-omics profiling (genomics, transcriptomics, proteomics, metabolomics) are revolutionizing how we study plant stress responses [25]. Phenomics enables the rapid measurement of plant traits (morphology, physiology, performance) at scale using sensors and imaging, providing rich datasets on how different genotypes perform under stress in realistic environments. Meanwhile, omics technologies allow dissection of the genetic and biochemical underpinnings of stress tolerance by surveying the genome and downstream molecules (RNA, proteins, metabolites) comprehensively. Integrating phenomics and omics information—effectively connecting observable traits to specific genes, proteins, and metabolites—offers a powerful approach to identify stress-adaptive mechanisms and candidate tolerance genes. For example, high-throughput field imaging can pinpoint genotypes with cooler canopies or sustained greenness under drought, and genome-wide association studies (GWAS) can then map genomic loci linked to those phenotypic traits [26]. Likewise, transcriptomic and proteomic analyses of stress-exposed plants reveal the activation of protective pathways (like the ABA–SnRK2 or HSF–HSP modules), which can be correlated with observed stress tolerance phenotypes. By connecting these scales—from satellite pixels capturing field performance down to proteins and genes governing stress responses—researchers are developing holistic strategies to breed and manage crops for improved drought–heat resilience. This review provides an overview of these advances in multi-scale phenomics and multi-omics integration and discusses how they are being translated into climate-smart agriculture practices (Figure 1). Figure 1 operationalizes the "pixels-to-proteins" pipeline: spectral/thermal signals map to field-level traits (e.g., CTD, PRI, SIF), align with tolerance phenotypes, and connect—via GWAS/TWAS and targeted metabolomics—to causal pathways (ABA/HSF–HSP/ROS), closing the loop to breeding priorities and decision thresholds. This schematically illustrates how remote sensing and omics layers converge from canopy to molecular scale, emphasizing the integrative nature of this review.

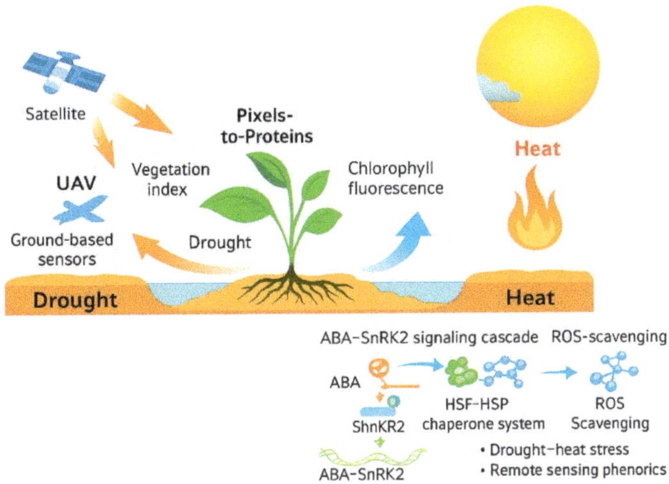

Figure 1. Integrating Remote Sensing and Multi-Omics to Dissect Crop Tolerance to Combined Drought and Heat.

2. Multi-Scale Remote-Sensing Phenomics

High-throughput plant phenomics leverages remote sensing technologies across multiple scales—from satellites imaging entire regions to drones and ground-based sensors imaging individual plots or plants—to capture crop performance under stress in a rapid, quantitative manner. In the context of drought and heat stress tolerance, multi-scale remote-sensing phenomics provides invaluable traits such as canopy temperature, vegetation indices, and fluorescence signals that serve as proxies for plant health and stress levels [27,28]. This section details the major phenotyping platforms and sensors, the key indicators they measure, and the role of advanced data analytics (AI and machine learning) in extracting stress-relevant information. Schematic of a phenotyping pipeline that integrates satellite, drone, and ground-based imaging to quantify crop performance under drought and heat. Key indicators—canopy temperature, vegetation indices, and fluorescence signals—are extracted and analyzed with AI/ML to derive stress-relevant traits (Figure 2).

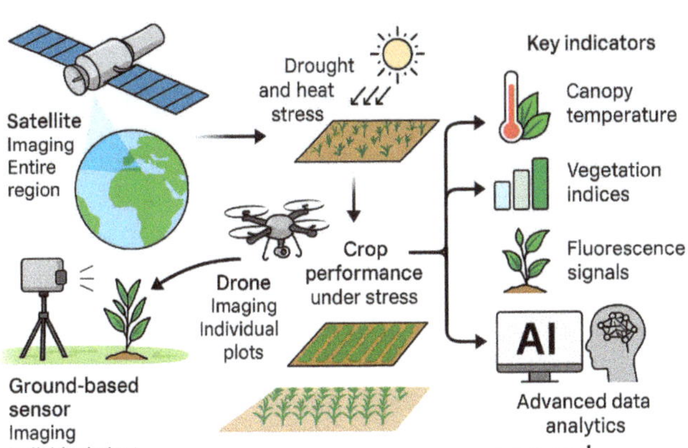

Figure 2. Satellite-to-Ground Phenotyping Workflow for Heat and Drought Monitoring.

2.1. Phenotyping Platforms: Satellite, Airborne, UAV, and Ground-Based

Satellite remote sensing offers a macro-scale perspective on crop performance, proving essential for regional drought monitoring and yield forecasting. Operational sensors such as MODIS, Landsat, and Sentinel-2 capture vegetation health through multispectral imagery and thermal infrared measurements. For example, satellite-derived drought indices have monitored severity and crop impacts over multi-decade periods [29,30]. A study from 2001 to 2020 utilized MODIS data to map drought stress in Sichuan Province croplands via vegetation indices, revealing spatiotemporal patterns of impacts [31]. Satellites' key advantage is comprehensive coverage, regularly scanning vast agricultural regions to detect stress hotspots in real fields and bridge-controlled experiments with farm-scale outcomes. Limitations include moderate spatial resolution (10–30 m for Sentinel/Landsat, coarser for MODIS) and reliance on clear skies, as clouds obstruct optical sensors. Nevertheless, satellites are increasingly paired with field trials to assess genotype performance under authentic stress, with emerging radar technologies (e.g., Sentinel-1) addressing cloud issues. Typical GSD is 10–30 m (S2/L8) with 5–16 d revisit; dominant failure modes include cloud contamination and mixed pixels. BRDF correction and cross-dates normalization are recommended for time-series analyses.

Airborne platforms, including manned aircraft or helicopters with sensors, provide an intermediate scale with sub-meter resolution over trial fields, covering areas beyond ground systems' reach. They have aided breeding programs by phenotyping drought responses in large plots of wheat or beans [32]. One investigation deployed both unmanned aerial systems and high-altitude aircraft to acquire high-resolution multispectral imagery for dry bean phenotyping, correlating canopy cover and temperature with drought tolerance rankings [33]. Airborne methods support custom sensor configurations, such as hyperspectral imagers for detailed stress detection, and have been valuable for capturing data in research stations, especially pre-drone era. However, they are logistically demanding, expensive, and less flexible in timing compared to UAVs, requiring specialized operations.

Unmanned aerial vehicles (UAVs), or drones, have become central to field phenomics due to their high flexibility, fine resolution (cm-level), and relative affordability. UAVs accommodate multispectral cameras, thermal sensors, LiDAR, and hyperspectral imagers, allowing on-demand flights to target growth stages or stress onsets. In drought and heat studies, they measure NDVI, canopy temperature, plant height, and biomass across numerous plots per flight [8,34]. For wheat under drought, UAV NDVI and temperature imagery identified stay-green phenotypes and cooler canopies linked to better yields [35,36]. Studies show UAVs achieve similar or superior precision to manual measurements for NDVI, capturing whole-plot variation [37]. Under heat, thermal UAVs pinpoint efficient transpirational cooling, suggesting deeper roots [38]. Applications extend to rice, where drone RGB tracked senescence dynamics tied to genetics [39], and sorghum, where biomass data informed GWAS for resilience loci [40]. UAVs democratize phenomics, enabling small programs to collect high-throughput data, though regulatory and data processing challenges persist.

Ground-based phenotyping systems, ranging from handheld devices and portable carts to fixed installations, deliver the highest precision and support sensors unsuitable for aerial use, like chlorophyll fluorometers or gas-exchange analyzers. Phenomobiles (sensor-mounted vehicles) scan plots for reflectance, fluorescence, or architecture; one equipped with multispectral and thermal sensors identified drought-tolerant wheat via cooler temperatures [41]. In controlled environments, automated imaging captures wilting kinetics or growth under imposed stresses [42]. A low-cost greenhouse system with ML analyzed drought tolerance in grasses [43]. Ground measurements validate remote indices, e.g., correlating NDVI to leaf water potential or photosynthesis [44]. Despite lower throughput and labor needs, they complement by providing calibration for multi-scale integration.

Collectively, these platforms synergize satellites for breadth, airborne/UAVs for detail, ground for validation, fostering holistic crop assessments under drought and heat. As illustrated in Figure 1, multi-scale phenotyping platforms enable complementary data acquisition for comprehensive crop stress assessment (Figure 3) [45]. Practical examples highlight the power of these platforms. For instance, MODIS time-series vegetation indices were applied to monitor stay-green expression on wheat, revealing delayed senescence phenotypes that aligned with yield stability under terminal drought [46,47]. Such coarse-resolution monitoring provides a regional context and identifies hotspots that can later be examined with UAV flights. UAV thermal and hyperspectral imaging then resolve canopy-level responses—such as midday canopy cooling (CTD) and pigment dynamics (PRI/red-edge)—at the plot scale [48], while ground-based measurements provide calibration/validation (Cal/Val) through direct assessments of leaf water potential, stomatal conductance, or chlorophyll fluorescence (Fv/Fm) [49]. Together, these platforms enable a multi-scale pipeline: satellites for breadth and trend detection, UAVs for high-resolution trait dissection, and ground systems for mechanistic fidelity, bridging discovery from landscape variability to genotype-level selection.

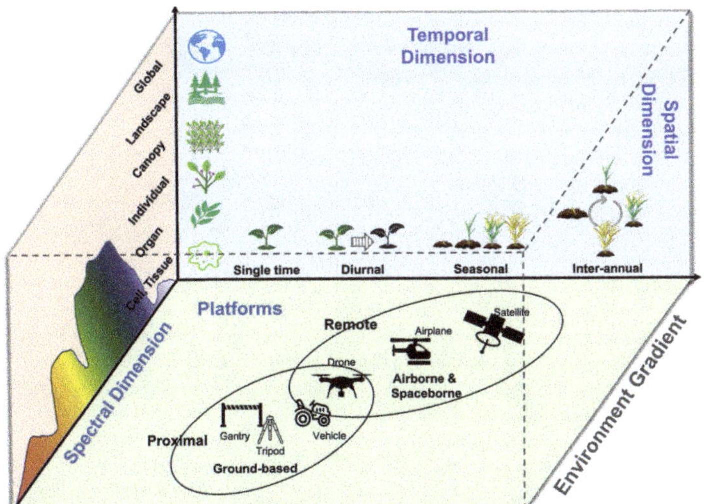

Figure 3. Schematic overview of proximal and remote sensing platforms in plant phenomics, including satellite, airborne, UAV, and ground-based systems [45]. Copyright 2022 Elsevier.

2.2. *Key Phenotypic Indicators of Drought and Heat Stress*

Remote-sensing phenomics relies on indicators that signal plant physiological status under stress, enabling early detection and genotypic screening. These metrics—spectral, thermal, and structural—proxy traits like water-use efficiency and photosynthesis. Multi-indicator integration enhances accuracy, as no single metric captures complexity. Below, we outline key indicators, their basis, applications, and genetics.

Normalized Difference Vegetation Index (NDVI), (NIR − Red)/(NIR + Red), correlates with biomass and chlorophyll. Under drought/heat, NDVI drops from senescence; "stay-green" genotypes sustain it, linking to yields in cereals [50,51]. NDVI proxies adaptations like rooting; QTLs control it in wheat/maize under drought [52]. Derivable across platforms, it is foundational but saturates in dense canopies.

Canopy temperature (CT), via thermal sensors, reflects transpirational cooling. Drought closes stomata, raising CT; cooler canopies indicate tolerance [53]. UAV mapping in wheat ties lower CT to roots/yields [54]. Canopy temperature depression (CTD) normalizes; CIMMYT selects via infrared for stressed environments [55]. Sensitive to microclimate, CT benefits contextual fusion.

Chlorophyll fluorescence (Fv/Fm = (Fm − F0)/Fm) assesses PSII yield, declining from photoinhibition [56]. Fluorometry identifies heat-tolerant wheat sustaining Fv/Fm [57]. Solar-induced fluorescence (SIF) proxies canopy photosynthesis, dropping in drought.

Advanced indices target specifics: Photochemical Reflectance Index (PRI = (R531 − R570)/(R531 + R570)) detects xanthophyll shifts for efficiency; NDWI estimates hydration. Red-edge senses chlorophyll early. Hyperspectral in stressed wheat outperformed NDVI for yields [58]. Table 1 lists some commonly used indices (e.g., NDVI, PRI, Normalized Difference Water Index) along with their formulae and what aspect of stress they reflect. By using hyperspectral sensors (which capture continuous spectra), researchers can calculate dozens of indices or even apply machine learning to find the most predictive wavelength features for stress tolerance [59]. For instance, in maize, machine learning on hyperspectral data identified specific spectral bands linked to drought susceptibility, which aligned with known pigment and water absorption features. Table 1 provides an overview of common platforms and sensor types, their typical spatial/temporal resolution, and example stress indicators obtainable from each. Table 1 contrasts major platforms (satellite, UAV, ground) by resolution/revisit, strengths, limitations, and Cal/Val essentials; Table 2 lists representative indicators and what aspect of stress they reflect.

Table 1. Comparative overview of phenotyping platforms for drought–heat detection.

Platform	Resolution	Key Indicators (Examples)	Strengths	Limitations	Representative Applications in Drought–Heat Detection
Satellite (e.g., Sentinel-2, MODIS, Landsat)	Low (10–500 m); revisit 5–16 d	NDVI, LST, NDWI	Regional coverage; multi-decadal time series; free/low cost	Cloud dependence; coarse resolution; limited genotype inference	Sentinel-2 NDVI & LST mapped regional drought severity and yield decline in wheat [60]; MODIS time-series revealed stay-green phenotypes in Sichuan wheat [31]
UAV/Drones	High (cm-level, on-demand)	CTD, PRI, hyperspectral reflectance, RGB/LiDAR	Plot-level precision; captures transient stress	Flight regulations; short battery life; heavy data pipelines	UAV NDVI + canopy temperature identified drought-tolerant wheat genotypes with cooler canopies [61]; UAV hyperspectral (red-edge, PRI) detected pigment dynamics under heat stress in maize [62]
Ground-based sensors (tripods, gantries, handhelds)	Very high (mm–cm)	Fv/Fm, PRI, leaf water potential	Highest mechanistic fidelity; essential for Cal/Val	Labor-intensive; very limited area	Porometry + leaf water potential validated UAV CTD signals in wheat [63]; chlorophyll fluorescence (Fv/Fm) distinguished heat-tolerant rice cultivars [49]
Proximal phenotyping vehicles (e.g., carts, tractor-based)	Medium (m-level)	Multispectral/LiDAR imaging	Higher throughput than handheld; bridges plots to fields	Logistic constraints; less scalable than UAVs or satellites	Phenomobile multispectral imaging quantified soybean canopy traits under drought, supporting QTL mapping [64]

Table 1 highlights the coverage–precision–scalability trade-off across platforms. Satellites provide unmatched coverage for early warning but lack genotype-level detail. UAVs capture fine-scale, transient stress signals yet face regulatory and processing limits. Ground systems offer mechanistic fidelity and serve as the Cal/Val standard but cover small areas, while proximal vehicles bridge UAV and ground scales. In practice, integrating satellites for breadth, UAVs for plot-level traits, and ground/proximal sensors for validation provides a robust pipeline for drought–heat phenotyping.

Where appropriate, satellite-derived indices can be downscaled and calibrated to plot scale using UAV subsets, enabling trial-wide monitoring when flight capacity is limited. For biomass and senescence traits, UAV RGB/LiDAR quantifies fractional cover and canopy height, and stress accelerates senescence. In wheat, image segmentation produced indices associated with heat QTLs [59], while rice time-series dynamics revealed drought-related genetic effects [65]. Combining spectral greenness with canopy temperature—e.g., NDVI + canopy temperature depression (CTD)—improves screening accuracy [66,67]. As illustrated in Figure 4, joint temporal trajectories of canopy temperature and greenness distinguish drought-avoidance and stay-green strategies in wheat [68].

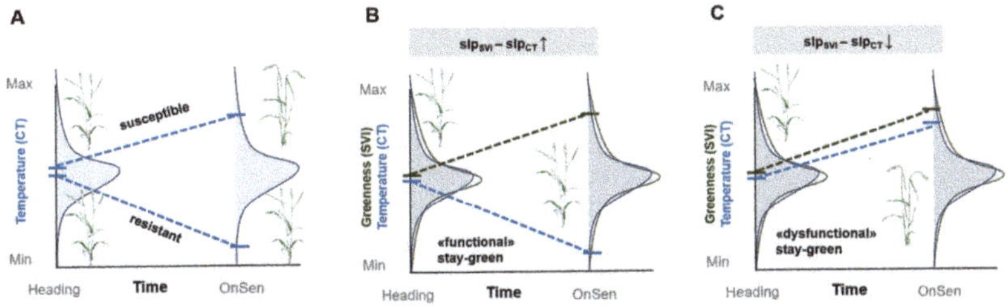

Figure 4. Contrasting stay-green strategies under drought–heat stress: effects on canopy temperature, photosynthesis, and chlorophyll dynamics [68]. Copyright 2021 Elsevier.

Figure 4 illustrates contrasting physiological strategies influencing drought–heat resilience. Panel A shows that tolerant genotypes often maintain lower canopy temperatures (more negative canopy temperature depression, CTD) due to sustained transpiration, whereas susceptible lines exhibit warmer canopies and accelerated stress onset. Panel B depicts differences in photosynthetic efficiency: stay-green lines sustain higher rates over time, while sensitive lines experience sharp declines during heading and grain filling. Panel C highlights chlorophyll dynamics, where delayed senescence ("stay-green") preserves green leaf area and supports assimilate supply under stress, in contrast to rapid yellowing in susceptible lines. Collectively, Figure 4 underscores how canopy cooling, photosynthetic stability, and chlorophyll retention serve as complementary mechanisms underpinning stress tolerance, and how these can be captured through integrated remote-sensing phenomics. This figure emphasizes that no single trait ensures resilience: canopy cooling, photosynthetic stability, and delayed senescence interact to determine yield outcomes. Importantly, each trait also has trade-offs—prolonged stay-green may delay remobilization under mild stress, while high transpirational cooling can increase water demand.

2.3. *AI and Machine Learning for Trait Extraction and Sensor Integration*

The surge in high-throughput phenomics produces expansive datasets, encompassing hyperspectral imagery, temporal sequences, and outputs from diverse sensors, demanding advanced tools for insightful analysis [69,70]. Artificial intelligence (AI) and machine learning (ML) have become essential in deriving valuable insights from these intricate datasets, enabling the identification of stress-associated traits in crops facing drought and heat challenges. Spanning supervised and unsupervised frameworks, these approaches range from traditional regression techniques to cutting-edge deep learning models, effectively tackling core phenotyping hurdles.

For trait extraction and image processing, convolutional neural networks (CNNs) stand out in handling visual inputs to measure morphological and physiological attributes. CNNs proficiently segment plant structures, identify senescence indicators, or categorize stress manifestations with superior accuracy. One exemplary case utilized a deep learning framework on UAV-captured maize leaf images to enumerate microscopic bulliform cells, facilitating large-scale evaluation of drought-adaptive features that were formerly confined to labor-intensive microscopy [71]. Likewise, ML-based segmentation in heat-stressed wheat quantified foliar senescence progression, linking these indices to yield resilience and aiding genomic breeding strategies [72].

In predictive analytics, ML harnesses sensor features to anticipate stress effects and yield projections. Methods like random forests and partial least squares regression dissect hyperspectral profiles to infer biochemical markers, such as chlorophyll and hydration levels, serving as surrogates for stress dynamics [73]. Ensemble frameworks in wheat under drought, amalgamating various spectral metrics, forecasted grain output with high fidelity, offering noninvasive aids for preliminary selection in breeding programs [74]. Moreover, classification models delineate stress categories—e.g., differentiating drought from nutritional deficits—via integrated thermal and spectral data [75].

ML enhances sensor fusion by merging disparate data modalities, boosting trait precision. Multivariate algorithms blend hyperspectral reflectance (reflecting canopy vitality) with thermal scans (depicting evaporative cooling) to estimate traits like stomatal conductance in drought-affected wheat [76,77]. Such integration delineates comprehensive stress profiles, surpassing isolated sensor utilities. Temporal scrutiny, via recurrent neural networks, illuminates dynamic patterns, including NDVI degradation post-drought initiation, unveiling attributes like recuperative capacity.

Beyond data handling, ML reveals emergent markers; for instance, importance assessments in soybean hyperspectral analyses pinpointed mid-infrared wavelengths tied to foliar hydration, refining stress metrics [74]. Holistic workflows, where AI-derived traits inform genome-wide association studies (GWAS) in sorghum, expedite resilience gene identification [78].

Ultimately, AI and ML convert unprocessed phenomics inputs into biologically meaningful traits, amplifying stress phenotyping for breeding resilient varieties. Progress in interpretable AI will better synchronize models with underlying biology, mitigating opacity issues and broadening utility in variable field settings.

2.4. Sensor Integration and High-Throughput Stress Phenotyping

An important trend in field phenotyping for climate resilience is the integration of multiple sensor modalities to capture the different dimensions of plant stress response. No single measurement can fully characterize a complex trait like drought tolerance; thus, researchers increasingly combine, for example, spectral and thermal imaging, or imaging and soil sensors, to obtain a holistic view. A prime illustration is the integration of hyperspectral and thermal UAV imagery in wheat: by simultaneously measuring canopy reflectance (which can indicate chlorophyll content, nitrogen status, etc.) and canopy temperature (indicating water status), one study was able to identify genotypes that maintained both high greenness and low temperature under drought—a desirable combination [79]. This dual-sensor approach led to the discovery of genomic loci associated with those combined traits, underscoring how sensor integration can sharpen the phenotype definition for genetic analysis [80].

Another example is combining fluorescence with imaging. New field phenotyping rigs can measure pulse-modulated chlorophyll fluorescence alongside spectral reflectance. In a drought experiment on maize, simultaneous recording of fluorescence (Fv/Fm) and

NDVI provided insight that some genotypes had high NDVI (stayed green) but low Fv/Fm (photoinhibition) under stress, whereas a few maintained both—the latter were truly tolerant [81]. Without measuring fluorescence, one might wrongly assume all green genotypes are fine; the integrated sensing revealed subtler differences in photosynthetic function.

Beyond plant-mounted sensors, integration often includes environmental sensors (weather stations, soil moisture probes) to precisely characterize the stress each plot experiences. This contextual data is crucial: for instance, linking a plot's canopy temperature to its soil moisture and atmospheric vapor pressure deficit helps differentiate whether a high temperature is due to genotype or simply due to variation in micro-environment. In phenomics experiments, networks of soil moisture sensors or IoT-based systems (like the iPOT system for automated drought imposition [82]) ensure each genotype's stress level is known, allowing more accurate association of phenotypes with genetic differences.

The concept of sensor fusion extends to merging data across scales. A notable approach is calibrating coarser-resolution satellite indices using finer drone data. For instance, if a breeding program can only obtain satellite NDVI but at 10 m resolution (covering many plots), they might use a drone on a subset of fields to develop a calibration model translating satellite signals into plot-level values or stress ratings. This multi-scale synergy can enable larger-scale monitoring of trials when drone coverage is limited.

Integration also implies combining aboveground and belowground phenotyping. While remote sensing focuses on aboveground traits (canopy, leaves), root traits are pivotal in drought tolerance. Some studies have integrated shovelomics or root imaging (e.g., via soil coring or ground-penetrating radar) with aboveground phenotypes. One study mapped root architectural traits in wheat and compared them with UAV-based drought indices, finding that deeper roots were associated with cooler canopies and higher yield in drought, tying the belowground mechanism to the aboveground signature [83,84]. Although root phenotyping remains challenging, these integrated efforts highlight how multi-faceted true drought tolerance is.

In summary, multi-scale remote-sensing phenomics is most powerful when multiple complementary sensors are used and their data combined. By capturing various physiological signals (greenness, temperature, fluorescence, structure) and integrating them with environmental measurements, researchers can obtain a comprehensive stress phenotype for each genotype. This improves the identification of truly resilient genotypes and the genetic loci underlying complex traits, as demonstrated in numerous high-throughput studies across crops [85,86]. The next sections will transition from phenotypes to the multi-omics underpinnings of drought–heat tolerance and then discuss how we integrate across genotype (G), environment (E), and phenotype (P) in emerging "pixels-to-proteins" frameworks.

Table 2 compares widely used indicators—NDVI, canopy temperature depression (CTD), Fv/Fm, plant height, and stay-green index—covering their physiological basis, measurement modality, and typical correlations with yield under combined drought–heat stress. Across crops, NDVI and CTD show consistently strong associations with yield maintenance, making them robust, scalable proxies for field screening, whereas Fv/Fm provides an early mechanistic signal at smaller plot scales. These patterns, drawn from multiple studies, offer practical guidance for trait prioritization.

Case study—compound drought–heat detection using the platforms in Table 1. Design: split-plot (well-watered vs. deficit) × passive heating (+2–4 °C at midday), $n \approx 24$ plots (≥4 genotypes). Sensing: UAV thermal to compute canopy temperature depression ($CTD = T_{air} - T_{canopy}$) and RGB/hyperspectral for structure/greenness at solar noon and late afternoon on stress days; co-sample leaf water potential and stomatal conductance within ±2 h. Thresholds: flag compound stress when CTD falls to ≤+1 °C at noon and the stay-green/greenness signal declines on ≥2 flights; confirm with $\Psi_{leaf} < -1.5$ MPa. Outputs: (i) a selection index combining CTD with greenness dynamics; (ii) prescriptions (pre-irrigation or temporary shading at booting–flowering); (iii) validation with yield components and membrane stability. (Thresholds require site-year calibration.) Threshold and flight-time choices follow recent UAV thermal drought-monitoring practice and calibrated ΔT/CWSI ranges in winter wheat; midday/early-afternoon flights maximize discriminatory power [46,87].

Table 2. Phenotypic indicators of drought–heat tolerance: physiology, measurement, correlation with yield (R), and representative applications. CTD = $T_{air} - T_{canopy}$ (larger CTD = cooler canopy).

Indicator	Physiological Basis	Measurement Type	Correlation with Yield Under Stress	Crops Reported	Representative Findings
NDVI (Normalized Difference Vegetation Index)	Canopy greenness/biomass	Multispectral reflectance	High (R = 0.6–0.8)	Wheat, Maize, Rice	NDVI decline correlated with yield loss in maize under South Asian drought trials [88]
CTD (Canopy Temperature Depression = $T_{air} - T_{canopy}$)	Transpirational cooling	Thermal imaging	High (R = 0.6–0.8)	Wheat, Sorghum	Cooler canopies identified tolerant wheat lines across multi-location trials [68]
Fv/Fm (maximum quantum efficiency of PSII)	Photosystem II efficiency	Chlorophyll fluorescence sensor	Moderate (R = 0.4–0.6)	Rice, Soybean	Heat-tolerant genotypes maintain higher Fv/Fm under stress [89]
Plant height	Growth/vigor proxy	RGB or LiDAR	Variable (context-dependent)	Maize, Sorghum	Height alone inconsistent; combined with NDVI improves prediction [90]
Stay-green index	Delayed senescence	NDVI time-series	Moderate to High (R = 0.5–0.7)	Wheat, Pearl Millet	Stay-green genotypes showed improved grain filling under heat stress [89]

Note: R ranges are indicative and context dependent. We use CTD = $T_{air} - T_{canopy}$ (larger CTD = cooler canopy → typically positive correlation with yield). Thermal flights should be made near solar noon; NDVI can saturate at high LAI; thresholds require site-year calibration.

3. Multi-Omics Dissection of Drought–Heat Tolerance

Understanding and improving crop tolerance to drought and heat requires delving into the biological mechanisms at the genomic, transcriptomic, proteomic, and metabolomic levels. High-throughput "omics" approaches enable comprehensive dissection of these mechanisms by identifying the genes, gene expression changes, proteins, and metabolites that differentiate stress-tolerant vs. susceptible genotypes. In recent years, multi-omics studies have greatly advanced our knowledge of drought–heat stress responses, often revealing complex networks of signaling pathways and protective molecules involved in tolerance [91,92]. This section summarizes key contributions of each omics level—genomics, transcriptomics, proteomics, and metabolomics—to unraveling drought/heat tolerance, with emphasis on some well-known pathways like ABA–SnRK2, HSP–HSF, and ROS scavenging, and highlights how these layers interact (Figure 5).

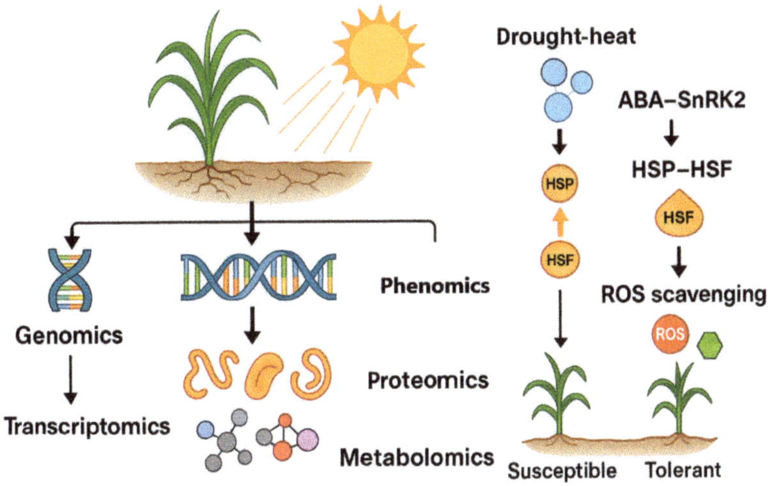

Figure 5. Multi-Omics Framework for Drought–Heat Tolerance: From Genomes to Metabolites and Key Protective Pathways.

3.1. *Genomic Insights: QTL Mapping and GWAS for Stress Tolerance Genes*

Genomics serves as the foundational blueprint for elucidating genetic variation underpinning crop tolerance to drought and heat stresses, enabling the identification of key loci through quantitative trait locus (QTL) mapping in biparental populations and genome-wide association studies (GWAS) in diverse germplasm panels [93]. These methodologies have systematically uncovered genomic regions governing yield stability and physiological traits under abiotic constraints across major crops, facilitating targeted breeding for enhanced resilience.

In wheat, QTL analyses have delineated multiple loci influencing grain yield under combined drought and heat, such as a prominent QTL on chromosome 4B that bolsters photosynthetic efficiency and root architecture [94]. Introgression of drought-tolerant QTLs from wild emmer relatives has validated their efficacy in elite cultivars [95], while GWAS incorporating phenomics traits like normalized difference vegetation index (NDVI) and canopy temperature has pinpointed markers on chromosomes 2A and 3B associated with cooler canopies, aligning with established drought-responsive genes [96].

Rice genomics highlights *qDTY* QTLs, including *qDTY12.1* on chromosome 12, which exert substantial effects on yield under upland drought and have been deployed via marker-assisted selection [97]. Recent GWAS leveraging UAV-derived dynamic drought responses has revealed novel loci modulating senescence kinetics [39]. For heat tolerance, genes like *TT1* (*thermotolerance1*) maintain pollen viability at elevated temperatures.

Maize studies, including large-scale GWAS, emphasize loci on chromosomes 7 and 8 enriched in dehydration-responsive element-binding (DREB) transcription factors and heat shock proteins in tolerant lines [98]. QTLs for anthesis-silking interval co-localize with abscisic acid (ABA) biosynthesis genes, mitigating reproductive delays under stress.

Sorghum, a resilient C4 crop, features cloned stay-green QTLs (*Stg1*–*Stg4*), with *Stg3* encoding an NAC transcription factor that delays senescence [99]. GWAS on diverse panels using drone phenotyping identified over 200 loci linked to biomass traits under drought.

In legumes like soybean and chickpea, GWAS has associated markers with canopy wilting and root traits, implicating stomatal regulation and ABA signaling genes [100]. Cotton heat tolerance QTLs near small heat shock proteins (HSPs) and transcription factors underscore fiber yield stability [101].

Collectively, these genomic investigations recurrently implicate core pathways: ABA signaling components (e.g., PYR/PYL receptors, SnRK2 kinases), HSP chaperones, transcription factor families (DREB, NF-Y, WRKY, NAC), and antioxidant enzymes [102,103]. Functional validation, such as overexpressing ABA receptors within QTL intervals, confirms enhanced drought survival [104].

Advancements extend to genomic selection (GS), which utilizes genome-wide markers to predict breeding values for polygenic traits. Integrating phenomics data into GS models, as in wheat yield forecasting under stress, improves predictive accuracy by capturing intermediate phenotypes [105]. Thus, genomics yields candidate genes and markers, bridging observable traits to molecular mechanisms, though functional elucidation necessitates complementary omics layers.

3.2. Transcriptomic Responses: Gene Expression Networks Under Stress

Transcriptomics, through RNA sequencing (RNA-Seq) or microarray profiling of mRNA expression, elucidates how plants reprogram gene expression in response to drought and heat stresses, highlighting differences between tolerant and susceptible genotypes. Extensive transcriptome studies across crops have illuminated regulatory networks and key genes activated during these stresses.

Under drought, tolerant genotypes typically exhibit rapid and robust induction of genes encoding osmoprotectants, late embryogenesis abundant (LEA) proteins, dehydrins, detoxification enzymes, and abscisic acid (ABA) signaling components compared to susceptible ones. For instance, in drought-tolerant wheat, ABA-responsive genes (e.g., *RAB18, RD29A*) and protective enzymes (e.g., glutathione S-transferases) are upregulated more swiftly [106,107]. In maize roots, tolerant lines uniquely induce genes for root elongation and aquaporins to sustain water uptake.

Heat stress induces widespread upregulation of heat shock proteins (HSPs) and their regulators, heat shock factors (HSFs). Tolerant rice varieties show elevated baseline and induced expression of small HSPs, chaperonins, and membrane-stabilizing genes, such as lipid desaturases. Transcriptomes of heat-stressed cotton anthers reveal sustained metabolic genes (e.g., sugar and energy pathways) correlating with pollen viability in resilient cultivars [108,109]. Heat also engages hormonal crosstalk, upregulating ABA-responsive transcription factors (e.g., ABFs/AREBs) and genes related to ethylene or brassinosteroids.

Combined drought–heat stress elicits unique expression profiles absent in individual stresses [15]. In barley flag leaves, this synergy induces ~1000 novel genes involved in protein repair and ubiquitination, emphasizing proteostasis, while amplifying HSF/HSP expression and suppressing photosynthesis-related genes more severely [110].

Key regulators include transcription factor families: DREB/CBFs for dehydration-inducible genes, bZIPs (e.g., ABFs) linking ABA to effectors, NACs and WRKYs for stress and senescence modulation, and HSFs for heat responses. Time-course RNA-Seq shows early activation (1–3 h) of regulators, followed by effectors like LEAs or HSPs [111,112]. A sunflower WRKY factor, upregulated only under combined stress, suggests mediation of drought–heat crosstalk [113].

Differential expression between genotypes reveals tolerance mechanisms: tolerant rice sustains *HSFA2* and HSPs during heat for enhanced protein and membrane stability [114,115]; drought-tolerant chickpea maintains proline (P5CS) and ABA (NCED) biosynthesis genes for osmoprotectant accumulation. High-resolution analyses uncover tissue specificity, with maize roots favoring ABA/cytokinin pathways for growth and leaves prioritizing HSF/HSPs and antioxidants.

Transcriptomic data enable co-expression networks identifying hubs, such as a soybean NAC factor enhancing drought tolerance upon overexpression [116]. They also refine QTL candidates: a rice drought-yield QTL gene (Dehydration Responsive Factor X) showed differential induction, validated transgenically.

In summary, transcriptomics delineates dynamic responses, confirming transcriptional regulation of ABA–SnRK2 and HSF–HSP pathways (e.g., *DREB2A* by ABA; *HSFA1/HSFA2* by heat) [117,118], while revealing crosstalk (e.g., *RD29A* in combined stress) [119] and ROS-scavenging genes in tolerant lines. This yields candidates for protein and metabolite-level scrutiny.

3.3. Proteomics: Stress Proteins and Post-Translational Adjustments

Proteomic profiling quantifies the proteome, the functional effectors of stress responses, offering insights into post-transcriptional and post-translational regulation where protein abundance diverges from transcript levels. Under drought and heat stress, proteomic analyses delineate differential protein accumulation in tolerant versus sensitive genotypes.

A hallmark of drought tolerance is the upregulation of late embryogenesis abundant (LEA) proteins and dehydrins, which function as molecular chaperones or hydrophilins to stabilize cellular structures during desiccation. For instance, comparative proteomics of drought-stressed barley revealed elevated dehydrin isoforms in leaves and crowns of tolerant cultivars, uncorrelated with mRNA induction, implicating translational or stability controls [120].

Heat stress elicits robust induction of heat shock proteins (HSPs), corroborating transcriptomic data. In heat-exposed wheat and rice flag leaves, tolerant lines exhibit pronounced accumulation of small HSPs (e.g., HSP17, HSP23), HSP70, and HSP90, mitigating protein aggregation compared to sensitive counterparts [121]. Proteomics further unveils post-translational modifications (PTMs), such as phosphorylation or oligomerization of HSPs, enhancing chaperone efficacy in tolerant genotypes.

Oxidative stress from drought and heat prompts elevated antioxidant and detoxification enzymes, including superoxide dismutase (SOD), ascorbate peroxidase (APX), glutathione peroxidase, catalase, and glutathione-ascorbate cycle components, constitutively or inducibly higher in tolerant plants [122]. In tomato under combined heat–drought, a tolerant accession sustained Cu/Zn-SOD and chloroplastic APX, correlating with reduced reactive oxygen species (ROS) [123]. Repair enzymes, such as DNA glycosylases and protein disulfide isomerases, also increase to ameliorate oxidative damage.

Metabolic reprogramming is evident: enzymes for osmolyte synthesis (e.g., betaine aldehyde dehydrogenase for glycine betaine, Δ1-pyrroline-5-carboxylate synthetase [P5CS] for proline) are upregulated, while energy-demanding pathways (e.g., Calvin cycle, nitrogen assimilation) are downregulated in sensitive genotypes. Tolerant rice maintains Rubisco activate and oxygen-evolving complex proteins under heat, prolonging photosynthesis [121,124].

Signaling and regulatory proteins, often invariant at transcript level, display PTM differences; abscisic acid (ABA) signaling kinases (SnRK2, CDPKs) and phosphatases exhibit enhanced phosphorylation in tolerant maize under drought, bolstering ABA transduction. Drought also induces alternative splicing, yielding distinct isoforms in tolerant lines [125].

For combined stresses, proteomics identifies multifunctional proteins: HSP101 accumulates exclusively in tolerant wheat, interfacing heat refolding with dehydration protection; thioredoxins and cyclophilins, elevated in tolerant chickpea, facilitate protein folding amid dual stressors.

The ABA-SnRK2 pathway manifests at protein level: PYR/PYL receptors accumulate or diversify in drought-tolerant canola. Phosphoproteomics demonstrates rapid PTM dynamics, e.g., heat-induced dephosphorylation of photosynthetic proteins and phosphorylation of chaperones/transcription factors in rice; LEA phosphorylation modulates interactions; ubiquitination escalates for proteasome-mediated clearance of misfolded proteins, being more efficient in tolerant Arabidopsis [126].

Proteomic discoveries inform breeding: a drought-associated LEA in peanut suggests enhancement targets; transgenic overexpression of LEAs/dehydrins confers drought resilience; a novel heat-induced enzyme in tolerant sorghum stabilizes thylakoids [127].

In summary, proteomics validates transcriptomic predictions, elucidates protein-centric regulations, and underscores chaperone/antioxidant accumulation as tolerance signatures, intertwined with ROS homeostasis. This transitions to metabolomics, examining downstream small-molecule effectors.

3.4. Metabolomics: Metabolic Reprogramming Under Drought and Heat

Metabolomics profiles small molecules that act as osmolytes, antioxidants, and signaling intermediates—the final effectors of stress tolerance. Under drought, tolerant genotypes accumulate compatible solutes such as proline, glycine betaine, trehalose, mannitol, and soluble sugars (glucose, fructose, sucrose) to maintain osmotic balance and stabilize macromolecules. For example, tolerant tomato accessions accumulate proline and raffinose to sustain turgor, while chickpea lines enrich sugars and sorbitol in stems [128]. Heat stress similarly promotes solute accumulation and lipid remodeling that preserve membrane integrity; tolerant wheat maintains polyunsaturated fatty acids and sterols, while grapevine increases flavanols and anthocyanins that scavenge reactive oxygen species (ROS) [129]. Under combined drought–heat, maize elevates γ-aminobutyric acid (GABA) and amino derivatives to buffer carbon and nitrogen metabolism, while sunflower boosts sucrose and malate for respiratory support. Recurrent tolerance-associated metabolites include proline, polyols (mannitol/sorbitol), glycine betaine, and raffinose, all repeatedly linked to resilience [130]. Adjustments in energy metabolism (malate, fumarate, citrate in the tricarboxylic acid, TCA cycle) and hormonal metabolites (abscisic acid, ABA; ethylene precursor 1-aminocyclopropane-1-carboxylic acid, ACC; cytokinins sustaining root growth) further differentiate tolerant from sensitive backgrounds [131,132]. Polyamines (putrescine, spermidine) and non-enzymatic antioxidants (ascorbate, glutathione, tocopherols, flavonoids) complement this defense, as observed in cotton and cucumber under combined stress [133,134].

To convert metabolite profiles into field-ready indicators, targeted and untargeted LC/GC–MS analyses should be paired with plot-level imaging—canopy temperature depression ($CTD = T_{air} - T_{canopy}$), Photochemical Reflectance Index (PRI), and red-edge indices—across stress time courses. Hyperspectral features in the near-infrared (NIR), short-wave infrared (SWIR), and red-edge regions can act as spectral proxies for key metabolites when calibrated with parsimonious models (partial least-squares regression, regularized ensembles) and externally validated across sites and seasons. A pragmatic workflow is to (i) use satellite or unmanned aerial vehicle (UAV) series to flag stress windows; (ii) sample a subset of genotypes for leaf metabolomics at standardized times of day; (iii) fit and validate spectral–metabolite models; (iv) track yield components to quantify decision value [135]. When metabolite patterns consistently co-vary with yield stability, they constitute selection biomarkers. Mapping metabolite quantitative trait loci (mQTLs) and integrating with GWAS/TWAS links these profiles to genetic regulators, enabling metabolite-aware indices to be embedded into $G \times E \times P$ pipelines for predictive modeling of drought–heat resilience. A comparative summary of key metabolite classes, functions, spectral proxies, and representative crops is provided in Table 3.

Table 3. Metabolite signatures and spectral proxies linked to drought–heat resilience.

Metabolite Class	Function In Stress Tolerance	Spectral Proxy (Indicative Features)	Representative Crops/Examples
Compatible solutes (proline, glycine betaine, trehalose, mannitol/sorbitol)	Osmotic adjustment; macromolecule stabilization; partial ROS buffering	NIR/SWIR water-sensitive bands (~1450–1510 nm, ~1940 nm); red-edge shifts; PRI context	Tomato, chickpea, maize; engineered glycine betaine in non-accumulators [128–130]
Antioxidants (ascorbate, glutathione, tocopherols; flavonols/anthocyanins)	ROS scavenging; membrane protection	UV–blue for flavonoids; red-edge position; PRI dynamics	Grapevine, cotton; tolerant lines sustain pools [129,135]
TCA/GABA shunts (malate, fumarate, citrate; GABA)	Respiratory priming; carbon/nitrogen buffering under combined stress	SWIR water/leaf chemistry features; time-series red-edge	Rice (TCA priming), maize (GABA surge) [131]
Polyamines (putrescine, spermidine)	Membrane stabilization; ROS mitigation; stress signaling	Indirect proxies via pigment/water-related bands; requires calibration/validation (Cal/Val)	Cucumber; exogenous application improves heat resilience [135]
Hormonal metabolites (ABA, ACC; cytokinins)	Stomatal closure (ABA); heat/ethylene signaling (ACC); root maintenance (cytokinins)	Indirect via CTD/PRI/red-edge combined with environmental covariates	ABA increase under drought; ACC under heat; tolerant lines maintain cytokinins [132]

3.5. *Key Pathways: ABA–SnRK2, HSF–HSP, and ROS Crosstalk*

Multi-omics convergence highlights interconnected pathways underpinning drought–heat tolerance.

- ABA–SnRK2 pathway: Drought sentinel; ABA-receptor (PYR/PYL) binding relieves *PP2C* inhibition, activating SnRK2 kinases to phosphorylate ABF transcription factors, inducing responsive genes. Omics corroborate genomic variants in receptors/SnRK2; transcript induction of biosynthesis/ABFs; proteomic phosphorylation/accumulation; metabolic ABA surge. Extends to heat: ABA-deficient mutants vulnerable via stomatal dysregulation; *HSFA6b* bridges ABA-heat. Agricultural exploitation: ABA agonists preemptively activate responses, efficacious in fields albeit cost-limited [136].
- HSF–HSP network: Heat master regulators; HSFA1 trimerizes post-HSP90 release, amplifying HSFA2/HSP cascade for protein refolding/membrane protection. Multi-omics: Elevated baseline/inducible HSF/HSP in tolerant crops; sHSPs safeguard organelles. Drought synergy: HSP70 osmotic induction; PTMs (phosphorylation/oligomerization) modulate. Intersects ABA/ethylene/calcium; calmodulin activates HSFs; TOR influences HSP translation. Transgenics overexpressing HSP/HSF confer thermotolerance, tempered by growth penalties [137].
- ROS management: Drought/heat disrupt electron transport, generating superoxide/H_2O_2, damaging biomolecules. Tolerance via enzymatic (SOD/catalase/peroxidases) and non-enzymatic (ascorbate/glutathione/flavonoids) scavenging. ROS dual role: Low levels signal defenses (H_2O_2 activates TFs/ABA); tolerant balance signaling/excess. Crosstalk: ABA employs ROS in stomatal closure; chloroplast singlet oxygen triggers nuclear acclimation; HSPs stabilize scavengers/induce antioxidants [138].
- Ancillary: SnRK1 integrates sugar/energy; Ca^{2+} spikes activate CDPKs for ROS/ABA targets; lipid peroxides signal; MAPKs cascade to HSFs; SnRK2 osmotic sans ABA; *HSFA6b* ABA-responsive [139,140].

Holistic network where ABA orchestrates dehydration tolerance, HSF–HSP cellular integrity, and ROS redox homeostasis, supported by metabolic/redox feedback. Crop enhancement: Select for ABA sensitivity/chaperone induction without yield trade-offs; physiological screens (canopy temperature/membrane stability) [139].

4. G × E × P Integration and "Pixels-to-Proteins" Framework

Advancing plant breeding requires G × E × P pipelines that integrate genotype (G)—markers, QTLs, GWAS/TWAS hits—environment (E)—weather, soil, and management—and phenotype (P)—multi-scale imaging and plot-level traits—across time and scale. Figure 6 illustrates a pixels-to-proteins workflow in which canopy-level indices—canopy temperature depression, Photochemical Reflectance Index (PRI), red-edge and related spectral features—are co-collected with transcriptomic, proteomic, and metabolomic signatures. Standardized metadata and calibration/validation (Cal/Val) align sensor-derived signals with physiological ground truth. Integrated phenomic–omic features feed explainable machine learning models with uncertainty quantification to capture genotype × environment interactions. Model outputs inform three decision layers: breeding (selection indices; probability of superiority across target environments), agronomy (irrigation, nutrient, or shading prescriptions with confidence intervals), and research (hypothesis generation and active sampling). A digital-twin module simulates crop trajectories under alternative interventions and seasons and closes the loop through field validation and iterative model retraining.

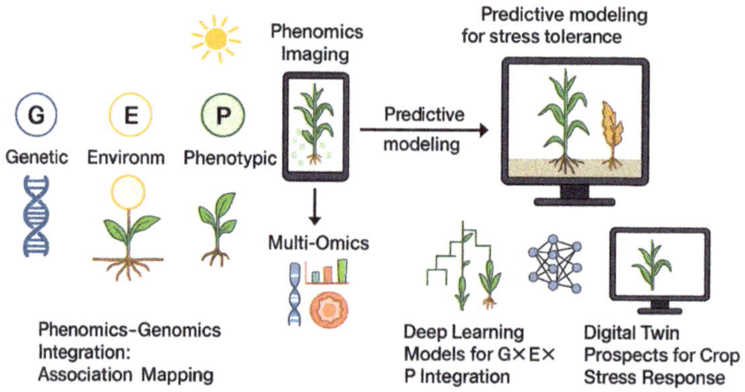

Figure 6. G × E × P integration via a pixels-to-proteins pipeline: harmonizing phenomics and multi-omics with Cal/Val and explainable AI to generate breeding and agronomy decisions.

4.1. Phenomics–Genomics Integration: Association Mapping

High-throughput field phenomics provides quantitative, heritable traits that power association analyses and expose loci governing drought–heat responses. In durum wheat, NDVI time-series derived from drone and ground imaging resolved stay-green QTL on chromosomes 2B/7A, co-localizing with chlorophyll-catabolism genes [141]. In rice, a UAV-derived drought-response index enabled GWAS that pinpointed recovery loci harboring ABA-signaling genes, linking canopy dynamics to regulatory variation [39].

Thermal and structural phenotyping further strengthens genetic dissection. In cotton and sorghum, drone thermography associated canopy temperature and biomass with genetic variation, delivering associations in sorghum [142]. For predictive breeding, incorporating UAV-based canopy temperature features increased genomic prediction accuracy for wheat yield under drought, demonstrating the value of phenomics priors in statistical models [143]. In parallel, envirotyping that uses remotely sensed weather and soil layers improved G × E modeling and genotype–environment matching across sites and seasons [144].

Conceptually, pixels-derived phenomics traits feed statistical genomics—phenomics-assisted genomics—to nominate markers and genes, followed by targeted validation and deployment. Consistent with this workflow, CRISPR/Cas9 editing of a UAV-GWAS candidate kinase in maize verified its causal role in drought tolerance, closing the loop from trait signal to gene function [145]. Best practice includes multi-environment replication with Cal/Val against physiological references (e.g., CTD, Fv/Fm, Ψ_{leaf}), explicit reporting of stress windows, and uncertainty estimates to ensure transferability of associations into breeding decisions.

4.2. Multi-Omics + Phenomics Case Studies

Holistic G × E × P integration—combining multi-scale phenomics with genomics, transcriptomics, proteomics and metabolomics—illuminates mechanisms that underlie field performance. In a potato diversity panel, high-throughput stress phenotyping (imaging) was paired with transcriptomics and metabolomics, and correlation networks linked ABA-biosynthesis genes together with proline and raffinose accumulation to osmotic adjustment and delayed wilting under water deficit [146].

In wheat introgression lines, spectral traits and yield were integrated with GWAS and RNA-seq, revealing a drought-associated QTL that encodes a detoxification enzyme; this gene was overexpressed in tolerant lines and coincided with reduced ROS, providing a mechanistic bridge from locus to phenotype [147]. Proteome-level validation further tightens this link: GWAS candidates in the HSP family showed concordant abundance changes, and a kinase within a sorghum stay-green QTL exhibited stress-responsive phosphorylation in tolerant genotypes, consistent with enhanced chaperone signaling [148].

Model-based integration delivers actionable prediction. In maize, a hybrid model that combined CNN (convolutional neural network) features from UAV imagery with genomic markers improved genomic prediction of biomass under drought compared with either data stream alone [145]. A UAV–GWAS–transcriptome workflow that sampled canopy-temperature extremes mapped loci and recovered DEGs (differentially expressed genes), implicating aquaporins in sustaining transpiration and canopy cooling [149]. More generally, Bayesian networks, multi-layer machine learning models, and knowledge graphs provide scalable frameworks to assimilate heterogeneous data, linking field traits to metabolites, genes, and QTL while preserving causal hypotheses for downstream testing [150]. Across these case studies, best practice includes explicit Cal/Val against physiological references, standardized reporting of stress windows, and external validation to ensure that discoveries translate across sites and seasons.

4.3. Deep Learning Models for G × E × P Integration

Deep learning (DL) provides a flexible framework to integrate phenomics, genomics, and environmental variables by modeling complex non-linear interactions. In image–genomic prediction, convolutional neural networks (CNNs) extract canopy and architectural features from phenomics imagery (P), while dense layers encode genomic markers (G), and combined outputs predict yield or resilience. Environmental signals are implicitly captured within imagery, thereby embedding the G × E component without the need for explicit covariates. These architectures already improve drought-yield predictions by coupling spectral indices with marker-based models.

Beyond image-based pipelines, DL enables omics–phenomics bridging and multimodal learning. Neural networks have been trained to predict drought-responsive transcript levels directly from hyperspectral reflectance, linking canopy spectra with molecular regulation [151]. Similarly, transcriptome-derived features at the seedling stage can be classified by neural nets to distinguish tolerant versus sensitive genotypes, supporting earlier and less invasive selection. More advanced frameworks aggregate inputs across domains—weather, soil, images, and genomic profiles—for trait or stress projections. For example, time-series UAV imagery combined with daily environmental records improves yield forecasts, while fusing soil-moisture signals with canopy images enhances drought detection.

A persistent challenge is the interpretability of deep networks. Emerging solutions include saliency maps, which highlight influential image regions, and attention mechanisms, which weight genomic loci or environmental variables. These techniques not only increase user trust but also uncover novel interactions—such as specific spectral band × weather variable combinations that forecast stress before visible symptoms, linked to membrane stability under vapor pressure deficit (VPD). Such explainable DL approaches help translate black-box models into biologically meaningful hypotheses, strengthening their role in breeding, agronomy, and decision-support systems.

4.4. Digital Twin Prospects for Crop Stress Response

Digital twins provide a virtual representation of crops or fields that update in near real time through the assimilation of imaging, sensor data, and simulations [152,153]. In principle, they embody G × E × P integration in silico, enabling stress forecasting and scenario testing. At the field scale, digital twins ingest satellite and UAV imagery (P) together with IoT environmental sensors (E) and genotype-parameterized crop models to predict yield and stress outcomes [154]. At the plant scale, twins can incorporate omics-derived physiological modules, such as ABA-mediated stomatal conductance regulation or HSP-driven canopy temperature control, linked to crop growth processes (water and carbon allocation) and gene networks underlying stress-induced senescence [155].

Although implementation remains nascent, prototypes illustrate the concept. Greenhouse-based digital twins have coupled imaging data with control systems for dynamic irrigation and climate regulation [156]. In open-field contexts, established crop models such as APSIM or DSSAT are being extended with real-time remote sensing inputs and genotype-specific coefficients, providing a steppingstone toward full digital twins. Such systems can simulate stress trajectories, explore irrigation scheduling, or assess the impact of heatwave onset on yield stability. They also support breeding ideotype design, for instance by evaluating how enhanced root hydraulics or cooler canopy traits would perform across environments.

Future development requires broader integration of remote sensing and weather records for establishing reliable baselines, together with extensions to predict extreme events. Ultimately, digital twins complement the progression from phenomics–genomic synergies for locus discovery, to multi-omics for trait dissection, to deep learning for data fusion, by adding a simulation layer that closes the loop between prediction and intervention. Early exemplars—such as UAV-led gene identification or omics-informed phenotypic modeling—underscore the potential of digital twins to foster proactive breeding and management strategies. This trajectory lays the foundation for truly data-enabled climate-smart agriculture (Table 4).

Table 4. Case Studies of Phenomics-Multi-Omics Integration in Crop Breeding.

Crop	Integration Approach	Key Findings I Outcome	Outcome
Rice	UAV NDVI + Transcriptomics	Identified ABA loci for yield stability	Improved varieties
Maize	Satellite LST + Genomics	QTLs for G × E interactions	Enhanced prediction models
Wheat	Hyperspectral + Proteomics	HSP markers for heat tolerance	Breeding acceleration
Soybean	AI fusion (pixels-to-proteins)	Metabolite predictions	Precision management

5. Translational Applications for Climate-Smart Agriculture

Advances in remote-sensing phenomics and multi-omics are moving beyond discovery to enable climate-smart agriculture (CSA). These integrated approaches support three main applications: (a) crop breeding for drought–heat tolerance, where trait-linked genomic and metabolomic signatures accelerate selection; (b) field management and precision agronomy, where canopy indices (NDVI, CTD, PRI) inform irrigation, fertilization, and shading; (c) predictive modeling and decision-support systems (DSS) that combine phenomics, omics, environmental, and forecast data to guide proactive interventions. Such tools deliver tangible benefits including more resilient varieties, optimized resource use, and early stress warnings. Building on the methodological framework in Figure 6, which integrates phenomics, omics, and explainable AI, Figure 7 illustrates how these pipelines translate into actionable breeding programs, precision agronomy, and farm-level DSS.

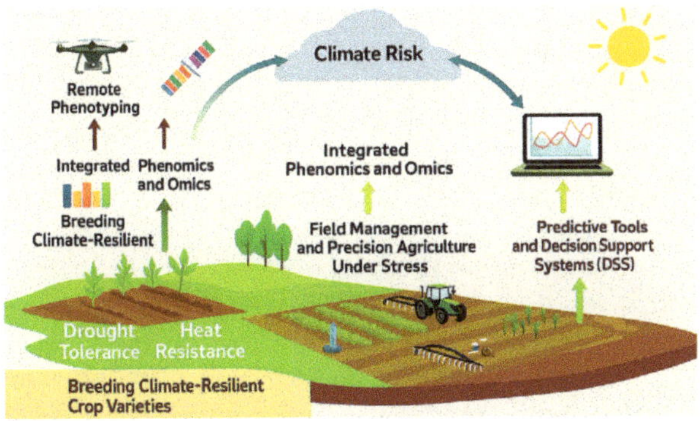

Figure 7. Phenomics–omics translation into breeding, field management, and decision-support systems (DSS).

Figure 7 shows the translation pathway from trait discovery to field applications. In practice, DSS integrate satellite and UAV indices with environmental and genetic layers to generate site-specific prescriptions, enabling farmers to act on stress forecasts and reduce yield losses. Effective deployment, however, requires calibrated thresholds, interoperability standards, and user-friendly delivery (e.g., mobile or SMS-based platforms for smallholders).

5.1. Breeding Climate-Resilient Crop Varieties

Breeding for drought and heat resilient cereals has long been impeded by the low heritability of yield under stress and the polygenic nature of the underlying physiology. The convergence of high-throughput phenotyping, genomics and accelerated generation advance is now delivering measurable genetic gain. Remote and proximal sensing of canopy temperature and normalized difference vegetation index (NDVI) have been fully embedded in the International Maize and Wheat Improvement Center (CIMMYT) wheat pipelines; selecting lines that maintain cooler canopies and higher NDVI in managed stress environments yields significantly greater stability than selection on grain yield alone [157,158].

Speed-breeding protocols that extend photoperiod in controlled environments allow five to six generations per year. When coupled with automated imaging, breeders can impose precise drought treatments on early generations, quantify shoot and root traits, and recycle only superior progeny, thus shortening the selection cycle by 30–40% [159]. Markers derived from phenomics genome-wide association studies (GWAS) and quantitative trait loci (QTL) mapping are now standard in rice and maize. Introgression of *qDTY* loci into

Asian mega-varieties confers a 0.5–1.0 t ha^{-1} advantage in on-farm drought trials [160]. Genomic selection models trained with grain yield plus physiological covariates (e.g., canopy temperature, hyperspectral indices) increase predictive ability for stress-year yield by up to 30% and permit off-site selection, further compressing the breeding cycle [161].

High-resolution phenotyping also accelerates pre-breeding. In sorghum, imaging and genomics disentangled wild alleles for early-season chilling tolerance from linkage drag, enabling their introgression into tropical backgrounds [162]. Precision phenomics confirmed the positive effect of wild emmer segments on water-limited grain filling in elite wheat without a yield penalty [163]. The same platforms are uncovering novel adaptive traits. Thermal imaging combined with gas-exchange assays revealed genotypic variation in night-time stomatal conductance; lines that close stomata overnight conserve soil moisture and achieve superior daytime cooling. Hyperspectral estimation of stem carbohydrate dynamics identified genotypes able to remobilize stem reserves during terminal stress, sustaining kernel weight [164].

Multi-omics has pinpointed central regulators now being manipulated by transgenics and gene editing. Overexpression of *DREB1A* or HSP101 enhances drought or heat tolerance but may incur yield costs in benign seasons. CRISPR knockout of a *PP2C* negative regulator of ABA signaling in rice hastens stomatal closure and improves survival under acute drought [165]. Multiplex editing of heat-shock factor networks or promoter engineering for stress-inducible expression is feasible, as exemplified by ongoing edits of *GhHB17* in cotton to safeguard reproductive tissues during heat waves [166,167].

Because drought seldom occurs in isolation, breeding programs increasingly test under combined water and thermal stress at "climate-analog" sites. In pearl millet and maize, selection in high-temperature drought nurseries has produced hybrids that outperform older checks across erratic Sahelian seasons. Similar gains are reported for heat-tolerant spring wheat in South Asia and rice on the Indo-Gangetic Plain [168].

Collectively, integrating phenomics with genomic tools is transforming climate adaptation breeding from empirical selection to data-driven precision, delivering cultivars that secure 20–30% yield advantages in the world's most vulnerable production zones.

Commercially deployed examples enabled by phenomics and multi-omics. Multiple programs have translated high-throughput phenotyping and omics-informed discovery into released, farmer-adopted varieties. In rice, drought-tolerant lines originating from marker/QTL pipelines have been deployed across Eastern India and South Asia through national systems, with on-farm advantages over local checks under water-limited conditions [65,160,168]. In wheat, canopy-temperature depression, hyperspectral traits, and thermal imaging have been used as indirect selection criteria to assemble heat- and drought-tolerant germplasm that national programs subsequently released, improving stability under late-season heat while maintaining performance in favorable seasons [53–55,66,67]. In sorghum, stay-green donor introgressions and canopy-development ideotypes—identified and tracked with field phenotyping—have underpinned drought-resilient materials used by breeding programs in semi-arid environments [40,50,51]. In maize, industry–public pipelines combining managed-stress trials, genomic selection, and field-based high-throughput phenotyping have delivered drought-tolerant hybrids that maintain yield under stress without sacrificing performance under optimal management [145,161]. Together, these cases illustrate a repeatable path from sensor-derived traits and omics loci to released varieties adopted at scale.

5.2. Field Management and Precision Agriculture Under Stress

Remote-sensing phenomics is now integral to climate-smart agronomy, translating sensor data into dynamic prescriptions that curb drought- and heat-induced yield loss [169].

Thermal and multispectral imagery captured by drones, satellites or pivot-mounted sensors drives variable-rate irrigation (VRI): canopy-temperature or crop water stress anomalies are mapped within seconds, so sprinklers boost flow only in water-deficient zones. Infrared-guided VRI in maize has saved ≈20% irrigation water without compromising yield, and NDVI-based prescription maps have been validated across the south-eastern United States [170].

Frequent Sentinel-2 revisits (≈5 d) enable early detection of incipient stress at landscape scale. Declines in NDVI, the Normalized Difference Moisture Index or solar-induced chlorophyll fluorescence trigger advisories that prompt irrigation, misting or shading before irreversible damage, with machine learning frameworks now delivering field-scale alerts several days ahead of visual wilting [171].

Sensor-guided variable-rate application extends beyond water: spectral or thermal cues delineate zones requiring foliar micronutrients, anti-transpirants, or targeted pesticides, reducing inputs while sustaining physiological activity during hot, dry spells. In orchards and vineyards, high-resolution thermal maps inform selective pruning or shadecloth deployment and automate mist-fan initiation, balancing transpiration with fruit quality. Remote-sensing metrics also underpin financial risk management. Index-based drought insurance already uses regional NDVI or rainfall anomalies to trigger payouts, enabling farmers to re-sow or purchase fodder after severe events; integration of satellite soil-moisture products refines the correlation between index and yield loss, cutting basis risk and improving actuarial performance [172].

Decision-support systems fuse these datasets with weather forecasts and crop models. When a forecast heatwave coincides with maize anthesis, the system may recommend pre-irrigation or foliar boron to safeguard pollen viability, actions verified against sensor-derived growth stage and stress intensity. At watershed scale, evapotranspiration maps guide reservoir releases, while drought hot spots identified via vegetation indices focus relief efforts. Technologies are also migrating directly into farmers' hands: smartphone-mounted thermal cameras and off-the-shelf NDVI drones enable rapid diagnosis of blocked drip lines or compaction-induced stress, allowing same-day remediation.

The synergy between management and genetics is clear—stress-tolerant cultivars maintain higher NDVI and cooler canopies longer, giving managers a wider decision window; conversely, sensing platforms expose thresholds at which even resilient genotypes need intervention. Together, precision phenomics and improved germplasm promise markedly better water productivity and yield stability in a warming, water-scarce world [172].

5.3. Predictive Tools and Decision-Support Systems (DSS)

Climate-smart agriculture is shifting from reactive responses to proactive, data-driven management. Multi-year satellite observations, weather forecasts, and phenomics now feed yield and stress forecasting that can flag risks (e.g., heat during flowering) weeks in advance, enabling pre-emptive actions such as targeted irrigation, stress-mitigation advisories, or import planning. On-farm DSS apps increasingly integrate forecast weather (E), crop genetics (G), and in-field or drone sensors (P) to tailor recommendations—adjusting irrigation cadence, foliar nutrition, and timing of inputs around heat waves to reduce stress [173].

Because heat and drought predispose crops to pests and diseases (e.g., spider mites, aflatoxin), modern DSS link abiotic stress maps with biotic risk alerts, directing scouting or prophylactic treatment to high-risk zones. At policy scales, integrated indicators (meteorological drought, soil moisture, crop condition) now improve drought declarations and disaster response, decreasing both missed events and false alarms. For long-term climate adaptation, G × E × P modeling and climate-scenario simulations identify trait packages and varieties suited to future heat and water regimes; climate-analog testing accelerates validation by trialing today in places that mirror tomorrow's climate [174].

Digital twins—already maturing in controlled environments—predict short-term stress from VPD and radiation, triggering pre-emptive irrigation or misting. Field-scale twins that pair hydrological and crop growth models can forecast soil-moisture drawdown and phenology, scheduling irrigation precisely when it has the greatest yield impact [154]. In resource-limited settings, community DSS deliver simple, timely advice via SMS or radio, while experts fuse satellite vegetation indices, weather outlooks, and crowdsourced observations to power drought early warning and cropping-strategy pivots (e.g., shifting planting dates or varieties) [174]. Evidence from farms and national systems shows that these tools save water, reduce heat damage, and buffer food systems by improving reserve planning. Ultimately, adoption hinges on usability and demonstrated benefit; as data pipelines and models improve, predictive DSS will become a central pillar of climate resilience—sustaining yields amid rising drought and heat variability.

Conclusions and actionable priorities. (1) Standardization and benchmarks: publish open Cal/Val protocols, multi-site reference datasets, and robust metrics penalizing domain-shift failure. (2) Scale-aware causal inference: combine plot-level trials with landscape replication; use causal graphs/negative controls to separate exposure from tolerance. (3) Compound-stress realism: synchronize thermal peaks with water deficits at reproduction; report stage-specific thresholds and uncertainty. (4) Explainable decision systems: co-develop interpretable ML and fail-safes, embedding prescription rules auditable by agronomists. For breeders: prioritize CTD×PRI stability during flowering; verify with GWAS/TWAS-backed candidates. For agronomists: implement two time-of-day UAV thermal flights on stress days and act on pre-defined CTD/PRI thresholds. For data scientists: use time-blocked external validation and physics-constrained features.

6. Knowledge Gaps and Future Directions

Despite substantial progress, three priority gaps continue to constrain the impact of phenomics–omics integration for multi-stress resilience.

(1) Standards and shared benchmarks. Heterogeneous metadata, trait vocabularies, and file formats still impede cross-study synthesis, while limited pairing of imaging and molecular measurements hampers reuse at scale. Wider adoption of community standards, explicit Cal/Val protocols, and AI-assisted harmonization are necessary, together with paired phenomics–omics acquisitions that are versioned and citable. The immediate deliverable is an open, MIAPPE-compliant benchmark—combining imaging and omics with persistent DOIs and versioned labels—to enable fair method comparisons and reproducibility across sites and years.

(2) Climate-realistic, multi-scale modeling with explainability. Field crops experience compound, non-additive stresses whose dynamics span organs, canopies, and fields; yet many pipelines remain single-stress and site-specific, and state-of-the-art models are often opaque. Future work should couple multi-site trials that impose realistic drought–heat regimes with models that link organ-level physiology to canopy signals and field performance, while embedding biophysical constraints and reporting transparent attributions. The concrete deliverable is a multi-site compound-stress dataset aligned across organ–canopy–field scales, plus openly evaluated models that pass time-blocked external validation and publish attention/SHAP attributions keyed to biological priors [152,175].

(3) The "E" in G × E × P must move beyond coarse labels to stage-aware thresholds that are portable across climates and actionable in farms. Trials should quantify timing,

intensity, and co-variation in stresses, and distill them into rules that breeders and agronomists can operationalize through decision support. The practical deliverable is a set of stage-specific stress thresholds validated in ≥3 contrasting climates, together with farmer-facing DSS pilots that report usability metrics and return-on-investment under commercial management [175].

(4) Regulatory and policy barriers. Field phenotyping increasingly relies on UAS/multi-sensor imaging, yet operations and data governance (ownership, privacy, cross-site sharing) impede multi-location trials and interoperability [48,64,74,142]. Varietal release and seed systems remain slow; aligning DUS/performance testing with climate-resilience traits and strengthening delivery pathways are emphasized in adaptation guidance [1,5]. Environmental permits and community consent can constrain heating/irrigation manipulations, and digital decision-support requires uncertainty disclosure and auditability. Targeted actions—enabling compliant UAS use, interoperable data standards, and streamlined release for resilience traits—would accelerate equitable adoption [74].

Despite significant progress, several knowledge gaps and challenges remain at the intersection of phenomics, omics, and stress tolerance, which guide future research directions. Addressing these will be crucial to fully realize the potential of integrated approaches for multi-stress resilience (Table 5).

Table 5. Knowledge Gaps and Future Directions in Phenomics–Multi-Omics Integration.

Gap	Description	Proposed Direction	Potential Impact
Data Standardization	Lack of uniform formats across platforms	Develop ontologies and AI harmonization	Improved cross-study comparisons
Combined Stress Models	Underrepresentation of multifactorial stresses	Multi-omics field trials with DL	Better real-world predictions
Accessibility in Developing Regions	High costs limit adoption	Low-cost sensors and open-source tools	Global equity in breeding
Digital Twins Implementation	Limited validation in diverse environments	Simulation-based breeding pipelines	Accelerated variety development

By addressing these gaps, we will be better poised to develop the next generation of crops and tools that can withstand the increasingly erratic climate. The vision is that perhaps in a decade, breeders routinely use integrated phenomics-genomic analyses to create multi-stress tolerant crops, farmers have AI-driven apps that tell them exactly when and how to protect their crops from an impending stress, and policymakers have robust early warning to organize support—all pieces working in concert for climate-smart agriculture.

7. Conclusions

Climate change-driven drought and heat stress pose unprecedented risks to global crop production. The convergence of multi-scale remote-sensing phenomics with genomics, transcriptomics, proteomics, and metabolomics establishes a transformative "pixels-to-proteins" framework. By linking canopy-level indicators such as canopy temperature depression and hyperspectral indices to regulatory pathways and allelic variations, this integration enables both mechanistic understanding and practical prediction of crop resilience. Tangible progress has already been made: breeders are identifying stress-adaptive variants, agronomists are applying sensor-informed decision-support systems (DSS), and data scientists are developing models that bridge phenotypes and multi-omics signatures.

Three research priorities now stand out. First, the development of data standards and cross-platform interoperability is essential to harmonize satellites, UAVs, ground sensors, and omics datasets, under FAIR principles of sharing and reuse. Second, the design of spatiotemporal field trials that simulate realistic multi-stress environments is required to validate biomarkers and ensure translation beyond controlled conditions. Third, the field deployment of digital twins and AI-driven DSS must progress from greenhouse proof-of-concept to robust, scalable systems, with interpretability, uncertainty quantification, and participatory farmer validation at the core.

Stakeholder-specific pathways further illustrate the value of this integration. For breeders, omics-guided phenomics accelerates gene discovery, supports marker-assisted selection, and enhances genomic prediction accuracy under stress environments. For agronomists, real-time canopy indices and DSS inform irrigation scheduling, nutrient optimization, and adaptive stress management at the field scale. For data scientists, the abundance of multi-modal datasets opens new opportunities for interpretable machine learning and digital-twin frameworks tailored to agriculture.

Despite these advances, caution is warranted. Digital twins and AI models remain largely confined to controlled or simulated contexts, and their effectiveness under heterogeneous field conditions is unproven. Achieving climate-smart agriculture will require coordinated international collaboration, robust data integration pipelines, and large-scale participatory validation in farmers' fields. Only by uniting breeders, agronomists, and data scientists within an interoperable, field-ready ecosystem can the promise of phenomics–omics integration be realized to safeguard crop productivity and global food security in an increasingly variable climate.

Author Contributions: Conceptualization, X.L. and S.Y.; Methodology, X.L.; Software, X.L.; Validation, X.L., S.Y., and Y.J.; Formal Analysis, X.L.; Investigation, X.L.; Resources, X.L.; Data Curation, X.L.; Writing—Original Draft Preparation, X.L.; Writing—Review and Editing, X.L., S.Y., Y.J., Y.W., and D.Y.; Visualization, X.L.; Supervision, S.Y.; Project Administration, S.Y.; Funding Acquisition, Y.J. All authors have read and agreed to the published version of the manuscript.

Funding: This article was funded by the Heilongjiang Provincial Natural Science Foundation of China (LH2023C066).

Data Availability Statement: Data are available through request to the corresponding author.

Conflicts of Interest: The authors declare no conflicts of interest.

Abbreviations

The following abbreviations are used in this manuscript:

ABA	Abscisic Acid
ACC	1-Aminocyclopropane-1-carboxylic Acid (ethylene precursor)
APX	Ascorbate Peroxidase
BRDF	Bidirectional Reflectance Distribution Function
Cal/Val	Calibration/Validation
CAT	Catalase
CDPK	Calcium-Dependent Protein Kinase
CT	Canopy Temperature
CTD	Canopy Temperature Depression
CWSI	Crop Water Stress Index
DSS	Decision-Support System
DT	Digital Twin
EVI	Enhanced Vegetation Index

Fv/Fm	Maximal Quantum Yield of PSII (Photosystem II)
g_s	Stomatal Conductance
GABA	γ-Aminobutyric Acid
GSD	Ground Sampling Distance
GWAS	Genome-wide Association Study
G × E × P	Genotype × Environment × Phenotype
HSF	Heat Shock Factor
HSP	Heat Shock Protein
iWUE	Intrinsic Water-Use Efficiency
JA	Jasmonic Acid/Jasmonates
LAI	Leaf Area Index
LEA	Late Embryogenesis Abundant proteins
LiDAR	Light Detection and Ranging
LST	Land Surface Temperature
ML	Machine Learning
mQTL	Metabolite Quantitative Trait Loci
NDRE	Normalized Difference Red-Edge Index
NDVI	Normalized Difference Vegetation Index
NDWI	Normalized Difference Water Index
NIR	Near-Infrared
OEC	Oxygen-Evolving Complex
P5CS	Δ^1-Pyrroline-5-Carboxylate Synthetase
PRI	Photochemical Reflectance Index
PTM	Post-Translational Modification
QTL	Quantitative Trait Locus/Loci
RGB	Red–Green–Blue imaging
ROS	Reactive Oxygen Species
Rubisco	Ribulose-1,5-bisphosphate Carboxylase/Oxygenase
SA	Salicylic Acid
SIF	Solar-Induced Fluorescence
SnRK2	Sucrose Non-Fermenting-1-Related Protein Kinase 2
SOD	Superoxide Dismutase
SWIR	Short-Wave Infrared
T_{air}	Air Temperature
T_{canopy}	Canopy Temperature
TCA	Tricarboxylic Acid cycle
TPE	Target Population of Environments
TWAS	Transcriptome-Wide Association Study
UAS	Unmanned Aircraft System
UAV	Unmanned Aerial Vehicle
UQ	Uncertainty Quantification
VI	Vegetation Index
VPD	Vapor Pressure Deficit
WUE	Water-Use Efficiency
Ψ_{leaf}	Leaf Water Potential

References

1. McCarthy, J.J.; Canziani, O.F.; Leary, N.A.; Dokken, D.J.; White, K.S. *Climate Change 2001: Impacts, Adaptation, and Vulnerability: Contribution of Working Group II to the Third Assessment Report of the Intergovernmental Panel on Climate Change*; Cambridge University Press: Cambridge, UK, 2001; Volume 2.
2. Mittler, R. Abiotic stress, the field environment and stress combination. *Trends Plant Sci.* **2006**, *11*, 15–19. [CrossRef]
3. Bita, C.E.; Gerats, T. Plant tolerance to high temperature in a changing environment: Scientific fundamentals and production of heat stress-tolerant crops. *Front. Plant Sci.* **2013**, *4*, 273. [CrossRef]
4. Barnabás, B.; Jäger, K.; Fehér, A. The effect of drought and heat stress on reproductive processes in cereals. *Plant Cell Environ.* **2008**, *31*, 11–38. [CrossRef]

5. FAO of the UN. *The State of Food and Agriculture 2021: Making Agrifood Systems More Resilient to Shocks and Stresses*; FAO of the UN: Rome, Italy, 2021.
6. Lesk, C.; Rowhani, P.; Ramankutty, N. Influence of extreme weather disasters on global crop production. *Nature* **2016**, *529*, 84–87. [CrossRef] [PubMed]
7. Asseng, S.; Ewert, F.; Martre, P.; Rötter, R.P.; Lobell, D.B.; Cammarano, D.; Kimball, B.A.; Ottman, M.J.; Wall, G.W.; White, J.W.; et al. Rising temperatures reduce global wheat production. *Nat. Clim. Change* **2015**, *5*, 143–147. [CrossRef]
8. Yang, G.; Liu, J.; Zhao, C.; Li, Z.; Huang, Y.; Yu, H.; Xu, B.; Yang, X.; Zhu, D.; Zhang, X.; et al. Unmanned aerial vehicle remote sensing for field-based crop phenotyping: Current status and perspectives. *Front. Plant Sci.* **2017**, *8*, 1111. [CrossRef]
9. Farooq, M.; Wahid, A.; Kobayashi, N.; Fujita, D.; Basra, S.M. Plant drought stress: Effects, mechanisms and management. In *Sustainable Agriculture*; Springer: Berlin/Heidelberg, Germany, 2009; pp. 153–188.
10. Jaleel, C.A.; Manivannan, P.; Wahid, A.; Farooq, M.; Al-Juburi, H.J.; Somasundaram, R.; Panneerselvam, R. Drought stress in plants: A review on morphological characteristics and pigments composition. *Int. J. Agric. Biol.* **2009**, *11*, 100–105.
11. Wahid, A.; Gelani, S.; Ashraf, M.; Foolad, M.R. Heat tolerance in plants: An overview. *Environ. Exp. Bot.* **2007**, *61*, 199–223. [CrossRef]
12. Kotak, S.; Larkindale, J.; Lee, U.; von Koskull-Döring, P.; Vierling, E.; Scharf, K.-D. Complexity of the heat stress response in plants. *Curr. Opin. Plant Biol.* **2007**, *10*, 310–316. [CrossRef]
13. Rizhsky, L.; Liang, H.; Shuman, J.; Shulaev, V.; Davletova, S.; Mittler, R. When defense pathways collide. The response of Arabidopsis to a combination of drought and heat stress. *Plant Physiol.* **2004**, *134*, 1683–1696. [CrossRef]
14. Suzuki, N.; Rivero, R.M.; Shulaev, V.; Blumwald, E.; Mittler, R. Abiotic and biotic stress combinations. *New Phytol.* **2014**, *203*, 32–43. [CrossRef]
15. Zandalinas, S.I.; Mittler, R.; Balfagón, D.; Arbona, V.; Gómez-Cadenas, A. Plant adaptations to the combination of drought and high temperatures. *Physiol. Plant.* **2018**, *162*, 2–12. [CrossRef] [PubMed]
16. Mittler, R.; Blumwald, E. Genetic engineering for modern agriculture: Challenges and perspectives. *Annu. Rev. Plant Biol.* **2010**, *61*, 443–462. [CrossRef]
17. Cutler, S.R.; Rodriguez, P.L.; Finkelstein, R.R.; Abrams, S.R. Abscisic acid: Emergence of a core signaling network. *Annu. Rev. Plant Biol.* **2010**, *61*, 651–679. [CrossRef]
18. Fujita, Y.; Yoshida, T.; Yamaguchi-Shinozaki, K. Pivotal role of the AREB/ABF-SnRK2 pathway in ABRE-mediated transcription in response to osmotic stress in plants. *Physiol. Plant.* **2013**, *147*, 15–27. [CrossRef]
19. Scharf, K.-D.; Berberich, T.; Ebersberger, I.; Nover, L. The plant heat stress transcription factor (Hsf) family: Structure, function and evolution. *Biochim. Biophys. Acta (BBA) Gene Regul. Mech.* **2012**, *1819*, 104–119. [CrossRef] [PubMed]
20. Wang, W.; Vinocur, B.; Shoseyov, O.; Altman, A. Role of plant heat-shock proteins and molecular chaperones in the abiotic stress response. *Trends Plant Sci.* **2004**, *9*, 244–252. [CrossRef]
21. Burgess, P.; Huang, B. Mechanisms of hormone regulation for drought tolerance in plants. In *Drought Stress Tolerance in Plants, Vol 1: Physiology and Biochemistry*; Springer: Cham, Switzerland, 2016; pp. 45–75.
22. Ohama, N.; Kusakabe, K.; Mizoi, J.; Zhao, H.; Kidokoro, S.; Koizumi, S.; Takahashi, F.; Ishida, T.; Yanagisawa, S.; Shinozaki, K.; et al. The transcriptional cascade in the heat stress response of Arabidopsis is strictly regulated at the level of transcription factor expression. *Plant Cell* **2016**, *28*, 181–201. [CrossRef] [PubMed]
23. Miller, G.; Shulaev, V.; Mittler, R. Reactive oxygen signaling and abiotic stress. *Physiol. Plant.* **2008**, *133*, 481–489. [CrossRef] [PubMed]
24. Suzuki, N.; Koussevitzky, S.; Mittler, R.; Miller, G. ROS and redox signalling in the response of plants to abiotic stress. *Plant Cell Environ.* **2012**, *35*, 259–270. [CrossRef]
25. Furbank, R.T.; Tester, M. Phenomics–technologies to relieve the phenotyping bottleneck. *Trends Plant Sci.* **2011**, *16*, 635–644. [CrossRef]
26. Chen, D.; Neumann, K.; Friedel, S.; Kilian, B.; Chen, M.; Altmann, T.; Klukas, C. Dissecting the phenotypic components of crop plant growth and drought responses based on high-throughput image analysis. *Plant Cell* **2014**, *26*, 4636–4655. [CrossRef]
27. Tardieu, F.; Cabrera-Bosquet, L.; Pridmore, T.; Bennett, M. Plant phenomics, from sensors to knowledge. *Curr. Biol.* **2017**, *27*, R770–R783. [CrossRef]
28. Pieruschka, R.; Schurr, U. Plant phenotyping: Past, present, and future. *Plant Phenomics* **2019**, *2019*, 7507131. [CrossRef]
29. Wardlow, B.D.; Anderson, M.C.; Verdin, J.P. *Remote Sensing of Drought: Innovative Monitoring Approaches*; CRC Press: Boca Raton, FL, USA, 2012.
30. AghaKouchak, A.; Farahmand, A.; Melton, F.S.; Teixeira, J.; Anderson, M.C.; Wardlow, B.D.; Hain, C.R. Remote sensing of drought: Progress, challenges and opportunities. *Rev. Geophys.* **2015**, *53*, 452–480. [CrossRef]
31. Chen, Y.; Yang, J.; Xu, Y.; Zhang, W.; Wang, Y.; Wei, J.; Cheng, W. Remote-Sensing Drought Monitoring in Sichuan Province from 2001 to 2020 Based on MODIS Data. *Atmosphere* **2022**, *13*, 1970. [CrossRef]

32. Chapman, S.C.; Merz, T.; Chan, A.; Jackway, P.; Hrabar, S.; Dreccer, M.F.; Holland, E.; Zheng, B.; Ling, T.J.; Jimenez-Berni, J. Pheno-copter: A low-altitude, autonomous remote-sensing robotic helicopter for high-throughput field-based phenotyping. *Agronomy* **2014**, *4*, 279–301. [CrossRef]
33. Sankaran, S.; Zhou, J.; Khot, L.R.; Trapp, J.J.; Mndolwa, E.; Miklas, P.N. High-throughput field phenotyping in dry bean using small unmanned aerial vehicle based multispectral imagery. *Comput. Electron. Agric.* **2018**, *151*, 84–92. [CrossRef]
34. Hunt, E.R., Jr.; Daughtry, C.S. What good are unmanned aircraft systems for agricultural remote sensing and precision agriculture? *Int. J. Remote Sens.* **2018**, *39*, 5345–5376. [CrossRef]
35. Rutkoski, J.; Poland, J.; Mondal, S.; Autrique, E.; Pérez, L.G.; Crossa, J.; Reynolds, M.; Singh, R. Canopy temperature and vegetation indices from high-throughput phenotyping improve accuracy of pedigree and genomic selection for grain yield in wheat. *G3 Genes Genomes Genet.* **2016**, *6*, 2799–2808. [CrossRef]
36. Deery, D.M.; Rebetzke, G.J.; Jimenez-Berni, J.A.; James, R.A.; Condon, A.G.; Bovill, W.D.; Hutchinson, P.; Scarrow, J.; Davy, R.; Furbank, R.T. Methodology for high-throughput field phenotyping of canopy temperature using airborne thermography. *Front. Plant Sci.* **2016**, *7*, 1808. [CrossRef]
37. Haghighattalab, A.; González Pérez, L.; Mondal, S.; Singh, D.; Schinstock, D.; Rutkoski, J.; Ortiz-Monasterio, I.; Singh, R.P.; Goodin, D.; Poland, J. Application of unmanned aerial systems for high throughput phenotyping of large wheat breeding nurseries. *Plant Methods* **2016**, *12*, 35. [CrossRef]
38. Perich, G.; Hund, A.; Anderegg, J.; Roth, L.; Boer, M.P.; Walter, A.; Liebisch, F.; Aasen, H. Assessment of multi-image unmanned aerial vehicle based high-throughput field phenotyping of canopy temperature. *Front. Plant Sci.* **2020**, *11*, 150. [CrossRef] [PubMed]
39. Jiang, Z.; Tu, H.; Bai, B.; Yang, C.; Zhao, B.; Guo, Z.; Liu, Q.; Zhao, H.; Yang, W.; Xiong, L.; et al. Combining UAV-RGB high-throughput field phenotyping and genome-wide association study to reveal genetic variation of rice germplasms in dynamic response to drought stress. *New Phytol.* **2021**, *232*, 440–455. [CrossRef] [PubMed]
40. Spindel, J.E.; Dahlberg, J.; Colgan, M.; Hollingsworth, J.; Sievert, J.; Staggenborg, S.H.; Hutmacher, R.; Jansson, C.; Vogel, J.P. Association mapping by aerial drone reveals 213 genetic associations for Sorghum bicolor biomass traits under drought. *BMC Genom.* **2018**, *19*, 679. [CrossRef]
41. Jimenez-Berni, J.A.; Deery, D.M.; Rozas-Larraondo, P.; Condon, A.G.; Rebetzke, G.J.; James, R.A.; Bovill, W.D.; Furbank, R.T.; Sirault, X.R. High throughput determination of plant height, ground cover, and above-ground biomass in wheat with LiDAR. *Front. Plant Sci.* **2018**, *9*, 237. [CrossRef]
42. Chen, D.; Shi, R.; Pape, J.-M.; Neumann, K.; Arend, D.; Graner, A.; Chen, M.; Klukas, C. Predicting plant biomass accumulation from image-derived parameters. *GigaScience* **2018**, *7*, giy001. [CrossRef]
43. Kim, Y.; Barnaby, J.Y.; Warnke, S.E. Development of a low-cost automated greenhouse imaging system with machine learning-based processing for evaluating genetic performance of drought tolerance in a bentgrass hybrid population. *Comput. Electron. Agric.* **2024**, *224*, 108896. [CrossRef]
44. Walter, A.; Liebisch, F.; Hund, A. Plant phenotyping: From bean weighing to image analysis. *Plant Methods* **2015**, *11*, 14. [CrossRef]
45. Tao, H.; Xu, S.; Tian, Y.; Li, Z.; Ge, Y.; Zhang, J.; Wang, Y.; Zhou, G.; Deng, X.; Zhang, Z.; et al. Proximal and remote sensing in plant phenomics: 20 years of progress, challenges, and perspectives. *Plant Commun.* **2022**, *3*, 100344. [CrossRef] [PubMed]
46. Liu, H.; Song, W.; Lv, J.; Gui, R.; Shi, Y.; Lu, Y.; Li, M.; Chen, L.; Chen, X. Precise Drought Threshold Monitoring in Winter Wheat Using the Unmanned Aerial Vehicle Thermal Method. *Remote Sens.* **2024**, *16*, 710. [CrossRef]
47. Deng, Q.; Zhang, C.; Dong, J.; Li, Y.; Li, Y.; Huang, Y.; Zhang, H.; Fan, J. Variations over 20 Years in Vegetation Dynamics and Its Coupled Responses to Individual and Compound Meteorological Drivers in Sichuan Province, China. *Atmosphere* **2024**, *15*, 1384. [CrossRef]
48. Qin, W.; Wang, J.; Ma, L.; Wang, F.; Hu, N.; Yang, X.; Xiao, Y.; Zhang, Y.; Sun, Z.; Wang, Z.; et al. UAV-Based Multi-Temporal Thermal Imaging to Evaluate Wheat Drought Resistance in Different Deficit Irrigation Regimes. *Remote Sens.* **2022**, *14*, 5608. [CrossRef]
49. Araus, J.L.; Cairns, J.E. Field high-throughput phenotyping: The new crop breeding frontier. *Trends Plant Sci.* **2014**, *19*, 52–61. [CrossRef]
50. Lopes, M.S.; Reynolds, M.P. Stay-green in spring wheat can be determined by spectral reflectance measurements (normalized difference vegetation index) independently from phenology. *J. Exp. Bot.* **2012**, *63*, 3789–3798. [CrossRef]
51. Borrell, A.K.; Mullet, J.E.; George-Jaeggli, B.; van Oosterom, E.J.; Hammer, G.L.; Klein, P.E.; Jordan, D.R. Drought adaptation of stay-green sorghum is associated with canopy development, leaf anatomy, root growth, and water uptake. *J. Exp. Bot.* **2014**, *65*, 6251–6263. [CrossRef] [PubMed]
52. Pinto, R.S.; Reynolds, M.P.; Mathews, K.L.; McIntyre, C.L.; Olivares-Villegas, J.-J.; Chapman, S.C. Heat and drought adaptive QTL in a wheat population designed to minimize confounding agronomic effects. *Theor. Appl. Genet.* **2010**, *121*, 1001–1021. [CrossRef]
53. Idso, S.B. Non-water-stressed baselines: A key to measuring and interpreting plant water stress. *Agric. Meteorol.* **1982**, *27*, 59–70. [CrossRef]

54. Amani, I.; Fischer, R.; Reynolds, M. Canopy temperature depression association with yield of irrigated spring wheat cultivars in a hot climate. *J. Agron. Crop Sci.* **1996**, *176*, 119–129. [CrossRef]
55. Reynolds, M.; Balota, M.; Delgado, M.; Amani, I.; Fischer, R. Physiological and morphological traits associated with spring wheat yield under hot, irrigated conditions. *Funct. Plant Biol.* **1994**, *21*, 717–730. [CrossRef]
56. Maxwell, K.; Johnson, G.N. Chlorophyll fluorescence—A practical guide. *J. Exp. Bot.* **2000**, *51*, 659–668. [CrossRef]
57. Sharma, D.K.; Andersen, S.B.; Ottosen, C.O.; Rosenqvist, E. Wheat cultivars selected for high Fv/Fm under heat stress maintain high photosynthesis, total chlorophyll, stomatal conductance, transpiration and dry matter. *Physiol. Plant.* **2015**, *153*, 284–298. [CrossRef]
58. Zarco-Tejada, P.J.; González-Dugo, V.; Williams, L.E.; Suarez, L.; Berni, J.A.; Goldhamer, D.; Fereres, E. A PRI-based water stress index combining structural and chlorophyll effects: Assessment using diurnal narrow-band airborne imagery and the CWSI thermal index. *Remote Sens. Environ.* **2013**, *138*, 38–50. [CrossRef]
59. Christopher, J.T.; Christopher, M.J.; Borrell, A.K.; Fletcher, S.; Chenu, K. Stay-green traits to improve wheat adaptation in well-watered and water-limited environments. *J. Exp. Bot.* **2016**, *67*, 5159–5172. [CrossRef]
60. Furbank, R.T.; Jimenez-Berni, J.A.; George-Jaeggli, B.; Potgieter, A.B.; Deery, D.M. Field crop phenomics: Enabling breeding for radiation use efficiency and biomass in cereal crops. *New Phytol.* **2019**, *223*, 1714–1727. [CrossRef]
61. Zhang, L.; Niu, Y.; Zhang, H.; Han, W.; Li, G.; Tang, J.; Peng, X. Maize Canopy Temperature Extracted From UAV Thermal and RGB Imagery and Its Application in Water Stress Monitoring. *Front. Plant Sci.* **2019**, *10*, 1270. [CrossRef]
62. López-Pereira, M.; Connor, D.J.; Hall, A.J. Intercepted radiation and radiation-use efficiency in sunflower crops grown at conventional and wide inter-row spacings: Measurements and modeled estimates of intercepted radiation. *Field Crops Res.* **2020**, *246*, 107684. [CrossRef]
63. Kim, S.L.; Kim, N.; Lee, H.; Lee, E.; Cheon, K.-S.; Kim, M.; Baek, J.; Choi, I.; Ji, H.; Yoon, I.S.; et al. High-throughput phenotyping platform for analyzing drought tolerance in rice. *Planta* **2020**, *252*, 38. [CrossRef] [PubMed]
64. Zhou, J.; Mou, H.; Zhou, J.; Ali, M.L.; Ye, H.; Chen, P.; Nguyen, H.T. Qualification of Soybean Responses to Flooding Stress Using UAV-Based Imagery and Deep Learning. *Plant Phenomics* **2021**, *2021*, 9892570. [CrossRef] [PubMed]
65. Swain, P.; Raman, A.; Singh, S.; Kumar, A. Breeding drought tolerant rice for shallow rainfed ecosystem of eastern India. *Field Crops Res.* **2017**, *209*, 168–178. [CrossRef]
66. Meroni, M.; Rossini, M.; Guanter, L.; Alonso, L.; Rascher, U.; Colombo, R.; Moreno, J. Remote sensing of solar-induced chlorophyll fluorescence: Review of methods and applications. *Remote Sens. Environ.* **2009**, *113*, 2037–2051. [CrossRef]
67. Babar, M.; Reynolds, M.; Van Ginkel, M.; Klatt, A.; Raun, W.; Stone, M. Spectral reflectance indices as a potential indirect selection criteria for wheat yield under irrigation. *Crop Sci.* **2006**, *46*, 578–588. [CrossRef]
68. Anderegg, J.; Aasen, H.; Perich, G.; Roth, L.; Walter, A.; Hund, A. Temporal trends in canopy temperature and greenness are potential indicators of late-season drought avoidance and functional stay-green in wheat. *Field Crops Res.* **2021**, *274*, 108311. [CrossRef]
69. Singh, A.; Ganapathysubramanian, B.; Singh, A.K.; Sarkar, S. Machine learning for high-throughput stress phenotyping in plants. *Trends Plant Sci.* **2016**, *21*, 110–124. [CrossRef]
70. Pound, M.P.; Atkinson, J.A.; Wells, D.M.; Pridmore, T.P.; French, A.P. Deep learning for multi-task plant phenotyping. In Proceedings of the IEEE International Conference on Computer Vision Workshops, Venice, Italy, 22–29 October 2017; pp. 2055–2063.
71. Ubbens, J.R.; Stavness, I. Deep plant phenomics: A deep learning platform for complex plant phenotyping tasks. *Front. Plant Sci.* **2017**, *8*, 1190. [CrossRef] [PubMed]
72. Kamilaris, A.; Prenafeta-Boldú, F.X. Deep learning in agriculture: A survey. *Comput. Electron. Agric.* **2018**, *147*, 70–90. [CrossRef]
73. Yang, W.; Feng, H.; Zhang, X.; Zhang, J.; Doonan, J.H.; Batchelor, W.D.; Xiong, L.; Yan, J. Crop phenomics and high-throughput phenotyping: Past decades, current challenges, and future perspectives. *Mol. Plant* **2020**, *13*, 187–214. [CrossRef]
74. Abdulridha, J.; Ampatzidis, Y.; Kakarla, S.C.; Roberts, P. Detection of target spot and bacterial spot diseases in tomato using UAV-based and benchtop-based hyperspectral imaging techniques. *Precis. Agric.* **2020**, *21*, 955–978. [CrossRef]
75. Resquin, F.; Navarro-Cerrillo, R.M.; Rachid-Casnati, C.; Hirigoyen, A.; Carrasco-Letelier, L.; Duque-Lazo, J. Allometry, growth and survival of three eucalyptus species (*Eucalyptus benthamii* Maiden and Cambage, *E. dunnii* Maiden and *E. grandis* Hill ex Maiden) in high-density plantations in Uruguay. *Forests* **2018**, *9*, 745. [CrossRef]
76. Zhao, T.; Stark, B.; Chen, Y.; Ray, A.L.; Doll, D. A detailed field study of direct correlations between ground truth crop water stress and normalized difference vegetation index (NDVI) from small unmanned aerial system (sUAS). In Proceedings of the 2015 International Conference on Unmanned Aircraft Systems (ICUAS), Denver, CO, USA, 9–12 June 2015; pp. 520–525.
77. Gonzalez-Dugo, V.; Zarco-Tejada, P.; Nicolás, E.; Nortes, P.A.; Alarcón, J.; Intrigliolo, D.S.; Fereres, E. Using high resolution UAV thermal imagery to assess the variability in the water status of five fruit tree species within a commercial orchard. *Precis. Agric.* **2013**, *14*, 660–678. [CrossRef]

78. Roosjen, P.P.; Brede, B.; Suomalainen, J.M.; Bartholomeus, H.M.; Kooistra, L.; Clevers, J.G. Improved estimation of leaf area index and leaf chlorophyll content of a potato crop using multi-angle spectral data–potential of unmanned aerial vehicle imagery. *Int. J. Appl. Earth Obs. Geoinf.* **2018**, *66*, 14–26. [CrossRef]
79. Deery, D.M.; Rebetzke, G.J.; Jimenez-Berni, J.A.; Bovill, W.D.; James, R.A.; Condon, A.G.; Furbank, R.T.; Chapman, S.C.; Fischer, R.A. Evaluation of the phenotypic repeatability of canopy temperature in wheat using continuous-terrestrial and airborne measurements. *Front. Plant Sci.* **2019**, *10*, 875. [CrossRef]
80. Rebetzke, G.; Jimenez-Berni, J.; Fischer, R.; Deery, D.; Smith, D. High-throughput phenotyping to enhance the use of crop genetic resources. *Plant Sci.* **2019**, *282*, 40–48. [CrossRef]
81. McIntyre, C.L.; Seung, D.; Casu, R.E.; Rebetzke, G.J.; Shorter, R.; Xue, G.P. Genotypic variation in the accumulation of water soluble carbohydrates in wheat. *Funct. Plant Biol.* **2012**, *39*, 560–568. [CrossRef]
82. Chapuis, R.; Delluc, C.; Debeuf, R.; Tardieu, F.; Welcker, C. Resiliences to water deficit in a phenotyping platform and in the field: How related are they in maize? *Eur. J. Agron.* **2012**, *42*, 59–67. [CrossRef]
83. Tracy, S.R.; Nagel, K.A.; Postma, J.A.; Fassbender, H.; Wasson, A.; Watt, M. Crop improvement from phenotyping roots: Highlights reveal expanding opportunities. *Trends Plant Sci.* **2020**, *25*, 105–118. [CrossRef] [PubMed]
84. Lynch, J.P. Root phenes for enhanced soil exploration and phosphorus acquisition: Tools for future crops. *Plant Physiol.* **2011**, *156*, 1041–1049. [CrossRef] [PubMed]
85. White, J.W.; Andrade-Sanchez, P.; Gore, M.A.; Bronson, K.F.; Coffelt, T.A.; Conley, M.M.; Feldmann, K.A.; French, A.N.; Heun, J.T.; Hunsaker, D.J.; et al. Field-based phenomics for plant genetics research. *Field Crops Res.* **2012**, *133*, 101–112. [CrossRef]
86. Araus, J.L.; Kefauver, S.C.; Zaman-Allah, M.; Olsen, M.S.; Cairns, J.E. Translating high-throughput phenotyping into genetic gain. *Trends Plant Sci.* **2018**, *23*, 451–466. [CrossRef] [PubMed]
87. Nguyen, T.H.; Gaiser, T.; Vanderborght, J.; Schnepf, A.; Bauer, F.; Klotzsche, A.; Lärm, L.; Hüging, H.; Ewert, F. Responses of field-grown maize to different soil types, water regimes, and contrasting vapor pressure deficit. *Biogeosciences* **2024**, *21*, 5495–5515. [CrossRef]
88. Tamás, A.; Kovács, E.; Horváth, É.; Juhász, C.; Radócz, L.; Rátonyi, T.; Ragán, P. Assessment of NDVI Dynamics of Maize (*Zea mays* L.) and Its Relation to Grain Yield in a Polyfactorial Experiment Based on Remote Sensing. *Agriculture* **2023**, *13*, 689. [CrossRef]
89. Tiwari, R.; Singh, A.K.; Rajam, M.V. GmPARPs differentially regulate the drought and heat stress tolerance in soybean. *Plant Growth Regul.* **2023**, *101*, 643–661. [CrossRef]
90. Wen, X.; Li, H.-Y.; Song, Y.-L.; Zhang, P.-Y.; Zhang, Z.; Bu, H.-H.; Dong, C.-L.; Ren, Z.-Q.; Chang, J.-Z. Genome-wide association study for plant height and ear height in maize under well-watered and water-stressed conditions. *BMC Genom.* **2025**, *26*, 745. [CrossRef]
91. Urano, K.; Kurihara, Y.; Seki, M.; Shinozaki, K. 'Omics' analyses of regulatory networks in plant abiotic stress responses. *Curr. Opin. Plant Biol.* **2010**, *13*, 132–138. [CrossRef]
92. Jogaiah, S.; Govind, S.R.; Tran, L.-S.P. Systems biology-based approaches toward understanding drought tolerance in food crops. *Crit. Rev. Biotechnol.* **2013**, *33*, 23–39. [CrossRef]
93. Fleury, D.; Jefferies, S.; Kuchel, H.; Langridge, P. Genetic and genomic tools to improve drought tolerance in wheat. *J. Exp. Bot.* **2010**, *61*, 3211–3222. [CrossRef]
94. Broz, A.K.; Manter, D.K.; Callaway, R.M.; Paschke, M.W.; Vivanco, J.M. A molecular approach to understanding plant–plant interactions in the context of invasion biology. *Funct. Plant Biol.* **2008**, *35*, 1123–1134. [CrossRef] [PubMed]
95. Peleg, Z.; Fahima, T.; Krugman, T.; Abbo, S.; Yakir, D.; Korol, A.B.; Saranga, Y. Genomic dissection of drought resistance in durum wheat× wild emmer wheat recombinant inbreed line population. *Plant Cell Environ.* **2009**, *32*, 758–779. [CrossRef] [PubMed]
96. Bennett, D.; Reynolds, M.; Mullan, D.; Izanloo, A.; Kuchel, H.; Langridge, P.; Schnurbusch, T. Detection of two major grain yield QTL in bread wheat (*Triticum aestivum* L.) under heat, drought and high yield potential environments. *Theor. Appl. Genet.* **2012**, *125*, 1473–1485. [CrossRef] [PubMed]
97. Sylvester-Bradley, R.; Riffkin, P.; O'Leary, G. Designing resource-efficient ideotypes for new cropping conditions: Wheat (*Triticum aestivum* L.) in the High Rainfall Zone of southern Australia. *Field Crops Res.* **2012**, *125*, 69–82. [CrossRef]
98. Kelliher, T.; Egger, R.L.; Zhang, H.; Walbot, V. Unresolved issues in pre-meiotic anther development. *Front. Plant Sci.* **2014**, *5*, 347. [CrossRef]
99. Prasad, V.B.R.; Govindaraj, M.; Djanaguiraman, M.; Djalovic, I.; Shailani, A.; Rawat, N.; Singla-Pareek, S.L.; Pareek, A.; Prasad, P.V.V. Drought and High Temperature Stress in Sorghum: Physiological, Genetic, and Molecular Insights and Breeding Approaches. *Int. J. Mol. Sci.* **2021**, *22*, 9826. [CrossRef]
100. Varshney, R.K.; Thudi, M.; Nayak, S.N.; Gaur, P.M.; Kashiwagi, J.; Krishnamurthy, L.; Jaganathan, D.; Koppolu, J.; Bohra, A.; Tripathi, S.; et al. Genetic dissection of drought tolerance in chickpea (*Cicer arietinum* L.). *Theor. Appl. Genet.* **2014**, *127*, 445–462. [CrossRef]

101. Ma, Z.; He, S.; Wang, X.; Sun, J.; Zhang, Y.; Zhang, G.; Wu, L.; Li, Z.; Liu, Z.; Sun, G.; et al. Resequencing a core collection of upland cotton identifies genomic variation and loci influencing fiber quality and yield. *Nat. Genet.* **2018**, *50*, 803–813. [CrossRef]
102. Nakashima, K.; Yamaguchi-Shinozaki, K.; Shinozaki, K. The transcriptional regulatory network in the drought response and its crosstalk in abiotic stress responses including drought, cold, and heat. *Front. Plant Sci.* **2014**, *5*, 170. [CrossRef]
103. Zhu, J.-K. Abiotic stress signaling and responses in plants. *Cell* **2016**, *167*, 313–324. [CrossRef] [PubMed]
104. Maršálová, L.; Vítámvás, P.; Hynek, R.; Prášil, I.T.; Kosová, K. Proteomic response of *Hordeum vulgare* cv. Tadmor and *Hordeum marinum* to salinity stress: Similarities and differences between a glycophyte and a halophyte. *Front. Plant Sci.* **2016**, *7*, 1154. [CrossRef]
105. Juliana, P.; Poland, J.; Huerta-Espino, J.; Shrestha, S.; Crossa, J.; Crespo-Herrera, L.; Toledo, F.H.; Govindan, V.; Mondal, S.; Kumar, U.; et al. Improving grain yield, stress resilience and quality of bread wheat using large-scale genomics. *Nat. Genet.* **2019**, *51*, 1530–1539. [CrossRef] [PubMed]
106. Aprile, A.; Mastrangelo, A.M.; De Leonardis, A.M.; Galiba, G.; Roncaglia, E.; Ferrari, F.; De Bellis, L.; Turchi, L.; Giuliano, G.; Cattivelli, L. Transcriptional profiling in response to terminal drought stress reveals differential responses along the wheat genome. *BMC Genom.* **2009**, *10*, 279. [CrossRef]
107. Kumar, J.; Gunapati, S.; Kianian, S.; Singh, S. Comparative analysis of transcriptome in two wheat genotypes with contrasting levels of drought tolerance. *Protoplasma* **2018**, *255*, 1487–1504. [CrossRef] [PubMed]
108. Driedonks, N.; Rieu, I.; Vriezen, W.H. Breeding for plant heat tolerance at vegetative and reproductive stages. *Plant Reprod.* **2016**, *29*, 67–79. [CrossRef]
109. Min, L.; Li, Y.; Hu, Q.; Zhu, L.; Gao, W.; Wu, Y.; Ding, Y.; Liu, S.; Yang, X.; Zhang, X. Sugar and auxin signaling pathways respond to high-temperature stress during anther development as revealed by transcript profiling analysis in cotton. *Plant Physiol.* **2014**, *164*, 1293–1308. [CrossRef]
110. Rollins, J.; Habte, E.; Templer, S.E.; Colby, T.; Schmidt, J.; Von Korff, M. Leaf proteome alterations in the context of physiological and morphological responses to drought and heat stress in barley (*Hordeum vulgare* L.). *J. Exp. Bot.* **2013**, *64*, 3201–3212. [CrossRef]
111. Guo, M.; Liu, J.-H.; Ma, X.; Zhai, Y.-F.; Gong, Z.-H.; Lu, M.-H. Genome-wide analysis of the Hsp70 family genes in pepper (*Capsicum annuum* L.) and functional identification of CaHsp70-2 involvement in heat stress. *Plant Sci.* **2016**, *252*, 246–256. [CrossRef]
112. Yan, J.; Wang, S.; Xiao, D.; Yang, M.; Miao, B.; Niu, B.; Wang, J.; Wei, J.; Ma, F.; Xu, J. Comparative Transcriptomic Profiling Reveals Divergent Drought-Response Mechanisms Between Resistant and Susceptible Apple Genotype Roots. *Agronomy* **2025**, *15*, 748. [CrossRef]
113. Li, J.; Islam, F.; Huang, Q.; Wang, J.; Zhou, W.; Xu, L.; Yang, C. Genome-wide characterization of WRKY gene family in Helianthus annuus L. and their expression profiles under biotic and abiotic stresses. *PLoS ONE* **2020**, *15*, e0241965. [CrossRef]
114. Jung, H.-S.; Crisp, P.A.; Estavillo, G.M.; Cole, B.; Hong, F.; Mockler, T.C.; Pogson, B.J.; Chory, J. Subset of heat-shock transcription factors required for the early response of Arabidopsis to excess light. *Proc. Natl. Acad. Sci. USA* **2013**, *110*, 14474–14479. [CrossRef] [PubMed]
115. Liu, H.C.; Liao, H.T.; Charng, Y.Y. The role of class A1 heat shock factors (HSFA1s) in response to heat and other stresses in Arabidopsis. *Plant Cell Environ.* **2011**, *34*, 738–751. [CrossRef]
116. Yang, C.; Huang, Y.; Lv, P.; Antwi-Boasiako, A.; Begum, N.; Zhao, T.; Zhao, J. NAC Transcription Factor GmNAC12 Improved Drought Stress Tolerance in Soybean. *Int. J. Mol. Sci.* **2022**, *23*, 12029. [CrossRef] [PubMed]
117. Qin, F.; Shinozaki, K.; Yamaguchi-Shinozaki, K. Achievements and challenges in understanding plant abiotic stress responses and tolerance. *Plant Cell Physiol.* **2011**, *52*, 1569–1582. [CrossRef] [PubMed]
118. Ohama, N.; Sato, H.; Shinozaki, K.; Yamaguchi-Shinozaki, K. Transcriptional regulatory network of plant heat stress response. *Trends Plant Sci.* **2017**, *22*, 53–65. [CrossRef] [PubMed]
119. Ndathe, R.; Dale, R.; Kato, N. Dynamic modeling of ABA-dependent expression of the Arabidopsis RD29A gene. *Front. Plant Sci.* **2022**, *13*, 928718. [CrossRef]
120. Vollenweider, P.; Cosio, C.; Günthardt-Goerg, M.S.; Keller, C. Localization and effects of cadmium in leaves of a cadmium-tolerant willow (*Salix viminalis* L.): Part II Microlocalization and cellular effects of cadmium. *Environ. Exp. Bot.* **2006**, *58*, 25–40. [CrossRef]
121. Zhou, X.; Cooke, P.; Li, L. Eukaryotic release factor 1–2 affects Arabidopsis responses to glucose and phytohormones during germination and early seedling development. *J. Exp. Bot.* **2010**, *61*, 357–367. [CrossRef]
122. Ghatak, A.; Chaturvedi, P.; Weckwerth, W. Cereal crop proteomics: Systemic analysis of crop drought stress responses towards marker-assisted selection breeding. *Front. Plant Sci.* **2017**, *8*, 757. [CrossRef]
123. Sun, Y.; Gu, L.; Dickinson, R.E.; Pallardy, S.G.; Baker, J.; Cao, Y.; DaMatta, F.M.; Dong, X.; Ellsworth, D.; Van Goethem, D.; et al. Asymmetrical effects of mesophyll conductance on fundamental photosynthetic parameters and their relationships estimated from leaf gas exchange measurements. *Plant Cell Environ.* **2014**, *37*, 978–994. [CrossRef] [PubMed]

124. Li, X.; Lawas, L.M.; Malo, R.; Glaubitz, U.; Erban, A.; Mauleon, R.; Heuer, S.; Zuther, E.; Kopka, J.; Hincha, D.K.; et al. Metabolic and transcriptomic signatures of rice floral organs reveal sugar starvation as a factor in reproductive failure under heat and drought stress. *Plant Cell Environ.* **2015**, *38*, 2171–2192. [CrossRef] [PubMed]
125. Hu, X.; Li, N.; Wu, L.; Li, C.; Li, C.; Zhang, L.; Liu, T.; Wang, W. Quantitative iTRAQ-based proteomic analysis of phosphoproteins and ABA-regulated phosphoproteins in maize leaves under osmotic stress. *Sci. Rep.* **2015**, *5*, 15626. [CrossRef]
126. Zhang, Y.; Zhou, Y.; Zhang, D.; Tang, X.; Li, Z.; Shen, C.; Han, X.; Deng, W.; Yin, W.; Xia, X. PtrWRKY75 overexpression reduces stomatal aperture and improves drought tolerance by salicylic acid-induced reactive oxygen species accumulation in poplar. *Environ. Exp. Bot.* **2020**, *176*, 104117. [CrossRef]
127. Mielke, K.; Wagner, R.; Mishra, L.S.; Demir, F.; Perrar, A.; Huesgen, P.F.; Funk, C. Abundance of metalloprotease FtsH12 modulates chloroplast development in Arabidopsis thaliana. *J. Exp. Bot.* **2021**, *72*, 3455–3473. [CrossRef]
128. Obata, T.; Witt, S.; Lisec, J.; Palacios-Rojas, N.; Florez-Sarasa, I.; Yousfi, S.; Araus, J.L.; Cairns, J.E.; Fernie, A.R. Metabolite profiles of maize leaves in drought, heat, and combined stress field trials reveal the relationship between metabolism and grain yield. *Plant Physiol.* **2015**, *169*, 2665–2683. [CrossRef]
129. Lu-Irving, P.; Marx, H.E.; Dlugosch, K.M. Leveraging contemporary species introductions to test phylogenetic hypotheses of trait evolution. *Curr. Opin. Plant Biol.* **2018**, *42*, 95–102. [CrossRef] [PubMed]
130. Aljuaid, B.S.; Ashour, H. Exogenous γ-Aminobutyric Acid (GABA) Application Mitigates Salinity Stress in Maize Plants. *Life* **2022**, *12*, 1860. [CrossRef]
131. Zhang, Y.; Fernie, A.R. The Role of TCA Cycle Enzymes in Plants. *Adv. Biol.* **2023**, *7*, 2200238. [CrossRef]
132. Gupta, A.; Hisano, H.; Hojo, Y.; Matsuura, T.; Ikeda, Y.; Mori, I.C.; Senthil-Kumar, M. Global profiling of phytohormone dynamics during combined drought and pathogen stress in Arabidopsis thaliana reveals ABA and JA as major regulators. *Sci. Rep.* **2017**, *7*, 4017. [CrossRef]
133. Zandi, P.; Schnug, E. Reactive Oxygen Species, Antioxidant Responses and Implications from a Microbial Modulation Perspective. *Biology* **2022**, *11*, 155. [CrossRef]
134. Rao, M.J.; Duan, M.; Zhou, C.; Jiao, J.; Cheng, P.; Yang, L.; Wei, W.; Shen, Q.; Ji, P.; Yang, Y.; et al. Antioxidant Defense System in Plants: Reactive Oxygen Species Production, Signaling, and Scavenging During Abiotic Stress-Induced Oxidative Damage. *Horticulturae* **2025**, *11*, 477. [CrossRef]
135. Beegum, S.; Hassan, M.; Ramamoorthy, P.; Bheemanahalli, R.; Reddy, K.; Reddy, V.; Reddy, K. Hyperspectral Reflectance-Based High Throughput Phenotyping to Assess Water-Use Efficiency in Cotton. *Agriculture* **2024**, *14*, 1054. [CrossRef]
136. Huang, Y.-C.; Niu, C.-Y.; Yang, C.-R.; Jinn, T.-L. The heat stress factor HSFA6b connects ABA signaling and ABA-mediated heat responses. *Plant Physiol.* **2016**, *172*, 1182–1199. [CrossRef] [PubMed]
137. Bakery, A.; Vraggalas, S.; Shalha, B.; Chauhan, H.; Benhamed, M.; Fragkostefanakis, S. Heat stress transcription factors as the central molecular rheostat to optimize plant survival and recovery from heat stress. *New Phytol.* **2024**, *244*, 51–64. [CrossRef] [PubMed]
138. Singh, D. Juggling with reactive oxygen species and antioxidant defense system—A coping mechanism under salt stress. *Plant Stress* **2022**, *5*, 100093. [CrossRef]
139. Carianopol, C.S.; Chan, A.L.; Dong, S.; Provart, N.J.; Lumba, S.; Gazzarrini, S. An abscisic acid-responsive protein interaction network for sucrose non-fermenting related kinase1 in abiotic stress response. *Commun. Biol.* **2020**, *3*, 145. [CrossRef]
140. Kulik, A.; Wawer, I.; Krzywińska, E.; Bucholc, M.; Dobrowolska, G. SnRK2 protein kinases--key regulators of plant response to abiotic stresses. *Omics* **2011**, *15*, 859–872. [CrossRef]
141. Lauterberg, M.; Saranga, Y.; Deblieck, M.; Klukas, C.; Krugman, T.; Perovic, D.; Ordon, F.; Graner, A.; Neumann, K. Precision phenotyping across the life cycle to validate and decipher drought-adaptive QTLs of wild emmer wheat (*Triticum turgidum* ssp. dicoccoides) introduced into elite wheat varieties. *Front. Plant Sci.* **2022**, *13*, 965287. [PubMed]
142. Chang, A.; Jung, J.; Maeda, M.; Landivar, J.; Carvalho, H.; Yoem, J. Measurement of Cotton Canopy Temperature Using Radiometric Thermal Sensor Mounted on the Unmanned Aerial Vehicle (UAV). *J. Sens.* **2020**, *2020*, 8899325. [CrossRef]
143. Shahi, D.; Guo, J.; Pradhan, S.; Khan, J.; Avci, M.; Khan, N.; McBreen, J.; Bai, G.; Reynolds, M.; Foulkes, J.; et al. Multi-trait genomic prediction using in-season physiological parameters increases prediction accuracy of complex traits in US wheat. *BMC Genom.* **2022**, *23*, 298. [CrossRef] [PubMed]
144. Amin, A.; Christopher, J.; Cooper, M.; Collins, B.; Voss-Fels, K.; Hickey, L.; Chenu, K. Envirotyping Facilitates Understanding of Genotype × Environment Interactions and Highlights the Potential of Stay-green Traits in Wheat. *Field Crops Res.* **2025**, *331*, 109940. [CrossRef]
145. Adak, A.; Kang, M.; Anderson, S.L.; Murray, S.C.; Jarquin, D.; Wong, R.K.W.; Katzfuß, M. Phenomic data-driven biological prediction of maize through field-based high-throughput phenotyping integration with genomic data. *J. Exp. Bot.* **2023**, *74*, 5307–5326. [CrossRef]

146. Zagorscak, M.; Abdelhakim, L.; Rodriguez-Granados, N.; Široká, J.; Ghatak, A.; Bleker, C.; Blejec, A.; Zrimec, J.; Novak, O.; Pencík, A.; et al. Integration of multi-omics and deep phenotyping provides novel insights into multiple abiotic stress responses in potato. *Plant Physiol.* **2024**, *197*, kiaf126. [CrossRef]
147. Iquebal, M.A.; Sharma, P.; Jasrotia, R.S.; Jaiswal, S.; Kaur, A.; Saroha, M.; Angadi, U.B.; Sheoran, S.; Singh, R.; Singh, G.P.; et al. RNAseq analysis reveals drought-responsive molecular pathways with candidate genes and putative molecular markers in root tissue of wheat. *Sci. Rep.* **2019**, *9*, 13917. [CrossRef]
148. Tsehaye, Y.; Menamo, T.M.; Abay, F.; Tadesse, T.; Bantte, K. Multi-locus genome-wide association study for grain yield and drought tolerance indices in sorghum accessions. *Plant Genome* **2024**, *17*, e20505. [CrossRef]
149. Maistriaux, L.C.; Laurent, M.J.; Jeanguenin, L.; Prado, S.A.; Nader, J.; Welcker, C.; Charcosset, A.; Tardieu, F.; Nicolas, S.D.; Chaumont, F. Genetic variability of aquaporin expression in maize: From eQTLs to a MITE insertion regulating PIP2;5 expression. *Plant Physiol.* **2024**, *196*, 368–384. [CrossRef]
150. Cembrowska-Lech, D.; Krzemińska, A.; Miller, T.; Nowakowska, A.; Adamski, C.; Radaczyńska, M.; Mikiciuk, G.; Mikiciuk, M. An Integrated Multi-Omics and Artificial Intelligence Framework for Advance Plant Phenotyping in Horticulture. *Biology* **2023**, *12*, 1298. [CrossRef]
151. Chen, Y.; Monks, L.; Rubio, V.E.; Cox, A.J.; Swenson, N.G. Linking leaf hyperspectral reflectance to gene expression. *Commun. Earth Environ.* **2025**, *6*, 694. [CrossRef]
152. Purcell, W.; Neubauer, T. Digital Twins in Agriculture: A State-of-the-art review. *Smart Agric. Technol.* **2023**, *3*, 100094. [CrossRef]
153. Zhang, R.; Zhu, H.; Chang, Q.; Mao, Q. A Comprehensive Review of Digital Twins Technology in Agriculture. *Agriculture* **2025**, *15*, 903. [CrossRef]
154. Melesse, T.Y. Digital twin-based applications in crop monitoring. *Heliyon* **2025**, *11*, e42137. [CrossRef]
155. Ariesen-Verschuur, N.; Verdouw, C.; Tekinerdogan, B. Digital Twins in greenhouse horticulture: A review. *Comput. Electron. Agric.* **2022**, *199*, 107183. [CrossRef]
156. Soy, H.; Dilay, Y. Fog-Connected Digital Twin Implementation for Autonomous Greenhouse Management. In *Digital Twin Driven Intelligent Systems and Emerging Metaverse*; Karaarslan, E., Aydin, Ö., Cali, Ü., Challenger, M., Eds.; Springer Nature Singapore: Singapore, 2023; pp. 125–139.
157. Lipper, L.; Thornton, P.; Campbell, B.M.; Baedeker, T.; Braimoh, A.; Bwalya, M.; Caron, P.; Cattaneo, A.; Garrity, D.; Henry, K.; et al. Climate-smart agriculture for food security. *Nat. Clim. Change* **2014**, *4*, 1068–1072. [CrossRef]
158. Steenwerth, K.L.; Hodson, A.K.; Bloom, A.J.; Carter, M.R.; Cattaneo, A.; Chartres, C.J.; Hatfield, J.L.; Henry, K.; Hopmans, J.W.; Horwath, W.R.; et al. Climate-smart agriculture global research agenda: Scientific basis for action. *Agric. Food Secur.* **2014**, *3*, 11. [CrossRef]
159. Turner, N.C.; Wright, G.C.; Siddique, K. Adaptation of grain legumes (pulses) to water-limited environments. *Adv. Agron.* **2001**, *71*, 193–231.
160. Bernier, J.; Kumar, A.; Ramaiah, V.; Spaner, D.; Atlin, G. A large-effect QTL for grain yield under reproductive—Stage drought stress in upland rice. *Crop Sci.* **2007**, *47*, 507–516. [CrossRef]
161. Messina, C.D.; Podlich, D.; Dong, Z.; Samples, M.; Cooper, M. Yield–trait performance landscapes: From theory to application in breeding maize for drought tolerance. *J. Exp. Bot.* **2011**, *62*, 855–868. [CrossRef]
162. Hammer, G.L.; van Oosterom, E.; McLean, G.; Chapman, S.C.; Broad, I.; Harland, P.; Muchow, R.C. Adapting APSIM to model the physiology and genetics of complex adaptive traits in field crops. *J. Exp. Bot.* **2010**, *61*, 2185–2202. [CrossRef]
163. Nevo, E.; Chen, G. Drought and salt tolerances in wild relatives for wheat and barley improvement. *Plant Cell Environ.* **2010**, *33*, 670–685. [CrossRef]
164. Blum, A. Osmotic adjustment is a prime drought stress adaptive engine in support of plant production. *Plant Cell Environ.* **2017**, *40*, 4–10. [CrossRef]
165. Sadau, S.B.; Liu, Z.; Ninkuu, V.; Guan, L.; Sun, X. DREB transcription factors are crucial regulators of abiotic stress responses in *Gossypium* spp. *Plant Stress* **2024**, *11*, 100350. [CrossRef]
166. Luqman, T.; Hussain, M.; Ahmed, S.R.; Ijaz, I.; Maryum, Z.; Nadeem, S.; Khan, Z.; Khan, S.M.U.D.; Aslam, M.; Liu, Y.; et al. Cotton under heat stress: A comprehensive review of molecular breeding, genomics, and multi-omics strategies. *Front. Genet.* **2025**, *16*, 1553406. [CrossRef]
167. Lin, J.; Song, N.; Liu, D.; Liu, X.; Chu, W.; Li, J.; Chang, S.; Liu, Z.; Chen, Y.; Yang, Q.; et al. Histone acetyltransferase TaHAG1 interacts with TaNACL to promote heat stress tolerance in wheat. *Plant Biotechnol. J.* **2022**, *20*, 1645. [CrossRef] [PubMed]
168. Dar, M.H.; Waza, S.A.; Shukla, S.; Zaidi, N.W.; Nayak, S.; Hossain, M.; Kumar, A.; Ismail, A.M.; Singh, U.S. Drought tolerant rice for ensuring food security in Eastern India. *Sustainability* **2020**, *12*, 2214. [CrossRef]
169. Yari, A.; Madramootoo, C.A.; Woods, S.A.; Adamchuk, V.I. Performance evaluation of constant versus variable rate irrigation. *Irrig. Drain.* **2017**, *66*, 501–509. [CrossRef]
170. Stone, K.; Billman, E.D.; Bauer, P.J.; Sigua, G. Using NDVI for variable rate cotton irrigation prescriptions. *Appl. Eng. Agric.* **2022**, *38*, 787–795. [CrossRef]

171. Descals, A.; Torres, K.; Verger, A.; Peñuelas, J. Evaluating Sentinel-2 for Monitoring Drought-Induced Crop Failure in Winter Cereals. *Remote Sens.* **2025**, *17*, 340. [CrossRef]
172. Nguyen, T.T.; Mushtaq, S.; Kath, J.; Nguyen-Huy, T.; Reymondin, L. Satellite-based data for agricultural index insurance: A systematic quantitative literature review. *Nat. Hazards Earth Syst. Sci.* **2025**, *25*, 913–927. [CrossRef]
173. Zhu, T.; Lu, M.; Yang, J.; Bao, Q.; New, S.; Pan, Y.; Qu, A.; Feng, X.; Jian, J.; Hu, S.; et al. Enhancing Ready-to-Implementation subseasonal crop growth predictions in central Southwestern Asia: A machine learning-climate dynamical hybrid strategy. *Agric. For. Meteorol.* **2025**, *370*, 110582. [CrossRef]
174. Wang, Y.; Chen, Y.; Wen, J.; Wu, C.; Zhou, W.; Han, L.; Tang, X. Early warning of drought-induced vegetation stress using multiple satellite-based ecological indicators. *Ecol. Indic.* **2024**, *169*, 112857. [CrossRef]
175. Sato, H.; Mizoi, J.; Shinozaki, K.; Yamaguchi-Shinozaki, K. Complex plant responses to drought and heat stress under climate change. *Plant J.* **2024**, *117*, 1873–1892. [CrossRef]

plants

MDPI

Article

The *tae-miR164-TaNAC6A* Module from Winter Wheat Could Enhance Cold Tolerance in Transgenic *Arabidopsis thaliana*

Ziyao Dai †, Xiaoyan Yang †, Wenwang Shan, Yiou Hao, Da Zhang, Kankan Peng and Qinghua Xu *

College of Life Science, Northeast Agricultural University, Harbin 150030, China; daiziyao224776@163.com (Z.D.); yang20230223@163.com (X.Y.); shiq18279@gmail.com (W.S.); 17804620855@163.com (Y.H.); zhangda@neau.edu.cn (D.Z.); pengkankan@neau.edu.cn (K.P.)

* Correspondence: xuqinghua@neau.edu.cn

† These authors contributed equally to this work.

Abstract

Cold stress impedes the growth and development of wheat (*Triticum aestivum*) and other crops, ultimately reducing both yields and quality. Research indicates that non-coding RNAs (ncRNAs) play a crucial role in regulating plant stress responses and resistance. In a previous study, we observed that the expression of *tae-miR164* was inversely correlated with the expression of *TaNAC6A* in Dongnongdongmai 1 (Dn1), a winter wheat variety with high cold resistance, under cold-stress conditions. However, the molecular mechanism governing the cold responsiveness of the *tae-miR164-TaNAC6A* module was not fully understood. We found *that tae-miR164* and *TaNAC6A* were both induced to express in opposite trends, and TaNAC6A was located in the nucleus. We also discovered that the expression of *tae-miR164* and its target gene, *TaNAC6A*, was responsive to short-term freezing stress in transgenic Arabidopsis plants. Compared to wild-type (WT) Arabidopsis plants, OE-*tae-miR164* plants showed decreased cold tolerance, whereas those overexpressing *TaNAC6A* demonstrated increased tolerance. On average, the OE-*TaNAC6A* and STTM-*tea-miR164* plants exhibited fewer morphological abnormalities in response to cold stress, higher antioxidant enzyme activities and gene expression levels, lower levels of reactive oxygen species (ROS) and malondialdehyde (MDA), and higher expressions of *AtDREB1*, *AtDREB2*, and *AtABI5* in the cold-signaling pathway. Thus, the biological functions of *tae-miR164* and *TaNAC6A* were initially confirmed through heterologous expression strategies, and we have made the first attempt to elucidate its associated *tae-miR164-TaNAC6A* module of cold resistance. The findings of this research will support further investigations into the regulation of plant stress resistance by ncRNAs and will inform molecular module breeding strategies aimed at enhancing the cold tolerance of crop plants. Molecular module design breeding, as a significant breakthrough in modern biotechnology, is transforming traditional breeding models. Conventional hybrid breeding relies on empirical screening, which is time-consuming and subject to randomness. In contrast, molecular module breeding directly targets key genes and achieves precise regulation through technologies such as gene editing and synthetic biology.

Keywords: winter wheat; *tae-miR164-TaNAC6A* module; stress response mechanism; cold-stress tolerance

1. Introduction

Temperature is well-known to affect plant growth and development, particularly exposure at low temperatures [1]. Most plants are able to adapt to gradually decreasing temperatures by altering their cold-responsive gene expression to acclimate their physiology and biochemistry [2]. Frost damage occurs at critical temperatures when ice nucleation begins, resulting in structural and membrane damage. Freezing-induced increases in membrane permeability result in solute loss, ionic imbalance, and organelle damage [3]. It is therefore critical to understand the molecular mechanism underlying cold-stress resistance in order to develop cold-tolerant crops.

Adaptation to cold stress entails a complex array of physiological and biochemical mechanisms. In most plants, cold acclimation typically involves the synthesis and accumulation of proline (Pro) and other amino acids, proteins, sugars, and antioxidants [4]. Cold stress disrupts plant photosynthesis and leads to the accumulation of excessive ROS, resulting in lipid peroxidation, increased membrane permeability, cytoplasmic leakage, elevated relative conductivity, and inhibited development [5,6]. In response, plants express antioxidant enzymes, such as superoxide dismutase (SOD), catalase (CAT), ascorbate peroxidase (APX), and peroxidase (POD), to neutralize ROS [7]. For instance, Guo et al. [8] observed that, in both rice varieties LJ25 (low-temperature stress resistance) and LJ11 (low-temperature stress sensitivity), cold stress prompted the accumulation of Pro, soluble protein, glutathione (GSH), SOD, and APX. However, an opposite trend in ROS and malondialdehyde (MDA) was observed in both varieties. This may reflect differences in the antioxidant capacity and membrane lipid peroxidation levels between the two rice varieties when coping with low-temperature stress.

In plants, miRNAs have been identified as playing a core regulatory role in the response to cold stress [9]. The low-temperature responsiveness of miRNAs was first observed in Arabidopsis (*Arabidopsis thaliana*) [9], and it was subsequently reported in poplar (*Populus*) [10], rice (*Oryza sativa*) [11], wheat [12], tomato (*Solanum lycopersicum*) [13], grape (*Vitis vinifera*) [14], and cotton (*Gossypium hirsutum*) [15]. In tomato, the overexpression of *Sly-miR319d* enhances cold tolerance by inhibiting the expression of GAMYB-like1 [16]. In transgenic Arabidopsis plants, the overexpression of *tae-miR399* from wheat inhibited the expression of *AtUBC24*, which, in turn, inhibited the AtUBC24-mediated degradation of *AtICE1*, inducing the CBF signaling pathway, activating antioxidant defenses, and enhancing cold tolerance [17]. The highly conserved *miR164* has been found to regulate not only plant development but also defense responses to environmental and biological stress signals [18,19]. In tomato, *Sly-miR164A* regulated the fruit ripening and quality by modulating the expressions of *SlNAM2* and *SlNAM3*. *Sly-miR164A* was preferentially expressed during late fruit development, thereby regulating the fruit ripening and quality [20]. In wheat seedlings, the tae-*miR164-TaNAC14* module regulated root growth and development as well as stress (drought and salt) tolerance. Specifically, *TaNAC14* promoted root growth and development and enhanced drought tolerance, while *tae-miR164* inhibited root development and decreased drought and salt tolerance by downregulating the expression of *TaNAC14* [21].

NAC (NAM, ATAF, and CUC) transcription factors (TFs) are one of the largest TF families in plants and play important roles in their growth, development, and environmental stress response [22]. For instance, the overexpression of *CaNAC064* from pepper (*Piper nigrum*) in transgenic Arabidopsis plants has been shown to reduce the MDA content, cold-injury index, and relative electrolytic leakage while increasing the antioxidant activity in response to cold stress [23]. Similarly, the overexpression of *VvNAC17* from grape in transgenic Arabidopsis plants enhanced drought tolerance by upregulating the expressions of abscisic acid (ABA)- and stress-related genes [24]. Lastly, the overexpression of *ZmSNAC13* of maize (*Zea mays*) in transgenic Arabidopsis plants improved the tolerance to both drought and salt stresses. In this case, the primary stress tolerance mechanism was found to involve the ABA pathway and the MAPK signaling cascade [25].

Wheat is one of the most important cereal crops. Notably, Dongnongdongmai 1 (Dn1), a hexaploid (2n = 6x = 42, AABBDD) wheat, was reported to be the first winter wheat variety capable of safely overwintering in Heilongjiang, China, exhibiting a greening rate of more than 85% [26]. This remarkable attribute makes Dn1 a promising research candidate to understand the molecular mechanism of cold-stress tolerance in wheat. In a previous study, we observed that the expression of *tae-miR164* in Dn1 was negatively correlated with the expression of *TaNAC6A* (TraesCS5B02G054200, loc: chr5B (-) 59,165,237 -59170277) in the field (−25 °C). However, the regulatory mechanism underlying the *tae-miR164-TaNAC6A* module remained unclear. Therefore, we sought to reveal this molecular mechanism by characterizing the biological functions of the *tae-miR164-TaNAC6A* module in regulating cold resistance. This study clarifies the key molecular modules of cold resistance in Triticum aestivum and their mechanisms of action, promotes the exchange of superior cold-resistant genetic resources, facilitates the cultivation of cold-resistant crop varieties, breaks through breeding bottlenecks, and provides new impetus for the development of the *Triticum aestivum* seed industry.

2. Results

2.1. Analysis of Conserved Motifs and Gene Structures of tae-miR164 and TaNAC6A

Using RNAfold to predict the secondary structure, we found that pre-tae-miR164 exhibited a complex but stable stem–ring structure. According to WebLogo analysis, the mature miR164 sequence is highly conserved. In order to verify whether *tae-miR164* regulates the cold-stress response, several cis-acting elements within the pre-tae-miR164 promoter were identified using PlantCare, namely, LRE, Me-JAR, GARE, LTR, ABRE, and an enhancer (Figure S1).

TaNAC6A consists of 306 amino acids, with a relative molecular weight of 33.50 kDa. TaNAC6A contains a typical conserved NAM domain and is predicted to be localized to the nucleus. The *ProTaNAC6A* promoter contains a variety of cis-acting elements, including LTR, SARE, ABRE, and DRE, among others (Figure S2).

Bioinformatics predictions consistently indicate that the transcription of *tae-miR164* and *TaNAC6A* is induced by low temperature.

2.2. Expressions of tae-miR164 and TaNAC6A of Dn1 Under Cold Stress

RT-qPCR was conducted to assess the expression levels of *tae-miR164* and *TaNAC6A* in the tillering nodes and leaves of Dn1 under cold stress.

In tillering nodes, the *tae-miR164* expression peaked at 5 °C and was the lowest at −25 °C, decreasing with colder temperatures.

The *TaNAC6A* expression initially rose then fell, with a minimum at 5 °C and a maximum at −10 °C, showing a threefold increase. At −25 °C, the *TaNAC6A* expression was half that at −10 °C and 2.5 times higher than that at 5 °C (Figure 1A).

In leaves, the *tae-miR164* expression peaked at 0 °C, initially rising then falling. The TaNAC6A expression increased with lower temperatures, peaking at −25 °C (Figure 1B).

Both the *tae-miR164* and *TaNAC6A* levels were higher in leaves than in tillering nodes, with *tae-miR164* negatively regulating *TaNAC6A*.

This suggests that TaNAC6A's elevated expression may relate to winter wheat cold resistance (Dn1).

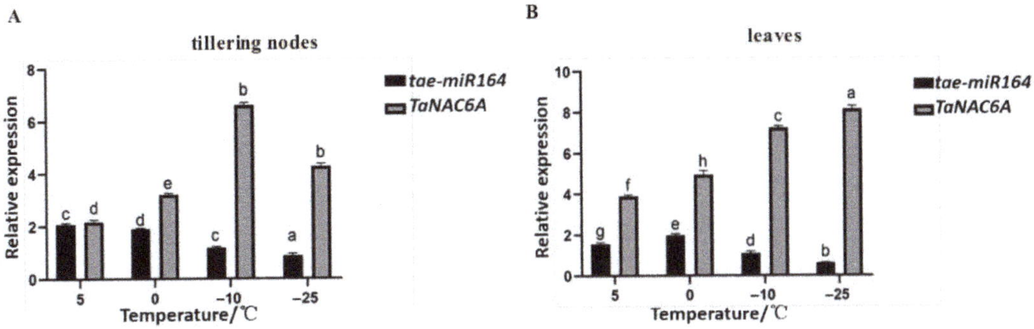

Figure 1. Expressions of *tae-miR164* and *TaNAC6A* in response to cold stress. (**A**) Expressions of *tae-miR164* and *TaNAC6A* in tillering nodes of Dn1. (**B**) Expressions of *tae-miR164* and *TaNAC6A* in leaves of Dn1. Values represent the means ± SDs (n = 10). Different lowercase letters indicate significant differences between treatments ($p < 0.05$), determined by two-way ANOVA. The reference gene used is *TaActin*.

2.3. Subcellular Localization of the TaNAC6A Protein

The pCAMBIA2300-*TaNAC6A-EGFP* fusion vector was developed for transient expression in tobacco to study the subcellular localization of TaNAC6A pr. A GFP signal was observed in the cell membranes, cytoplasm, and nuclei of tobacco leaves injected with *Agrobacterium tumefaciens* strain GV3101 carrying pCAMBIA2300-35S-*EGFP*. In contrast, green fluorescence was detected exclusively in the nuclei of tobacco leaves injected with *Agrobacterium tumefaciens* strain GV3101 carrying pCAMBIA2300-*TaNAC6A-EGFP* (Figure 2). This observation indicated that TaNAC6A is a protein located in the nucleus, and it may play a regulatory role.

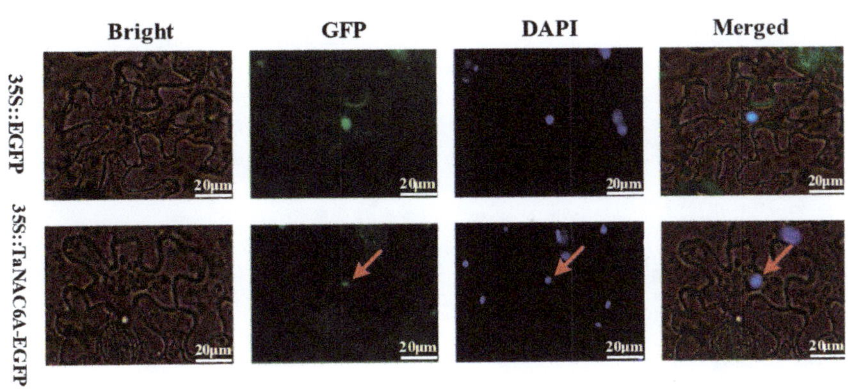

Figure 2. Subcellular localization of TaNAC6A protein. The Agrobacteria carrying 35 S:EGFP (control) and 35 S:*TaNAC6A-EGFP* vectors were transformed into the leaves of tobacco. The GFP signals of TaNAC6A-EGFP (**bottom**) and EGFP alone (**top**) were observed using a laser confocal microscope. DAPI was used as a nuclear marker in this experiment. Red arrow indicates the position of TaNAC6A-EGFP. Scale bars = 20 μm.

2.4. Verification of Low-Temperature Responsiveness of Pre-tae-miR164 and TaNAC6A

Biogenic analysis predicted the presence of low-temperature-responsive elements within the promoters of *pre-tae-miR164* and *TaNAC6A*. These predictions were substantiated by examining the promoter activities of these two genes under low-temperature conditions. The promoters of *tae-miR164* and *TaNAC6A* were capable of inducing GUS expression at normal temperatures (Figure 3A,B). However, after 4 h of exposure to 4 °C, the GUS staining appeared lighter in the *PromiR164* tobacco leaves and more intense in the *ProTaNAC6A* tobacco leaves (Figure 3C,D). These outcomes indicate that *tae-miR164* and *TaNAC6A* can respond to low temperature (4 °C), which might trigger the expression of key genes in the cold-signaling pathway.

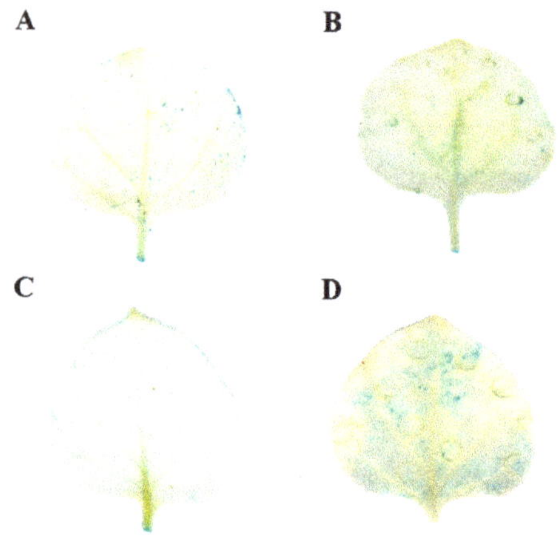

Figure 3. Analysis of promoter activity at different temperatures. GUS phenotype observed via histochemical staining. Leaves of tobacco separately injected with (**A**) 35S:*Pro-miR164-GUS* vector; (**B**) 35S:*Pro-TaNAC6A-GUS* vector; (**C**) 35S:*Pro-miR164-GUS* vector; (**D**) 35S:*Pro-TaNAC6A-GUS* vector after 4 h of exposure to 4 °C.

2.5. Validation of the tae-miR164-TaNAC6A Regulatory Relationship

To verify the relationship between *tae-miR164* and *TaNAC6A*, specific primers were designed at the 5′TUR of *TaNAC6A mRNA* (802 bp). PCR amplification using the 5′RACE technique yielded a single target band of 232 bp, corresponding to the size of the sequence after the first cleavage of *TaNAC6A* mRNA (Figure 4A). Sequencing of ten positive clones revealed that *tae-miR164* cleaved the target *mRNA* at a single site, located between the 10th and 11th bases from the 5′ UTR of *tae-miR164* (Figure 4B).

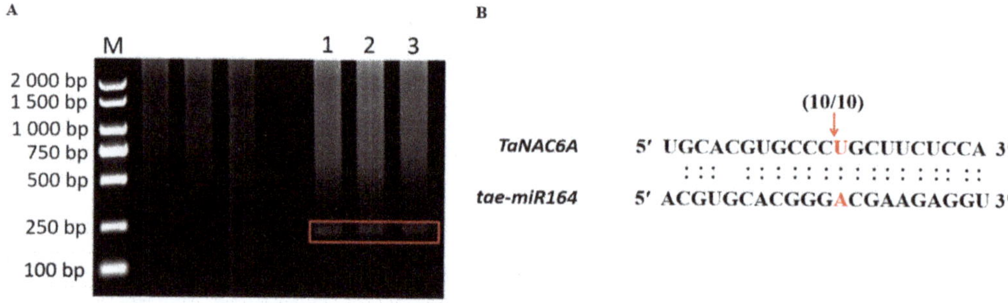

Figure 4. Interaction between *tae-miR164* and its target (*TaNAC6A*). (**A**) PCR amplification products after 5′RACE; red box indicates the amplified band. M: Marker 2000. Lanes 1, 2, and 3: cleaved *TaNAC6A* fragments; (**B**) model of the cDNA structure and mRNA cleavage site of *TaNAC6A*, as determined by 5′ RACE. Red letters show tae-miR164 complementary sites with the nucleotide positions of *TaNAC6A*. The expanded regions show the RNA sequence of each complementary site from 5′ to 3′ and the miRNA sequence from 3′ to 5′. The red arrow indicates a cleavage site verified by 5′-RACE, with the frequency of cloned PCR products shown above the alignment.

The relationship between *tae-miR164* and *TaNAC6A* was further investigated using transient co-transformation in tobacco. Four types of *Agrobacterium tumefaciens* strain GV3101 were prepared: the first with the pBI121-*GUS* construct, the second with the pBI121-*pre-tae-miR164* construct, the third with the pBI121-*TaNAC6A* construct, and the fourth consisting of a mixture containing both the pBI121-*pre-tae-miR164* and pBI121-*TaNAC6A* constructs. These types were individually infiltrated into tobacco leaves. The GUS phenotype observed in leaves inoculated with the pBI121-*GUS* empty vector (Figure 5A) was comparable to that seen in leaves treated with the pBI121-*TaNAC6A* strain (Figure 5C). Conversely, leaves infiltrated with the pBI121-*pre-tae-miR164* strain exhibited no GUS staining, as *pre-tae-miR164* suppressed the GUS gene expression (Figure 5B). Furthermore, leaves co-transformed with pBI121-*pre-tae-miR164* and pBI121-*TaNAC6A* displayed significantly reduced GUS-staining intensities (Figure 5D), suggesting that *tae-miR164* negatively regulates the expression of *TaNAC6A*.

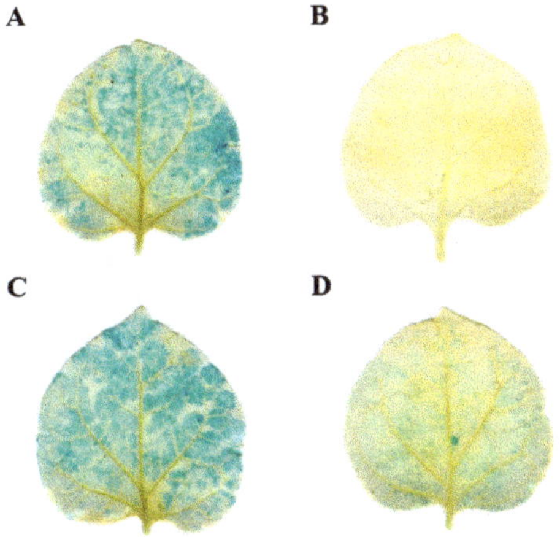

Figure 5. GUS phenotypes observed through histochemical staining. (**A**) Leaf injected with pBI121-*GUS* empty vector; (**B**) leaf injected with pBI121-*pre-tae-miR164* vector; (**C**) leaf injected with pBI121-*TaNAC6A* vector; (**D**) leaf injected with the mixture of pBI121-*pre-tae-miR164* and pBI121-*TaNAC6A* vectors.

The interaction between *tae-miR164* and *TaNAC6A* was validated once more using a dual-luciferase assay. Following the successful construction of the pGreenII-62-SK-*pre-tae-miR164* expression vector, the pGreenII 0800-tae-*miR164*-*TaNAC6A* and pGreenII 0800-*tae-miR164*-m*TaNAC6A* (with a mutation in the *TaNAC6A* cleavage site by *tae-miR164*) expression vectors were also successfully constructed. The constructed vectors were introduced into the *Agrobacterium tumefaciens* strain GV3101 and subsequently expressed in tobacco leaves. The fluorescence intensity was observed to determine whether a targeting relationship exists between *tae-miR164* and *TaNAC6A*. The fluorescence intensity was significantly reduced in tobacco leaves co-infiltrated with *Agrobacterium tumefaciens* strain GV3101 containing the SK-*tae-miR164* plasmid and the LUC-*TaNAC6A* plasmid. However, the fluorescence intensity was enhanced in tobacco leaves co-infiltrated with the SK-*tae-miR164* plasmid and the LUC-m*TaNAC6A* plasmid, which was introduced through an MRE (miRNA response element/miRNA binding site) mutation (Figure 6).

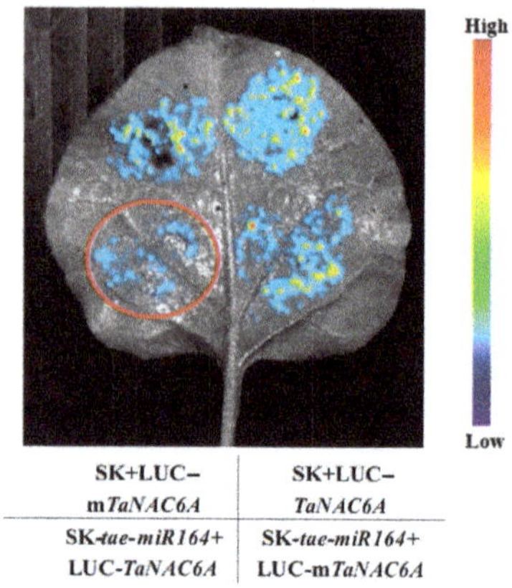

Figure 6. Dual-luciferase assay to verify the relationship between *tae-miR164* and *TaNAC6A*. Dual-luciferase assay showing the activation of SK-driven luciferase reporter gene in tobacco leaves. *Agrobacterium tumefaciens* strain GV3101 carrying different combinations of constructs was infiltrated into different regions of a tobacco leaf. The portion marked with a red circle shows that the luciferase activity was significantly inhibited. Images of luciferase activities were taken three days after the infiltration.

The three results collectively confirm that *tae-miR164* targets and negatively regulates the expression of *TaNAC6A*. And it was found that the 5′ base pairing is indeed essential for MRE to exert its crucial function. *tae-miR164* and *TaNAC6A* indeed exhibit an interaction relationship, and the *tae-miR164-TaNAC6A* module performed biological functions.

2.6. Phenotypic Changes in Transgenic Arabidopsis Plants

To investigate the functions of *tae-miR164* and *TaNAC6A* in relation to cold stress, we constructed overexpression vectors for 35S-*tae-miR164* and 35S-*TaNAC6A* and transformed them into Arabidopsis plants. From these transformations, we selected and obtained some OE-*tae-miR164* and OE-*TaNAC6A* plants; among them, the OE-*tae-miR164*-3, OE-*tae-miR164*-7, OE-*TaNAC6A*-4, and OE-*TaNAC6A*-8 plants were high-expression lines. Additionally, we constructed the STTM-*tae-miR164* (Short Targeted Tandem Motif knockout of *tae-miR164*) recombinant plasmid and transformed it into Arabidopsis plants. Subsequently, the third generation of genetically stable transgenic Arabidopsis plants, STTM-*tae-miR164* plants, were selected; among them, the STTM-*tae-miR164*-1 plants and STTM-*tae-miR164*-2 plants were the lines in which *tae-miR164* was most effectively knocked down. The *nac6* mutant (SALK K135734C, TARI) originated from the Columbia accession (Col-0), with the mutation site located in the homologous gene *AtNAC1* (AT1G56010) corresponding to *TaNAC6A* (Figure S3 and Table S5).

The phenotypes of five Arabidopsis plants (WT, OE-*tae-miR164*, OE-*TaNAC6A*, STTM-*tae-miR164, and nac6*) were observed under freezing stress (Figure 7A). No significant phenotypic differences were noted among the plants between 24 °C and 4 °C. However, at −10 °C, all plants displayed some degree of leaf curling or color deepening. Compared to the STTM-*tae-miR164* and OE-*TaNAC6A* plants, the WT, OE-*tae-miR164*, and *nac6* plants exhibited more pronounced color changes. Overall, the OE-*tae-miR164* plants appeared to be the most affected by low temperature. Following a 7-day recovery period, both the STTM-*tae-miR164* and OE-*TaNAC6A* plants returned to normal, whereas almost all the WT, OE-*tae-miR164*, and *nac6* plants perished. The survival rates of the STTM-*tae-miR164* and OE-*TaNAC6A* plants was about 80%, which were significantly higher than that of the WT, OE-*tae-miR164*, and *nac6* plants (Figure 7B).

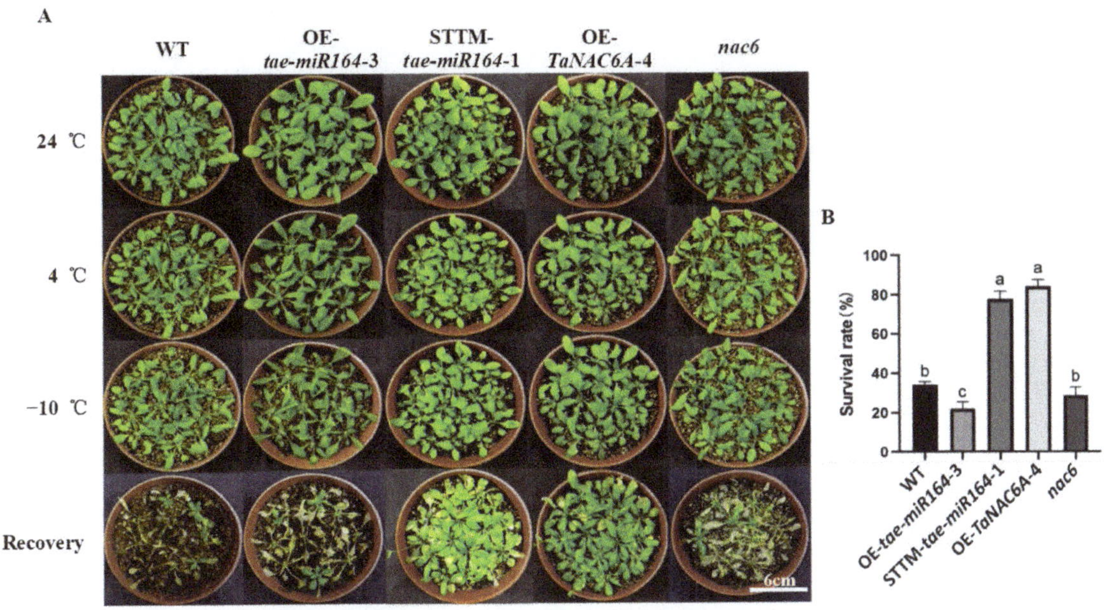

Figure 7. Phenotypes and survival rates of Arabidopsis plants under freezing stress. (**A**) Plant phenotypes; scale bars = 6 cm; (**B**) survival rates. Values represent the means ± SEs (n = 30). Different lowercase letters indicate significant differences between treatments ($p < 0.05$), determined by two-way ANOVA.

Direct observation of the leaves from various lines revealed distinct phenotypic differences (Figure 8A). Compared to the WT, OE-*tae-miR164*, and *nac6* plants, the leaves of the STTM-*tae-miR164* and OE-*TaNAC6A* plants exhibited increased serrations along the leaf margins. Additionally, the leaves of the OE-*tae-miR164* plants displayed the lowest trichome densities. *nac6* also had relatively low trichome densities, although they were not significantly different from those of the WT. Moreover, the trichomes of the WT and OE-*tae-miR164* plants typically had two to three branches, whereas those of the STTM-*tae-miR164* and OE-*TaNAC6A* plants generally had three to four branches (Figure 8).

Under normal growth conditions, the primary roots of the OE-*TaNAC6A* plants were significantly longer than those of the WT and *nac6* plants. Furthermore, the primary roots of the STTM-*tae-miR164* plants were significantly longer than those of the WT and OE-*tae-miR164* plants. Finally, the primary roots of the OE-tae-miR164 plants were much shorter, and they had fewer lateral roots compared to the other four types of plants (Figure 9).

Many plants possess an outer epidermal layer composed of fatty acids or monohydric alcohols, known as the cuticle. This protective structure acts as a barrier against abiotic stressors, such as radiation, low temperature, and drought. In this study, we utilized scanning electron microscopy (SEM) to observe alterations in the cuticles of leaves cultivated under low temperature. At 24 °C, the OE-*TaNAC6A* and STTM-*tae-miR164* plants exhibited thicker cuticles than those of the WT, OE-*tae-miR164*, and *nac6* plants (Figure 10A). Following 3 days of exposure to 4 °C, the average cuticle thickness increased in all four transgenic Arabidopsis plants, mirroring the WT. Notably, the wax contents in the leaves of the OE-*TaNAC6A* and STTM-*tae-miR164* plants were significantly greater than those of the other three Arabidopsis plants (Figure 10B).

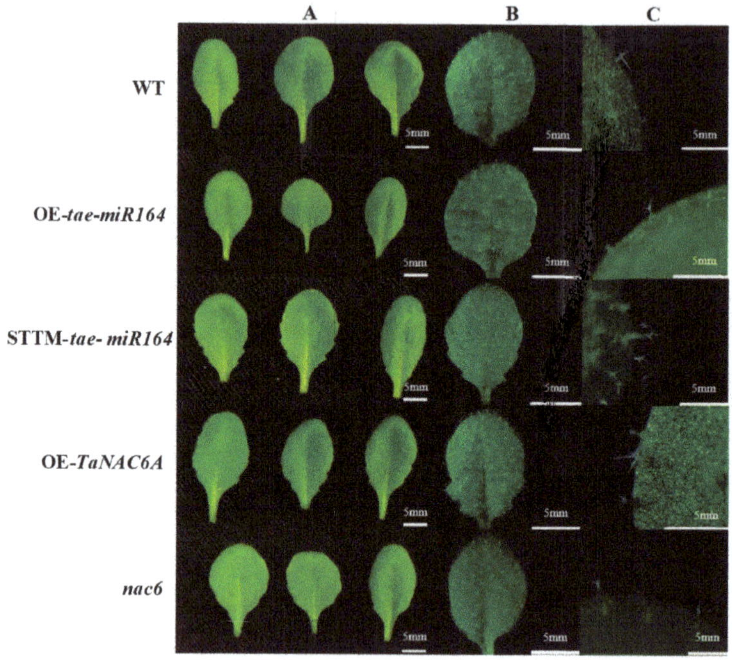

Figure 8. Changes in the leaf margins and trichomes of Arabidopsis plants. (**A**) Images of leaf margins from various lines; (**B**) increased trichome densities and (**C**) altered numbers of branches were associated with *tae-miR164* deletions and *TaNAC6A* overexpression. Images were taken from the third leaves of 14-day-old seedlings. OE-*tae-miR164* represents OE-*tae-miR164*-3 plants, STTM-*tae-miR164* represents STTM-*tae-miR164*-1 plants, and OE-*TaNAC6A* represents OE-*TaNAC6A*-4 plants. Scale bars = 5 mm.

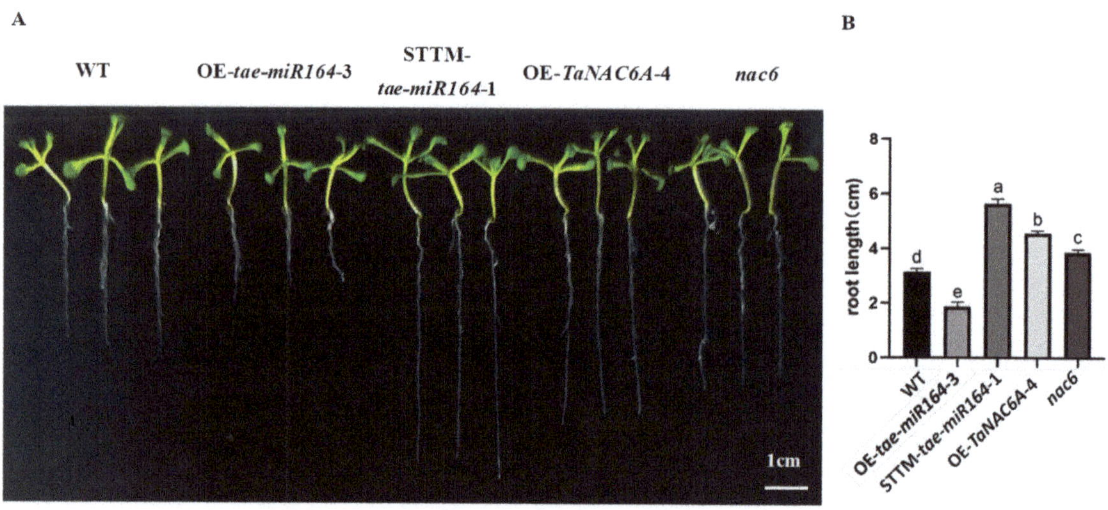

Figure 9. Root changes in Arabidopsis plants. (**A**) Lengths of primary roots at 10 days; scale bars = 1 cm; (**B**) statistical analysis of the lengths of the primary roots. Values represent the means ± SDs (n = 15). Different lowercase letters indicate significant differences between treatments ($p < 0.05$), determined by two-way ANOVA.

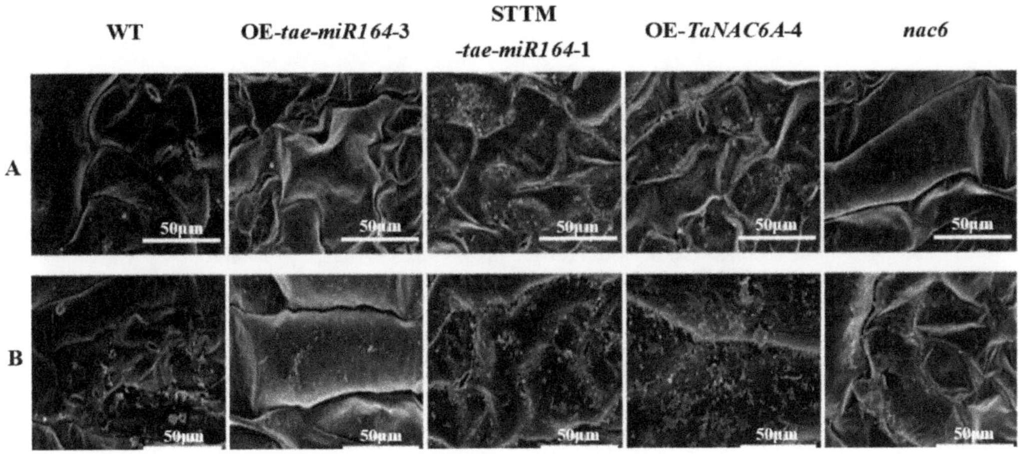

Figure 10. SEM micrographs of alterations in the cuticles of leaves of Arabidopsis plants. (**A**) OE-*TaNAC6A* and STTM-*tae-miR164* plants exhibited thicker cuticles at 24 °C; (**B**) wax contents of OE-*TaNAC6A* and STTM-*tae-miR164* plants were greater after 3 days of 4 °C cold stress. Scale bars = 50 μm.

Stomata regulate the exchange of gases between the plant and the external environment. SEM observations indicated that the stomata of all the Arabidopsis plants were fully open at 24 °C. Following three days of exposure to 4 °C, the stomata of all lines tended to close to some degree. Notably, the stomata of the OE-*TaNAC6A* and STTM-*tae-miR164* plants exhibited some degree of openness as well as wax buildup (Figure 11).

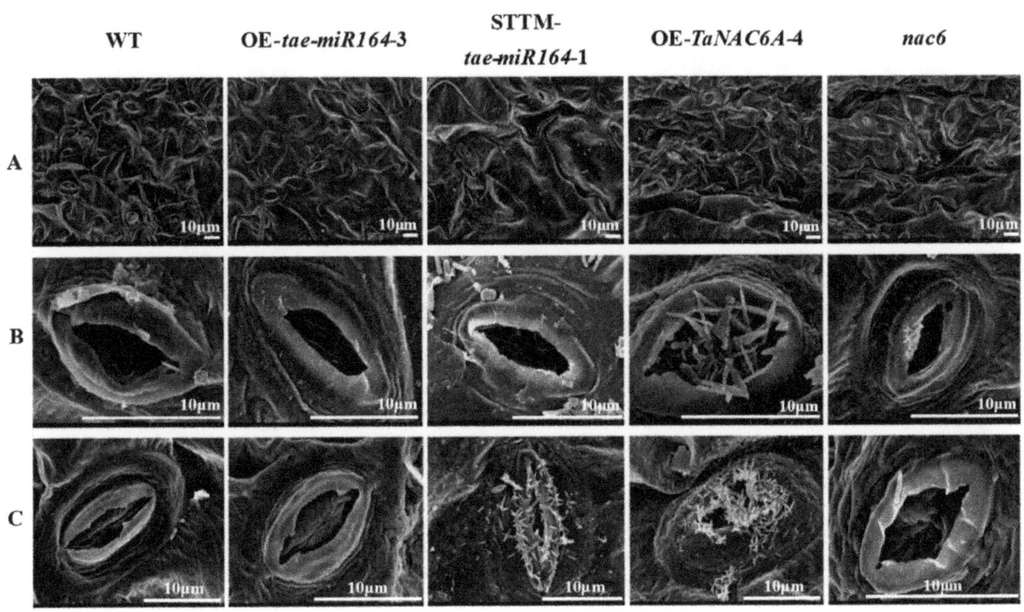

Figure 11. SEM micrographs of stomatal morphology changes in Arabidopsis plants. (**A**,**B**) Stomata of all Arabidopsis plants were fully open at 24 °C; (**C**) stomata of OE-*TaNAC6A* and STTM-*tae-miR164* plants still exhibited some degree of openness as well as wax buildup after 3 days of 4 °C cold stress. Scale bars = 10 μm.

The subcellular structures of four different Arabidopsis plants were observed using transmission electron microscopy (TEM). At 24 °C, the mesophyll cells of the four plants exhibited well-defined structures with clear organelle contours, with no significant differences observed. After 3 days of exposure to 4 °C, the mesophyll cells of the WT plants experienced severe plasmolysis and displayed abnormal nuclear and mitochondrial morphologies. Furthermore, their chloroplasts were deformed and disintegrating, with ruptured chloroplast membranes migrating towards the centers of the mesophyll cells. These deformities were markedly more severe in the OE-*tae-miR164* and *nac6* plants, exhibiting notable increases in both the numbers and sizes of starch granules. In contrast, the OE-*TaNAC6A* plants showed no significant changes, apart from a slight tendency for chloroplasts to move away from the edges of the mesophyll cells. Additionally, these plants had significantly reduced numbers of starch granules (Figure 12).

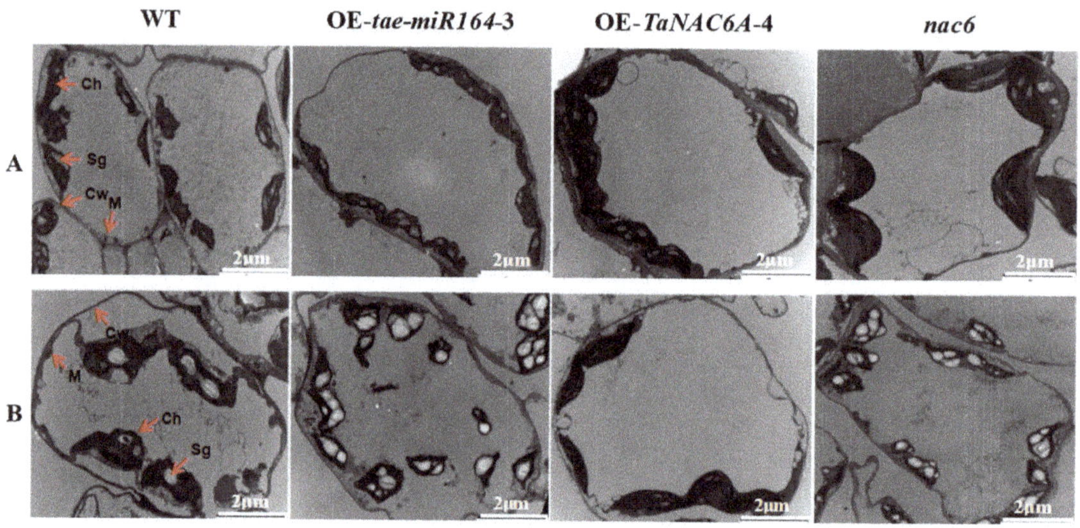

Figure 12. TEM micrographs of subcellular structure changes in mesophyll cells of Arabidopsis plants. (**A**) Cells of all plants exhibited well-defined structures with clear organelle contours at 24 °C; (**B**) all cells suffered varying degrees of damage; however, OE-*TaNAC6A* plants showed no significant changes after 3 days of 4 °C cold stress. Ch: chloroplast. Cw: cell wall. M: mitochondrion. Sg: starch granule. Scale bars = 2 μm.

At 24 °C, the grana thylakoid structures in four plant genotypes were relatively well-ordered. Additionally, their mitochondria were round or oval and typically located either at junctions between chloroplasts and the plasma membrane or between two adjacent chloroplasts. After 3 days of exposure to 4 °C, the organelles in the WT plants began to disintegrate alongside vacuolization. Furthermore, the chloroplasts in the WT plants started to swell and deform, including the enlargement of starch granules. The number of grana decreased, and the thylakoid layers began to loosen and become disorganized, with some layers fusing together, resulting in blurred or missing thylakoid structures. However, in comparison to the WT plants, the OE-*TaNAC6A* plants retained normal chloroplast and mitochondrial structures even after being exposed to 4 °C for 3 days. This was evidenced by the absence of cell membrane disintegration and starch granule formation, indicating that the mesophyll cells could still maintain normal photosynthetic activity at this stage. In contrast, the cellular structures of the mesophyll cells in the OE-*tae-miR164* and *nac6* plants were significantly more damaged, with markedly increased numbers of starch granules.

This suggested a substantial weakening in the ability to export photosynthetic products and an inability to perform normal photosynthesis. The numbers of starch granules increased in these plants, suggesting a weakened ability to export photosynthetic products (Figure 13).

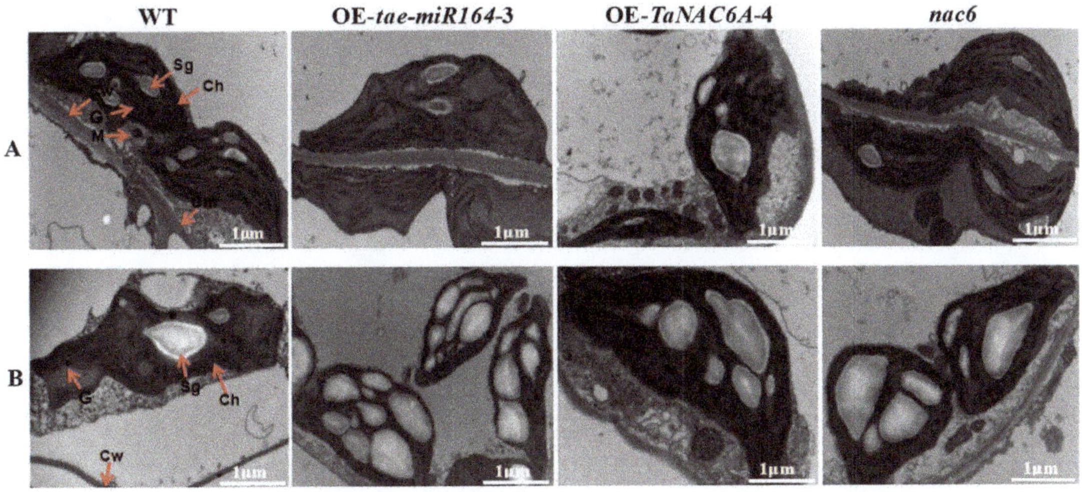

Figure 13. TEM micrographs of organellar morphologies undergoing changes in Arabidopsis plants. (**A**) Chloroplasts and mitochondria of all cells are normal at 24 °C; (**B**) chloroplasts and mitochondria of OE-*TaNAC6A* plants still remained normal after 3 days of 4 °C cold stress. Ch: chloroplast. Cw: cell wall. Cm: cell plasma membrane. M: mitochondrion. Sg: starch granule. G: osmiophilic body. Scale bars = 1 μm.

2.7. Physiological Markers of Cold Resistance in Transgenic Arabidopsis Plants

The MDA content is one of the important indicators for evaluating the degree of peroxidation in the plant cell plasma membrane. The higher the MDA content, the greater the degree of membrane lipid peroxidation, and the more severe the cellular damage. Under normal conditions, plants contain relatively low Pro contents. However, when subjected to adverse environmental stresses, such as drought or cold, Pro accumulates in large amounts. The variation in conductivity can reflect the extent of cellular damage. Generally speaking, under cold stress, the lower the relative levels of electrical conductivity and MDA content, and the higher the Pro content, the less damage plant cells suffer, indicating stronger cold resistance in the plant.

Consequently, in order to evaluate whether the overexpression of *tae-miR164* and *TaNAC6A* enhances cold tolerance in transgenic Arabidopsis plants, the electronic conductivities and the contents of MDA and Pro were measured.

Overall, the electric conductivities of the five types of Arabidopsis plants tended to increase with decreasing temperature. At 4 °C and −10 °C, the electric conductivities of the STTM-*tae-miR164* and OE-*TaNAC6A* plants were significantly lower than those of the WT, OE-*tae-miR164*, and *nac6* plants (Figure 14A), indicating that the OE-*tae-miR164* plants suffered more severe damage from cold stress. Conversely, the cold resistance of the transgenic Arabidopsis plants was improved, in which the expression level of *TaNAC6A* was relatively high.

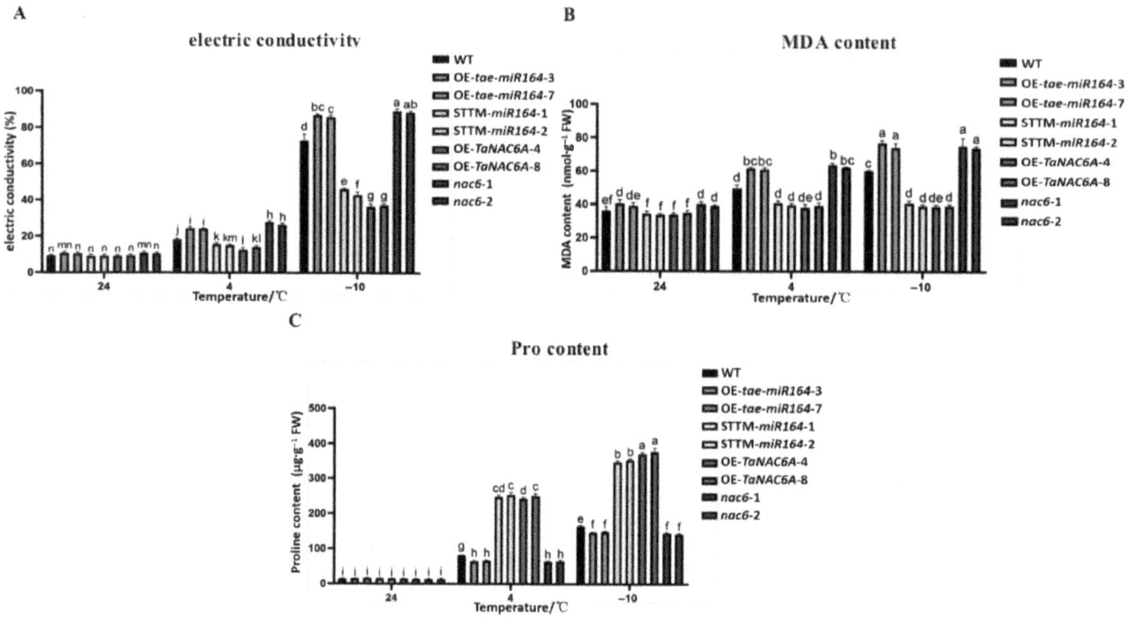

Figure 14. Physiological markers of cold tolerance in transgenic Arabidopsis plants. (**A**) Electric conductivity; (**B**) MDA content; (**C**) Pro content. Values represent the means ± SDs (n = 10). Different lowercase letters indicate significant differences between treatments ($p < 0.05$), determined by two-way ANOVA.

At 4 °C and −10 °C, the MDA contents in the Arabidopsis plants significantly increased compared to those at 24 °C, indicating that the cell membranes suffered varying degrees of oxidative damage. The MDA contents in the STTM-*tae-miR164* and OE-*TaNAC6A* plants were lower than those of the WT, OE-*tae-miR164*, and *nac6* plants at 24 °C, 4 °C, and −10 °C, although the difference was most pronounced at −10 °C (Figure 14B). Furthermore, the Pro contents in the STTM-*tae-miR164* and OE-*TaNAC6A* plants were higher than those of the WT, OE-*tae-miR164*, and *nac6* plants at 4 °C and −10 °C (Figure 14C). It appears that the overexpression of *tae-miR164* reduced the cold tolerance of the *OE-tae-miR164* plants, whereas the overexpression of *TaNAC6A* enhanced the cold tolerance of the OE-*TaNAC6A* plants.

2.8. Cold-Induced ROS Production in Transgenic Arabidopsis Plants

Cold stress frequently leads to the overproduction of reactive oxygen species (ROS), including hydrogen peroxide (H_2O_2) and superoxide radicals ($O_2^{\cdot-}$), among others. To ascertain whether the overexpression of *tae-miR164* and *TaNAC6A* influences the elimination of ROS, the levels of H_2O_2 and $O_2^{\cdot-}$ were quantified in cold-stressed Arabidopsis plants. The DAB and NBT chemical staining assays indicated that the WT plants exhibited more intense staining than both the OE-*TaNAC6A* and STTM-*tae-miR164* plants at cold stress (4 °C and −10 °C). However, the OE-*tae-miR164* and *nac6* plants displayed darker staining than the WT plants (Figure 15A,B). The levels of H_2O_2 and $O_2^{\cdot-}$ in the WT plants gradually increased as the temperature decreased, with significantly higher levels at cold stress compared to those at 24 °C. No significant differences were observed in the levels of H_2O_2 and $O_2^{\cdot-}$ in the OE-*TaTaNAC6A* and STTM-*tae-miR164* plants at 4 °C compared to at 24 °C. However, the levels of H_2O_2 and $O_2^{\cdot-}$ were higher at −10 °C than those at 24 °C. Furthermore, the levels of H_2O_2 and $O_2^{\cdot-}$ were lower in both the OE-*TaTaNAC6A* and

STTM-*tae-miR164* plants compared to those of the WT plants at both 4 °C and −10 °C. In contrast, the levels of H_2O_2 and $O_2^{\cdot-}$ in the OE-*tae-miR164* and *nac6* plants were higher than those in the WT plants (Figure 15C,D).

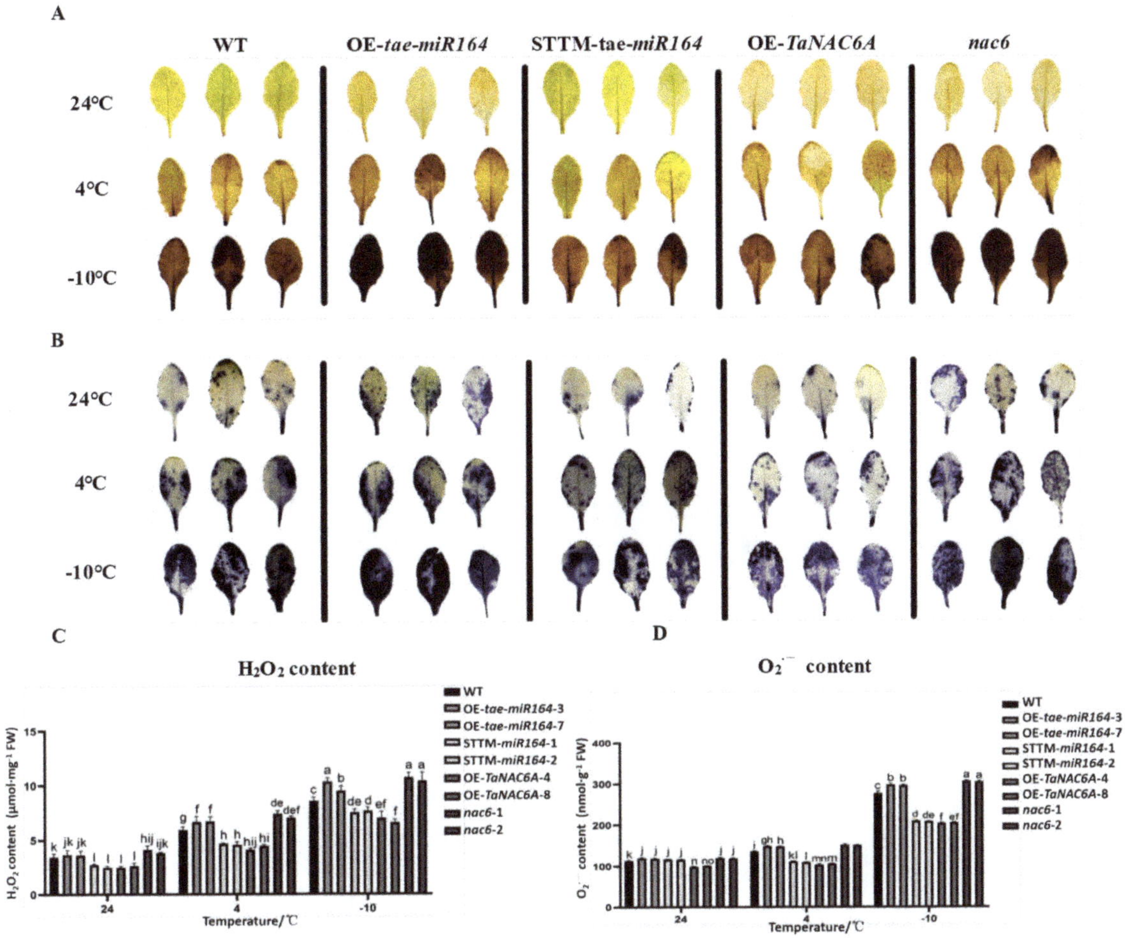

Figure 15. Changes in ROS levels in Arabidopsis plants under cold stress. (**A**) Leaves stained by DAB; (**B**) leaves stained by NBT; (**C**) H_2O_2 content; (**D**) $O_2^{\cdot-}$ content. Values represent the means ± SDs (n = 10). Different lowercase letters indicate significant differences between treatments ($p < 0.05$), determined by two-way ANOVA.

2.9. Changes in Antioxidant Enzyme Activities and Gene Expression Levels in Transgenic Arabidopsis Plants Under Cold Stress

The SOD, POD, and CAT activities of the OE-*TaNAC6A* plants were the strongest under cold stress, followed by those of the STTM-*tae-miR164* and WT plants, while the *nac6* plants exhibited the lowest activity (Figure 16A,D,F). The RT-qPCR results for antioxidase-related genes (*AtSOD1*, *AtSOD2*, *AtPER3*, *AtCAT1*, *AtCAT2*, and *AtCAT3*) indicated that the expression levels of the OE-*TaNAC6A* and STTM-*tae-miR164* plants were consistently higher than those of the WT and *nac6* plants under the same low-temperature conditions (Figure 16B,C,E,G–I).

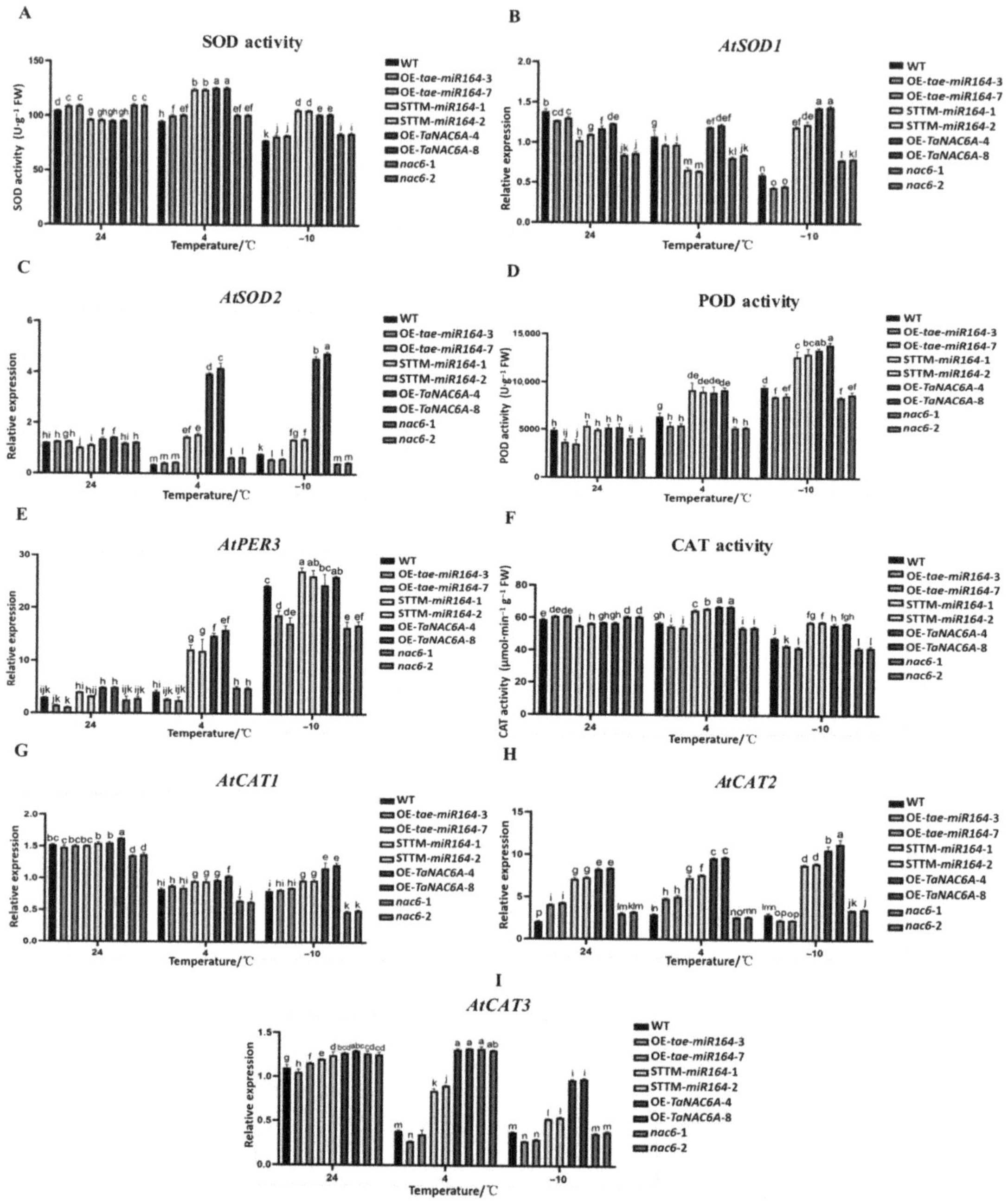

Figure 16. Changes in antioxidant enzyme activities and gene expression levels in Arabidopsis plants under cold stress. (**A**) SOD activity; (**B**) expression level of *AtSOD1*; (**C**) expression level of *AtSOD2*; (**D**) POD activity; (**E**) expression level of *AtPER3*; (**F**) CAT activity; (**G**) expression level of *AtCAT1*; (**H**) expression level of *AtCAT2*; (**I**) expression level of *AtCAT3*. Values represent the means ± SDs (n = 10). Different lowercase letters indicate significant differences between treatments ($p < 0.05$), determined by two-way ANOVA. The reference gene used is *AtActin*.

2.10. Alterations in the Expression of Cold-Signaling-Related Genes in OE-TaNAC6A Plants

Cold signaling is segregated into two major signaling pathways: ABA-independent and ABA-dependent pathways. In Arabidopsis, cold stress has been shown to induce the expression of ICE1, activating downstream DREB1/CBF members, which, in turn, induces an ABA-independent signaling pathway [27]. It is well-known that ABIs are key members in the ABA-dependent signaling pathway.

Therefore, we detected two genes (*AtDREB1*, *AtDREB2*) in the ABA-independent signaling pathway and three genes (*AtABI3*, *AtABI4*, and *AtABI5*) in the ABA-dependent signaling pathway by RT-qPCR. The results indicated that, as the temperature decreased, the expression levels of *AtDREB1*, *AtDREB2*, and *AtABI5* in the OE-*TaNAC6A* plants with high *TaNAC6A* expression exhibited a gradual increase. At 4 °C, the expression levels of *AtABI3* and *AtABI4* were both higher than those at 24 °C and −10 °C (Figure 17).

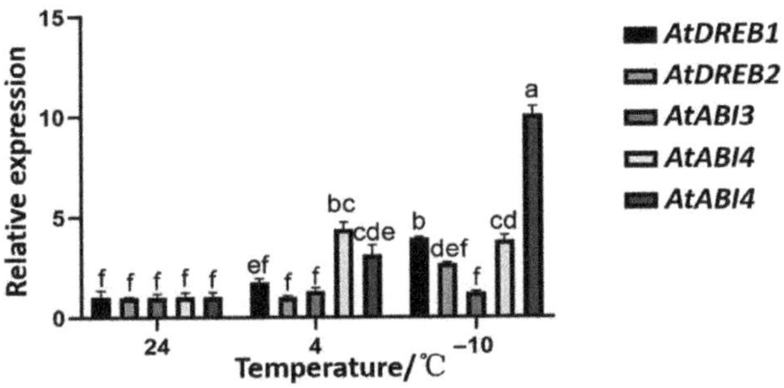

Figure 17. Relative expression levels of downstream target genes in Arabidopsis plants. Values represent the means ± SDs (n = 10). Different lowercase letters indicate significant differences between treatments ($p < 0.05$), determined by two-way ANOVA. The reference gene used is *AtAction*.

These results indicate that the overexpression of *TaNAC6A* enhanced the cold-signaling pathways in the Arabidopsis plants.

3. Discussion

An increasing body of evidence indicates that miRNAs are crucial in regulating plant growth, development, and stress tolerance [28]. In a previous study, high-throughput RNA sequencing libraries were constructed using tillering nodes from Dn1 as samples, collected at temperatures of 5 °C and −10 °C. Thirty miRNA families were identified in the library from samples at 5 °C, while 24 miRNA families were identified in the library from samples at −10 °C. A total of 53 genes were predicted to be targets of the 30 miRNA families in the 5 °C library [29]. To fully leverage Dn1's abundant genetic resources for cold resistance, it is essential to characterize the functions and regulatory mechanisms of the miRNAs associated with cold resistance. miR164 is one of the most conserved miRNAs and is involved in plant growth, development, and abiotic and biotic stress responses [21]. One miRNA molecule can control multiple target genes. Spychała et al. found that *tae-miR164* targeted *Lr46-RLK3*, a rapid responsive gene to *Puccinia triticina* infection [30]. In Wang et al.'s research, *TaMAPK4* was verified to be a target of *tae-miR164*, which played an important role in signaling during the wheat–*Pst* interaction [31].

Previous studies have also shown that *tae-miR164* targets plant-specific NAC transcription factors from the NAM subfamily of wheat. Feng et al. revealed that *TaNAC21/22*, a NAC transcription factor gene, was the target of *tae-miR164* and played an important role in regulating the resistance of host plants to stripe rust [32].

Specifically, *miR164* negatively regulats the expression of its target NAC transcription factor (TF). We identified *TaNAC6A mRNA* as the target gene of *tae-miR164* through psRNATarget analysis. Subsequently, this study confirmed that the *tae-miR164-TaNAC6A* module enhanced the cold resistance of Dn1, following a combination analysis of 5′RACE, tobacco transient co-transfection, and dual-luciferase assays.

Research into the stress-regulating functions of NAC transcription factors (TFs) has primarily focused on drought and salt stresses, with limited studies exploring signal transduction under low-temperature stress, particularly at sub-zero temperatures. Zhou et al. identified that the transient overexpression of *TaNAC6A/6B/6D* reduced the haustorium index of Yangmai158, and the stable transformation of *TaNAC6A* enhanced its resistance against *Bgt*, implying that TaNAC6s play important roles in basal resistance [33]. *AtNAC1* in Arabidopsis is the homologous gene of *TaNAC6A* in wheat, which was analysized by WheatOmics. Rodríguez-García et al. identified that AtNAC1 activates the expression of peptidase-encoding *AtCEPs* in roots to limit the root hair growth of Arabidopsis plants [34]. Xie et al.'s study provided mechanistic insights into how AtNAC1 regulates root ground tissue maturation by coordinating with the SCR/SHR-CYCD6;1 module in Arabidopsis plants [35]. Xie et al. even discovered the AtNAC1 transduced auxin signal downstream of TIR1 to promote lateral root development [36]. Our study is the first to validate the biological functions of *TaNAC6A* in wheat and *AtNAC1* in Arabidopsis plants in regulating plant cold tolerance.

Dong et al. discovered, in their research on tomato, that the *miR164a-NAM3* module induces ethylene synthesis by directly regulating the expressions of *SlACS1A*, *SlACS1B*, *SlACO1*, and *SlACO4*, thereby conferring cold tolerance to transgenic tomato plants [37].

Li et al. treated wheat with phenanthrene (polycyclic aromatic hydrocarbons, PAHs) and found that phenanthrene stress (ubiquitous organic pollutants in the environment) accelerated the senescence and death of wheat roots while stimulating the emergence of new roots. Phenanthrene stress upregulated the expression of *tae-miR164* and enhanced the silencing of *TaNAC1*, thereby inhibiting the occurrence of adventitious roots [38]. Our study found that the *tae-miR164-TaNAC6A* module can regulate cold tolerance in transgenic Arabidopsis plants, and we validated this at the phenotypic, physiological, and molecular levels.

Following recovery from exposure to −10 °C, the survival rates for the OE-*TaNAC6A* and STTM-*tae-miR164* plants were 84% and 77.6%, respectively. In contrast, the survival rates for the OE-*tae-miR164*, *nac6*, and WT plants were 22%, 28.4%, and 34%, respectively. The OE-*TaNAC6A* plants exhibited a higher survival rate than the STTM-*tae-miR164* plants, indicating significant cold tolerance. Additionally, under cold stress, the OE-*tae-miR164* plants showed higher MDA contents and electrical conductivities and lower Pro contents compared to the other four lines.

In general, extremely low temperatures often result in the excessive production of ROS, the accumulation of MDA, and increased conductivity—biomarkers indicative of cell membrane damage [39]. Plants strictly regulate their ROS levels by recruiting enzymatic and non-enzymatic antioxidants [40,41]. The activities of SOD, CAT, and POD in wheat leaves increased after the Methyl jasmonate (MJ) pretreatment of the plants subjected to hardening temperature, as compared to the control [42]. In our study, at 4 °C and −10 °C, the leaves of the OE-*tae-miR164* and *nac6* plants exhibited deeper staining by DAB and NBT than those of the other three lines. Moreover, the H_2O_2 and $O_2^{\cdot -}$ contents of the OE-*tae-miR164* and *nac6* plants increased with decreasing temperature. In addition, markers of membrane permeability and membrane lipid peroxidation increased in the OE-*tae-miR164* plants but decreased in the OE-*TaNAC6A* plants. The activities of SOD, POD, and CAT gradually decreased with decreasing temperature in the OE-*tae-miR164* and *nac6* plants. Finally, the expression levels of *AtSOD1/2*, *AtPER3*, and *AtCAT1/2/3* were lower in the

OE-*tae-miR164* and *nac6* plants than those in the other lines, consistent with the antioxidant activity assays.

DREB TFs regulate plant stress resistance. DREB1 and DREB2 regulate two different signaling pathways to resist stress caused by cold or dehydration [43]. ABA plays a crucial role in regulating plant growth during stress responses. By using molecular biology approaches, significant positive regulatory genes associated with ABA responses have been identified, including *ABI3*, *ABI4*, and *ABI5* [44]. Here, we evaluated the differential expressions of five cold-stress-related genes in transgenic OE-*TaNAC6A* plants. The expression levels of *AtDREB1/2* and *AtABI3/4/5* were significantly altered by exposure to cold temperatures. Together, the results suggest that *tae-miR164* negatively regulates cold tolerance while *TaNAC6A* promotes cold resistance.

The leaf morphology, size, and number are crucial for plant growth and development [45]. Compared to the WT plants, the *OE-tae-miR164* plants exhibited smoother leaf margins, whereas the STTM-*tae-miR164* plants showed increased serration. Notably, the trichome density in the OE-*tae-miR164* plants was lower than that in the other lines. Most plant leaves possess a waxy cuticle, composed of fatty acids or monoalcohols synthesized by their epidermal cells [46], which serves as a barrier against abiotic stressors such as radiation, low temperatures, and drought. SEM observations revealed that both the OE-*TaNAC6A* and STTM-*tae-miR164* plants had thicker, more waxy cuticles, even under normal temperature conditions. After exposure to 4 °C for three days, the stomata of all five lines tended to close, and their cuticles thickened. Nonetheless, the cuticles of the OE-*TaNAC6A* and STTM-*tae-miR164* plants remained thicker than those of the other lines.

Exposure to low temperatures often results in plasma vesiculation, enlargement of chloroplasts and mitochondria, and the blurring or disappearance of nuclear and plasma membranes, which can sometimes lead to complete cellular disintegration [47]. In this study, TEM observations showed normal cell structures in all five lines at room temperature, with the OE-*TaNC6A* plants having larger starch granules. After three days of exposure to 4 °C, the WT, OE-*tae-miR164*, and *nac6* plants displayed fusiform to circular chloroplast swelling, internal vacuolation, increases in osmophilic bodies, and expanded and/or disintegrated starch granules. In contrast, the OE-*TaNC6A* plants exhibited a relatively stable cell structure. Overall, the OE-*tae-miR164* plants showed poor organellar stability in response to cold stress. Conversely, plants overexpressing *TaNC6A* demonstrated stable cellular structures in response to low temperature, indicating improved cold tolerance.

Given that wheat is the allohexaploid progenitor, it is well-known that its genome is vast and complex, resulting in slow progress in gene cloning and functional studies. Therefore, this study selected Arabidopsis plants for heterologous validation, which also was the potential limitation to the in-depth analysis of the molecular mechanism of the *tae-miR164-TaNAC6A* module in wheat.

4. Materials and Methods

4.1. Plant Materials, Growth Conditions, and Treatments

Seeds of winter wheat, Dn1 (with strong cold resistance), provided by our laboratory, were sown on 12 September 2020. Approximately 200 seeds were planted per row at 0.2 m intervals and a depth of 5 cm within a 2 m planting area. The seeds were covered with soil to promote natural germination, and the plants were managed using conventional practices. Under natural cooling conditions in the field, tillering nodes and leaves were sampled when the ten-day-average minimum temperature reached about 5 °C (on 10 October 2020), 0 °C (on 29 October 2020), −10 °C (on 25 November 2020), and −25 °C (on 3 January 2021) [48].

Arabidopsis seeds, ecotype Col-0, as wild type (WT), provided by our laboratory, Arabidopsis mutant seeds *nac6* (SALK135734C, TAIR), purchased from The Arabidopsis Information Resource (TAIR), and the other three transgenic Arabidopsis seeds constructed by this study were cultivated under a light intensity of 120–150 μmolm^{-2} s^{-1}, 24 °C, a 16/8 h day/night photoperiod, and a relative humidity of 60%. After 28 days of cultivating, the Arabidopsis seedlings were cold-acclimated at 4 °C for 3 days and subsequently exposed to −10 °C for 3 h. Leaves were sampled at three distinct time points: before cold acclimation, after 3 days of cold acclimation, and after 3 h of exposure to −10 °C. Samples were wrapped in tin foil, immediately frozen in liquid nitrogen, and stored at −80 °C for future use [49]. Following the cold treatment, plants were grown at room temperature for an additional 7 days to calculate the survival rate. Plants that were dry and yellow and unable to regenerate were considered dead.

4.2. Gene Expression Analysis

Frozen tissues of winter wheat and fresh leaves of Arabidopsis plants were homogenized using liquid nitrogen. Total RNA was isolated from the plants using the Ultrapure RNA Kit (CW0581M, CWBIO, Suzhou, China). The cDNA was then synthesized from the mRNA with the HiScript III 1st Strand cDNA Synthesis Kit (R312, Vazyme, Beijing, China). RT-qPCR was employed to quantify the gene expression, following our previously published method [50]. The mRNA levels of the target genes were quantified with the $2^{-\Delta\Delta CT}$ method [51]. (Relevant RT-qPCR primers are listed in Table S1).

4.3. Bioinformatics Analysis

The amino acid sequence of TaNAC6A, along with the gene and promoter sequences of *tae-miR164* and *TaNAC6A*, were obtained from wheatOmics (http://202.194.139.32/, 20 December 2024) The precursor and mature sequences of *tae-miR164* were sourced from miRBase (https://www.mirbase.org/, 20 December 2024) and Dn1 miRNA libraries previously established by our lab. The base conservation in the mature miRNA was analyzed using WebLogo (http://weblogo.berkeley.edu/, accessed on 8 September 2025).

RNAfold (http://rna.tbi.univie.ac.at/cgi-bin/RNAWebSuite/RNAfold.cgi, 12 January 2025) was used to predict the stem–loop structure of the tae-miR164 precursor. ExPASy (http://expasy.org, 12 January 2025) was utilized to determine the physical and chemical properties of TaNAC6A, including the amino acid length, molecular weight, isoelectric point, and aliphatic index. PlantCare (http://bioinformatics.psb.ugent.be/webtools/plantcare/html, 12 January 2025) was employed to obtain the pre-*tae-miR164* and *TaNAC6A* promoter sequences for cis element analysis. The subcellular localization prediction website (http://www.csbio.sjtu.edu.cn/bioinf/Cell-PLoc-2, 12 January 2025) was used to forecast the subcellular localization of TaNAC6A.

4.4. Subcellular Localization Analysis

The coding sequence (CDS) of *TaNAC6A* was constructed into the pCAMBIA2300 -EGFP vector. Tobacco leaves (seeded 28 days prior) were injected with *Agrobacterium tumefaciens* strain GV3101 carrying pCAMBIA2300-*EGFP* and pCAMBIA2300-*TaNAC6A-EGFP*. Seventy-two hours after the injection, the leaves were harvested to observe the EGFP signals under a fluorescence microscope (SZX10, Olympus, Tokyo, Japan).

4.5. Transient Expression Assay in Tobacco

To analyze the promoter activity of pre-*tae-miR164* and *TaNAC6A* and verify the regulatory relationship between *tae-miR164* and its target gene (*TaNAC6A*), a transient expression assay was performed on tobacco (*Nicotiana tabacum*). The pre-*tae-miR164* and *TaNAC6A* promoter sequences (2000 bp) were constructed into the pBI121-GUS vector (relevant PCR primers are listed in Table S2). Tobacco leaves (seeded 28 days prior) were subsequently injected with *Agrobacterium tumefaciens* strain GV3101 carrying pBI121-*pre-tae-miR164-GUS* and pBI121-*ProTaNAC6A-GUS*. Seventy-two hours after the injection, the tobacco plants were exposed to 4 h at 4 °C. Subsequently, the leaves were harvested for histochemical staining.

4.6. 5′RACE Assay

The 5′RACE assay was conducted using a FirstChoice RLM-RACE Kit (Am1700, Thermo Fisher, Waltham, MA, USA), following the manufacturer's instructions. (The 5′ primers, including 5′ RACE inner, and 5′ RACE outer, were provided by the kit; the 3′ primers are listed in Table S3).

4.7. Dual-Luciferase Assay

The *pre-tae-miR164* sequence was constructed into the pGreenII-62-SK vector, and the upstream and downstream 100 bp sequences of the *tae-miR164* recognition site mutation of *TaNAC6A* were constructed into the pGreenII0800-miRNA vector. (Relevant PCR primers are listed in Table S4.) Tobacco leaves, aged 28 days, were subsequently injected with *Agrobacterium tumefaciens* strain GV3101 (pSoup), carrying both control and target vectors at a 1:1 (*v:v*) ratio. The tobacco plants were subsequently cultured in the dark for 1 day and in the light for 2 days. The abaxial surfaces of the leaves, facing upwards, were sprayed with a reaction solution containing 1 mM of D-luciferin and then placed in a dark environment for 15 min. Following treatment, the leaves were observed using a chemiluminescence/fluorescence imaging analysis system (5200Multi, Tanon, Shanghai, China) to detect the luminescence.

4.8. Overexpression Vector Construction and Arabidopsis Transformation

The ORFs of *TaNAC6A*, *pre-tae-miR164*, and STTM-*tae-miR164* were separately cloned into pCAMBIA2300, which was driven by the CaMV35S promoter and then transformed into the *Agrobacterium tumefaciens* strain GV3101. Particularly, the sequence was constructed using the mature *tae-miR164* sequence as a template, with three bases (CTA) inserted between complementary bases at positions 9 and 10 at the 5′ end of *tae-miR164*. STTM-*tae-miR164* forms a stem–loop structure following transcription, and the complementary portions of *tae-miR164* and STTM-*tae-miR164* exist in the form of single strands, binding complementarily with *tae-miR164*. The *Agrobacterium tumefaciens* strain GV3101 containing the target vector was individually transferred into Arabidopsis seedling ecotype Col-0 via the floral dip method. The positive transgenic lines were first screened on MS plates containing 50 $mg \cdot mL^{-1}$ kanamycin, and then the seedlings were detected by PCR.

4.9. Phenotyping and Physiological Index Assays

The seeds of Arabidopsis plants were sown on 1/2 MS solid medium, stratified for 2 days at 4 °C, and transferred to growth chambers (16 h day/8 h night at 24 °C). The length of the main root was measured by Image J 1.47 software. After 28-day-old Arabidopsis seedlings were cold-acclimated at 4 °C for 3 days, the morphologies of the leaves, stomata, and cellular ultrastructure were observed by a stereomicroscope (SZX10, Olympus, Tokyo, Japan), a scanning electron microscope (SU8010, Hitachi, Tokyo, Japan), and a transmission electron microscope (HT7800, Hitachi, Tokyo, Japan), respectively.

Leaf samples were collected after treatment. The electrical conductivity (relatively) was measured by a conductivity meter (DDS-307A, Leici, Shanghai, China). The relative H_2O_2 contents of the leaves were assessed by DAB staining, and the relative $O_2^{\cdot -}$ contents were assessed by NBT staining.

The contents of H_2O_2, $O_2^{\cdot-}$, Pro, and MDA were measured using one-to-one corresponding commercial kits from Comin Biotechnology (H_2O_2-2-Y, SA-2-G, PRO-2-Y, MDA-2-Y, Coming, Suzou, China), following the manufacturer's instructions. By using a visible spectrophotometer (721, Jinghua, Shanghai, China), the absorbance at 415 nm, 530 nm, 520 nm, and 532 nm was measured to calculate the contents of H_2O_2, $O_2^{\cdot-}$, Pro, and MDA.

The activities of SOD, POD, and CAT were measured using the relevant commercial kits from Comin Biotechnology (SOD-2-W, POD-2-Y, CAT-2-W, Comin, Suzhou, China), following the manufacturer's instructions. For the SOD, POD, or CAT activity determination, 0.1 g leaf samples was homogenized in 1 mL of PBS extraction buffer on ice. The absorbance at 560 nm, 470 nm, and 240 nm was measured to calculate the activities of SOD, POD, and CAT using a visible spectrophotometer (721, Jinghua, Shanghai, China). All experimental data were calculated according to the fresh weight.

4.10. Statistical Analysis

Three biological replicates were used for each experiment. Data are presented as the means of three biological replicates ± standard deviations (SDs), and data analyses were performed using IBM SPSS 26. All data were subjected to two-way analysis of variance (ANOVA). GraphPad Prism 8.0 was used for in-depth data analysis and for generating figures. A value of $p < 0.05$ was used to determine significance.

5. Conclusions

Taken together, our study suggests that the overexpression of *tae-miR164* significantly downregulated the expression of *TaNAC6A* and reduced the cold tolerance of Arabidopsis plants. We identified that tae-miR164 and TaNAC6A might be involved in the low-temperature response. The subcellular location indicated that it might function as a transcription factor in cold-stress tolerance. We confirmed a physical interaction between *tae-miR164* and *TaNAC6A*. Meanwhile, the results of the *TaNAC6A* and *tae-miR164* overexpression in Arabidopsis plants indicated that the *tae-miR164-TaNAC6A* module played a key role in regulating the cold tolerance in the transgenic Arabidopsis plants in the epidermal hair development and leaf margin phenotype and in regulating the balance of ROS, as well as in promoting the expressions of *AtDREB1*, *AtDREB2*, and *AtABI5*. Conversely, the overexpression of *TaNAC6A* enhances the cold tolerance of Arabidopsis plants. Our research findings reveal the physiological and molecular mechanisms of the *miR164-TaNAC6A* module in winter wheat responding to cold stress, providing new insights for in-depth studies on the ncRNA regulation of plant stress resistance. This has significant guiding implications for the molecular module breeding of cold resistance in other crops in order to create crop varieties that better meet the needs of humans. But bottlenecks still exist in molecular module breeding technology, such as high costs and an incomplete understanding of the interaction mechanisms between module molecules.

Supplementary Materials: The following supporting information can be downloaded at https://www.mdpi.com/article/10.3390/plants14182849/s1, Figure S1. Bioinformatics analysis of *tae-miR164* and its promoters. Figure S2. Bioinformatic analysis of TaNAC6A and the cis element in its promoter. Figure S3. Phylogenetic tree conducted by MEGA11 software. Table S1. Sequences of primers for qRT-PCR in wheat and Arabidopsis plants. Table S2. Sequences of primers for promoter amplification. Table S3. Sequences of primers for 5RACE. Table S4. Sequences of primers for dual-luciferase assay. Table S5. Sequences of primers for target gene clone.

Author Contributions: Q.X. designed the research; Z.D., X.Y., W.S. and Y.H. performed the experiments; Z.D. and W.S. analyzed the data; Q.X. and Z.D. wrote the manuscript; D.Z. and K.P. revised the manuscript. All authors have read and agreed to the published version of the manuscript.

Funding: This research was funded by the National Natural Science of China (31601236).

Data Availability Statement: The authors confirm that all data, tables, and figures in this manuscript are original.

Conflicts of Interest: The authors declare no conflicts of interest.

References

1. Ding, Y.; Shi, Y.; Yang, S. Molecular Regulation of Plant Responses to Environmental Temperatures. *Mol. Plant.* **2020**, *13*, 544–564. [CrossRef]
2. Soualiou, S.; Duan, F.; Li, X.; Zhou, W. Crop Production under Cold Stress: An Understanding of Plant Responses, Acclimation Processes, and Management Strategie. *Plant Physiol. Biochem.* **2022**, *190*, 47–61. [CrossRef] [PubMed]
3. Thomashow, M.F. Plant Cold Acclimation: Freezing Tolerance Genes and Regulatory Mechanisms. *Ann. Rev. Plant Physiol. Plant Mol. Biol.* **1999**, *50*, 571–599. [CrossRef] [PubMed]
4. John, R.; Anjum, N.A.; Sopory, S.K.; Akram, N.A.; Ashraf, M. Some Key Physiological and Molecular Processes of Cold Acclimation. *Biol. Plantarum.* **2016**, *60*, 603–618. [CrossRef]
5. Zhou, J.; Wang, J.; Shi, K.; Xia, X.J.; Zhou, Y.H.; Yu, J.Q. Hydrogen Peroxide is Involved in the Cold Acclimation Induced Chilling Tolerance of Tomato Plants. *Plant Physiol. Biochem.* **2012**, *60*, 141–149. [CrossRef]
6. Wang, Y.X.; Ya, H.U.; Chen, B.H.; Zhu, Y.F.; Dawuda, M.M.; Svetla, S. Physiological Mechanisms of Resistance to Cold Stress Associated with 10 Elite Apple Rootstocks. *J. Integr. Agric.* **2018**, *17*, 857–866. [CrossRef]
7. Hussain, N.; Yasmeen, A.; Afzal Muhammad, A. Exogenously Applied Growth Promoters Modulate the Antioxidant Enzyme System to Improve the Cotton Productivity under Water Stress Conditions. *Ital. J. Agron.* **2020**, *15*, 165. [CrossRef]
8. Guo, Z.; Cai, L.; Liu, C.; Chen, Z.; Guan, S.; Ma, W.; Pan, G. Low temperature Stress Affects Reactive Oxygen Species, Osmotic Adjustment Substances, and Antioxidants in Rice (*Oryza sativa* L.) at the Reproductive Stage. *Sci. Rep.* **2022**, *12*, 6224. [CrossRef]
9. Sun, X.; Lin, L.; Sui, N. Regulation Mechanism of MicroRNA in Plant Response to Abiotic Stress and Breeding. *Mol. Biol. Rep.* **2019**, *46*, 1447–1457. [CrossRef]
10. Lu, S.; Sun, Y.H.; Chiang, V.L. Stress-responsive microRNAs in Populus. *Plant J. Cell Mol. Biol.* **2008**, *5*, 131–151. [CrossRef] [PubMed]
11. Barrera-Figueroa, B.E.; Gao, L.; Wu, Z.; Zhou, X.; Zhu, J.; Jin, H.; Liu, R.; Zhu, J.-K. High Throughput Sequencing Reveals Novel and Abiotic Stress-Regulated MicroRNAs in the Inflorescences of Rice. *BMC Plant Biol.* **2012**, *12*, 132. [CrossRef] [PubMed]
12. Tang, Z.; Zhang, L.; Xu, C.; Yuan, S.; Zhang, F.; Zheng, Y.; Zhao, C. Uncovering Small RNA-mediated Responses to Cold Stress in a Wheat Thermosensitive Genic Male-Sterile Line by Deep Sequencing. *Plant Physiol.* **2012**, *159*, 721–738. [CrossRef]
13. Cao, X.; Wu, Z.; Jiang, F.; Zhou, R.; Yang, Z. Identification of Chilling Stress-Responsive Tomato MicroRNAs and Their Target Genes by High-Throughput Sequencing and Degradome Analysis. *BMC Genom.* **2014**, *15*, 1130. [CrossRef]
14. Wang, P.; Yang, Y.; Shi, H.; Wang, Y.; Ren, F. Small RNA and Degradome Deep Sequencing Reveal Respective Roles of Cold-Related MicroRNAs Across Chinese Wild Grapevine and Cultivated Grapevine. *BMC Genom.* **2019**, *20*, 740. [CrossRef]
15. Wang, Q.; Liu, N.; Yang, X.; Tu, L.; Zhang, X. Small RNA-Mediated Responses to Low and High-Temperature Stresses in Cotton. *Sci. Rep.* **2016**, *6*, 35558. [CrossRef]
16. Shi, X.; Jiang, F.; Wen, J.; Wu, Z. Overexpression of Solanum Habrochaites microRNA319d (sha-miR319d) Confers Chilling and Heat Stress Tolerance in Tomato (*S. lycopersicum*). *BMC Plant Biol.* **2019**, *19*, 214. [CrossRef]
17. Peng, K.; Tian, Y.; Sun, X.; Song, C.; Ren, Z.; Bao, Y.; Xing, J.; Li, Y.; Xu, Q.; Yu, J.; et al. Tae-miR399-UBC24 Module Enhances Freezing Tolerance in Winter Wheat via a CBF Signaling Pathway. *J. Agric. Food Chem.* **2021**, *69*, 13398–13415. [CrossRef] [PubMed]
18. Tsai, W.A.; Sung, P.H.; Kuo, Y.W.; Chen, M.-C.; Jeng, S.-T.; Lin, J.-S. Involvement of microRNA164 in Responses to Heat Stress in Arabidopsis. *Plant Sci.* **2023**, *329*, 111598. [CrossRef]
19. Fang, Y.; Xie, K.; Xiong, L. Conserved miR164-targeted NAC Genes Negatively Regulate Drought Resistance in Rice. *J. Exp. Bot.* **2014**, *65*, 2119–2135. [CrossRef] [PubMed]
20. Lin, D.; Zhu, X.; Qi, B.; Gao, Z.; Tian, P.; Li, Z.; Lin, Z.; Zhang, Y.; Huang, T. SlmiR164A Regulates Fruit Ripening and Quality by Controlling SlNAM2 and SlNAM3 in Tomato. *Plant Biotechnol. J.* **2022**, *20*, 1456–1469. [CrossRef]
21. Qing, C.; Du, L.; Wen, M.; Ruo, N.; Bao, W.; Li, G.; Meng, M.; Xiang, L.; Hui, Z. The MiR164-TaNAC14 Module Regulates Root Development and Abiotic-stress Tolerance of Wheat Seedlings. *J. Integr. Agric.* **2023**, *22*, 981–998.
22. Diao, P.; Chen, C.; Zhang, Y.; Meng, Q.; Lv, W.; Ma, N. The Role of NAC Transcription Tactor in Plant Cold Response. *Plant Signal. Behav.* **2020**, *15*, 1785668. [CrossRef]
23. Hou, X.M.; Zhang, H.F.; Liu, S.Y.; Wang, X.-K.; Zhang, Y.-M.; Meng, Y.-C.; Luo, D.; Chen, R.-G. The NAC Transcription Factor CaNAC064 is a Regulator of Cold Stress Tolerance in Peppers. *Plant Sci.* **2020**, *291*, 110346. [CrossRef]

24. Ju, Y.L.; Yue, X.F.; Min, Z.; Wang, X.-H.; Fang, Y.-L.; Zhang, J.-X. VvNAC17, a Novel Stress-responsive Grapevine (*Vitis vinifera* L.) NAC Transcription Factor, Increases Sensitivity to Abscisic Acid and Enhances Salinity, Freezing, and Drought Tolerance in Transgenic Arabidopsis. *Plant Physiol. Biochem.* **2020**, *146*, 98–111. [CrossRef]
25. Luo, P.; Chen, Y.; Rong, K.; Lu, Y.; Wang, N.; Xu, Z.; Pang, B.; Zhou, D.; Weng, J.; Li, M.; et al. ZmSNAC13, a Maize NAC Transcription Factor Conferring Enhanced Resistance to Multiple Abiotic Stresses in Transgenic Arabidopsis. *Plant Physiol. Biochem.* **2022**, *170*, 160–170. [CrossRef] [PubMed]
26. Liu, L.; Cang, J.; Yu, J.; Wang, X.; Huang, R.; Wang, J.; Lu, B. Effects of Exogenous Abscisic Acid on Carbohydrate Metabolism and the Expression Levels of Correlative Key Enzymes in Winter Wheat Under Low Temperature. *Biosci. Biotechnol. Biochem.* **2013**, *77*, 516–525. [CrossRef] [PubMed]
27. Abhinandan, K.; Skori, L.; Stanic, M.; Hickerson, N.M.N.; Jamshed, M.; Samuel, M.A. Abiotic Stress Signaling in Wheat-An Inclusive Overview of Hormonal Interactions During Abiotic Stress Responses in Wheat. *Front. Plant Sci.* **2018**, *9*, 734. [CrossRef]
28. Ma, X.; Zhao, F.; Zhou, B. The Characters of Non-Coding RNAs and Their Biological Roles in Plant Development and Abiotic Stress Response. *Int. J. Mol. Sci.* **2022**, *23*, 4124. [CrossRef]
29. Huang, R.; Cang, J.; Yu, J.; Lu, B.; Liu, L.; Wang, J.; Guo, R.; Xu, C. Solexa Sequencing and Bioinformatics Analysis of Small RNA in Winter Wheat. *Chin. Bull. Bot.* **2014**, *49*, 8–18. [CrossRef]
30. Spychała, J.; Tomkowiak, A.; Noweiska, A.; Bobrowska, R.; Bocianowski, J.; Sobiech, A.; Kwiatek, M.T. Diversity of Expression Patterns of Lr34, Lr67, and Candidate Genes towards Lr46 with Analysis of Associated miRNAs in Common Wheat Hybrids in Response to *Puccinia triticina* Fungus. *Curr. Issues Mol. Biol.* **2024**, *46*, 5511–5529. [CrossRef]
31. Wang, B.; Song, N.; Zhang, Q.; Wang, N.; Kang, Z. TaMAPK4 Acts as a Positive Regulator in Defense of Wheat Stripe-Rust Infection. *Front. Plant Sci.* **2018**, *15*, 152. [CrossRef]
32. Feng, H.; Duan, X.; Zhang, Q.; Li, X.; Wang, B.; Huang, L.; Wang, X.; Kang, Z. The Target Gene of tae-miR164, a Novel NAC Transcription Factor from the NAM Subfamily, Negatively Regulates Resistance of Wheat to Stripe rust. *Mol. Plant Pathol.* **2014**, *15*, 284–296. [CrossRef]
33. Zhou, W.; Qian, C.; Li, R.; Zhou, S.; Zhang, R.; Xiao, J.; Wang, X.; Zhang, S.; Xing, L.; Cao, A. TaNAC6s are Involved in the Basal and Broad-spectrum Resistance to Powdery Mildew in Wheat. *Plant Sci.* **2018**, *277*, 218–228. [CrossRef]
34. Rodríguez-García, D.R.; Rondón Guerrero, Y.D.C.; Ferrero, L.; Rossi, A.H.; Miglietta, E.A.; Aptekmann, A.A.; Marzol, E.; Pacheco, J.M.; Carignani, M.; Gabarain, V.B.; et al. Transcription Factor NAC1 Activates Expression of Peptidase-encoding AtCEPs in Roots to Limit Root Hair Growth. *Plant Physiol.* **2023**, *194*, 81–93. [CrossRef]
35. Xie, C.; Li, C.; Wang, F.; Zhang, F.; Liu, J.; Wang, J.; Zhang, X.; Kong, X.; Ding, Z. NAC1 Regulates Root Ground Tissue Maturation by Coordinating with the SCR/SHR-CYCD6;1 Module in Arabidopsis. *Mol. Plant* **2023**, *16*, 709–725. [CrossRef]
36. Xie, Q.; Frugis, G.; Colgan, D.; Chua, N.-H. Arabidopsis NAC1 Transduces Auxin Signal Downstream of TIR1 to Promote Lateral Root Development. *Genes Dev.* **2000**, *14*, 3024–3036. [CrossRef] [PubMed]
37. Dong, Y.; Tang, M.; Huang, Z.; Song, J.; Xu, J.; Ahammed, G.J.; Yu, J.; Zhou, Y. The miR164a-NAM3 Module Confers Cold Tolerance by Inducing Ethylene Production in Tomato. *Plant J.* **2022**, *111*, 440–456. [CrossRef]
38. Li, J.; Zhang, H.; Zhu, J.; Shen, Y.; Zeng, N.; Liu, S.; Wang, H.; Wang, J.; Zhan, X. Role of miR164 in the Growth of Wheat New Adventitious Roots Exposed to Phenanthrene. *Environ. Pollut.* **2021**, *284*, 117204. [CrossRef] [PubMed]
39. Agarwal, P.; Mitra, M.; Banerjee, S.; Roy, S. MYB4 Transcription Factor, a Member of R2R3-subfamily of MYB Domain Protein, Regulates Cadmium Tolerance via Enhanced Protection Against Oxidative Damage and Increases Expression of PCS1 and MT1C in Arabidopsis. *Plant Sci.* **2020**, *297*, 110501. [CrossRef] [PubMed]
40. Feng, S.; Wang, B.; Li, C.; Guo, H.; Bao, A.-K. Transcriptomic Analysis Provides Insight into the ROS Scavenging System and Regulatory Mechanisms in *Atriplex canescens* Response to Salinity. *Int. J. Mol. Sci.* **2022**, *24*, 242. [CrossRef]
41. Hasanuzzaman, M.; Bhuyan, M.; Zulfiqar, F.; Raza, A.; Mohsin, S.M.; Al Mahmud, J.; Fujita, M.; Fotopoulos, V. Reactive Oxygen Species and Antioxidant Defense in Plants under Abiotic Stress: Revisiting the Crucial Role of a Universal Defense Regulator. *Antioxidants* **2020**, *9*, 681. [CrossRef]
42. Repkina, N.; Ignatenko, A.; Holoptseva, E.; MiszalskI, Z.; Kaszycki, P.; Talanova, V. Exogenous Methyl Jasmonate Improves Cold Tolerance with Parallel Induction of Two Cold-Regulated (COR) Genes Expression in *Triticum aestivum* L. *Plants* **2021**, *10*, 1421. *Plants* **2021**, *10*, 1421. [CrossRef]
43. Reeves, W.M.; Lynch, T.J.; Mobin, R.; Finkelstein, R.R. Direct Targets of the Transcription Factors ABA-Insensitive (ABI4 and ABI5) Reveal Synergistic Action by ABI4 and Several bZIP ABA Response Factors. *Plant Mol. Biol.* **2011**, *75*, 347–363. [CrossRef]
44. Agarwal, P.K.; Agarwal, P.; Reddy, M.K.; Sopory, S.K. Role of DREB Transcription Factors in Abiotic and Biotic Stress Tolerance in Plants. *Plant Cell Rep.* **2006**, *25*, 1263–1274. [CrossRef]
45. Arvidsson, S.; Pérez-Rodríguez, P.; Mueller-Roeber, B. A Growth Phenotyping Pipeline for Arabidopsis Thaliana Integrating Image Analysis and Rosette Area Modeling for Robust Quantification of Genotype Effects. *New Phytol.* **2011**, *191*, 895–907. [CrossRef] [PubMed]

46. Riederer, M. Thermodynamics of the Water Permeability of Plant Cuticles: Characterization of the Polar Pathway. *J. Exp. Bot.* **2006**, *57*, 2937–2942. [CrossRef] [PubMed]
47. Wang, Y.M.; Zhou, D.M.; Yuan, X.Y.; Zhang, X.H.; Li, Y. Modeling the Interaction and Toxicity of Cu Cd Mixture to Wheat Roots Affected by Humic Acids, in Terms of Cell Membrane Surface Characteristics. *Chemosphere* **2018**, *199*, 76–83. [CrossRef] [PubMed]
48. Bao, Y.; Xing, J.; Liang, Y.; Ren, Z.; Fu, L.; Yu, J.; Wang, D.; Zhang, D.; Xu, Q.; Cang, J. Analysis of Overwintering Indexes of Winter Wheat in Alpine Regions and Establishment of A Cold Resistance Model. *Field Crops Res.* **2022**, *275*, 108347. [CrossRef]
49. Tian, Y.; Peng, K.; Bao, Y.; Zhang, D.; Meng, J.; Wang, D.; Wang, X.; Cang, J. Glucose-6-phosphate Dehydrogenase and 6-phosphogluconate Dehydrogenase Genes of Winter Wheat Enhance the Cold Tolerance of Transgenic Arabidopsis. *Plant Physiol. Biochem.* **2021**, *161*, 86–97. [CrossRef]
50. Peng, K.; Tian, Y.; Cang, J.; Yu, J. Overexpression of TaFBA-A10 from Winter Wheat Enhances Freezing Tolerance in Arabidopsis thaliana. *J. Plant Growth Regul.* **2022**, *41*, 314–326. [CrossRef]
51. Livak, K.J.; Schmittgen, T.D. Analysis of Relative Gene Expression Data Using Real-time Quantative PCR and the $2^{-\Delta\Delta CT}$ Method. *Methods* **2001**, *25*, 402–408. [CrossRef] [PubMed]

plants

MDPI

Article

Leveraging Transcriptome Insights and *GsHZ4* Gene Expression to Improve Alkaline Tolerance in *Lupinus angustifolius*

Jixiang Tang [1,†], Mengyu Liu [1,†], Yujing Liu [1], Xiaoyu Wang [1], Xinlei Du [1], Xinyao Liu [1], Mengyue Zhou [1], Honglin Mao [1], Yixuan Feng [1], Qing Gao [1], Junfeng Zhang [1,2] and Lei Cao [1,*]

[1] College of Horticulture and Landscape Architecture, Northeast Agricultural University, No. 600, Changjiang Road, Xiangfang District, Harbin 150030, China; jxtang1019@163.com (J.T.); liumm0730@163.com (M.L.); yujingliu0000@163.com (Y.L.); wxy766696114@163.com (X.W.); duxinlei0101@163.com (X.D.); liuxinyao000927@163.com (X.L.); 15995318092@163.com (M.Z.); 18345749688@163.com (H.M.); 18846043687@163.com (Y.F.); 13039763390@163.com (Q.G.); jfzhang1989@163.com (J.Z.)

[2] School of Geography and Tourism, Harbin University, No. 109, Zhongxing Avenue, Nangang District, Harbin 150076, China

[*] Correspondence: leicao@neau.edu.cn

[†] These authors contributed equally to this work.

Abstract

Soil alkalinity severely restricts the cultivation of *Lupinus angustifolius*, a valuable legume. Wild soybean (*Glycine soja*) is a leguminous plant with extremely strong alkaline resistance (pH 8.5). Transferring the alkali-tolerant genes from wild soybeans into lupinus can effectively enhance the alkali tolerance. In this study, we combined transcriptome profiling and genetic transformation to elucidate the molecular basis of alkaline stress response in lupinus. RNA-seq analysis of root tips under acid (HCl, pH 4.0) and alkali ($NaHCO_3$, pH 8.5) stress revealed 104,353 annotated unigenes, with differential expression patterns highlighting enrichment in cellular component, binding, and catalytic activity categories. KEGG pathway analysis indicated that early responses involved ribosome-related pathways, while later stages activated plant hormone signaling and MAPK pathways. Notably, no homeodomain-leucine zipper (HD-Zip) family genes were identified in the lupinus genome. Therefore, we transferred GsHZ4, an alkali-resistant HD-Zip transcription factor from wild soybean into lupinus hairy roots via *Agrobacterium rhizogenes*-mediated transformation. Overexpression of *GsHZ4* significantly enhanced antioxidant enzyme activities (CAT, POD, and SOD) and reduced malondialdehyde content under $NaHCO_3$ stress. Furthermore, the promoter of *GsHZ4* expression was strongly induced by indole-3-acetic acid (IAA). Key alkali-responsive genes (*LaKIN*, *LaMYB34*, *LaDnaJ1*, *LaDnaJ20*, *LaNAC22*, and *LaNAC35*) were upregulated in transgenic lines, suggesting that *GsHZ4* integrates into the endogenous stress-regulation network. Our findings demonstrate that heterologous expression of *GsHZ4* can enhance alkaline tolerance of lupinus, providing a novel strategy for breeding stress-resistant varieties and expanding lupinus cultivation in saline–alkali soils.

Keywords: *Lupinus angustifolius*; *Glycine soja*; RNA-seq; *GsHZ4*; alkaline stress

1. Introduction

Saline–alkali soils restrict agricultural production (crop growth and yields) and development in China, an issue that has been a focus of research [1]. Relative to salt stress, alkali stress involves higher level pH; it usually causes Na^+, $HCO_3{}^-$, $SO_3{}^{2-}$, and $CO_3{}^{2-}$ toxicity.

 https://doi.org/10.3390/plants14223408

However, the mechanism of plant alkali tolerance is not fully understood [2]. Lupinus (*Lupinus angustifolius)*, a member of the butterfly flower subfamily of the legume family, has palmate compound leaves with 5–9 leaflets and 5–10 cm-long terminal racemes. This colorful plant is used widely in landscaping. Lupinus leaves are rich in protein and are good source of essential amino acids for animals. They are used widely in chicken, duck, cow, rabbit, and pig feed to improve meat and milk production quality [3]. Lupinus roots could use solid phosphorus from the soil and fix nitrogen through rhizobia, improving soil nutrition. Thus, lupinus has high economic value. Compared to neutral or alkaline soil, lupinus is more suitable for growing in slightly acidic soil [4]. Adapting/Adjusting soil pH during planting is difficult, it often requires soil amendment by applying sulfur powder, which complicates the process and increases the costs of introduction and cultivation. This study was conducted to examine lupinus' alkaline resistance, with the aim of increasing the lupinus planting area and improving the utilization of saline–alkali soil.

The publication of the lupinus genome has facilitated molecular breeding efforts aimed at increasing the plant's alkaline resistance and reducing its reliance on acidic soil. In 2019, the lupinus genome was reported [5], enabling the small-scale adaptation of the species [6]. Lupinus transformation via *Agrobacterium rhizogenes* has been described [7]. Alkali stress inhibits lupinus growth and nutrient absorption, and the strong buffering capacity of HCO_3^- is a key factor determining the survival of lupinus in calcareous soil [4]. The treatment of farmland soil with filter mud improved the growth, seed alkaloid content, and yield of lupinus termis by upregulating the plant-water relationship [8]. Previous research had focused on the molecular mechanism of solid phosphorus utilization by lupinus.

Wild soybean (*Glycine soja*) has strong resistance to stress and can grow under extreme conditions [9]. Its genes have been used to improve the salt–alkali resistance of cultivated soybean [10,11]. Lupinus and wild soybean are leguminous plants. Thus, the transfer of wild soybean genes related to adaptation to specific environments into related flowering legumes and forage is an efficient and reliable means of improving these varieties. Exogenous genes can improve the abiotic stress tolerance of model plants. *GmWRKY16* enhances the abiotic tolerance of *Arabidopsis* [12]. The overexpression of *GmUBC2* enhances the drought tolerance of *Arabidopsis* [13]. The overexpression of the rice gene *SNAC1* improves the salt tolerance of transgenic cotton [14]. *ZmmiR156* improves the drought and salt resistance of tobacco [15]. Wild soybean genes can be transferred into lupinus to improve its alkaline stress tolerance. In this study, we transferred wild soybean genes into lupinus hairy roots, where they were expressed normally. These exogenous genes responded to stress and hormones by regulating peroxide metabolism and endogenous stress-tolerance genes in lupinus.

The HB family constitutes a superfamily of transcription factors characterized by a conserved homeodomain, playing pivotal roles in plant embryonic development, organogenesis, and cell differentiation [16]. Its subfamily, HD-Zip, defined by the distinctive homeodomain-leucine zipper (HD-LZ) motif, functions through dimerization to bind DNA. It is further categorized into four classes (I–IV), which specifically regulate diverse processes including stress responses, photomorphogenesis, vascular development, and epidermal differentiation, representing a unique and crucial group of plant-specific regulatory factors. The HB genes share a homologous protein domain with a conserved 60-amino-acid motif that always interacts specifically with DNA sequences from downstream genes' promoters. Members of the HD-Zip family have a leucine zipper (LZ) motif immediately downstream of the HD. The HD is responsible for specific binding to DNA, and the LZ acts as a dimerization motif. HD-Zip proteins bind to DNA as dimers, and the absence of an LZ abolishes their binding ability. HD-Zip genes participate in plant responses to salt stress by activating the antioxidant system, regulating osmotic homeostasis, and maintaining Na^+/K^+ homeostasis. For example, the HD-ZIP I gene *MdHB7*, found in apples, reduces the H_2O_2 and O_2^- concentrations and increases superoxide dismutase (SOD) and peroxidase (POD) activity, thereby reducing oxidative damage under salt stress [17]. The heterologous overexpression of the pepper gene *CaHB1* in tomato improves photosynthesis and stress resistance [18]. Lupinus was found to have some genes homologous to those of wild soybean, but to lack alkali tolerance genes such as *GsHZ4*. This genomic difference may be attributable to evolution and may underlie the difference in tolerance. The *GsHZ4* gene was derived from wild soybean strain *G07256*, which has strong saline–alkali resistance. Expression pattern and tissue and subcellular localization analyses and yeast one-hybrid assays indicated that *GsHZ4* is an HD-Zip transcription factor that binds to the *GsNAC019* promoter and improves plants' alkali resistance [9,19]. It improved this resistance in lupinus, and this approach represents a new strategy enabling the expansion of lupinus cultivation in saline–alkali soil.

2. Results

2.1. Lupinus Responses to Acid and Alkali Treatments

The DNBSEQ platform was used to measure a total of 95.77 Gb of data. After assembly and redundancy removal, 104,353 unigenes were obtained, with a total length, average length, N50, and GC content of 166,491,728 bp, 1595 bp, 2183 bp, and 40.45%, respectively (Table 1). Quality statistics for the filtered reads and the results of reference genome alignment are provided in Tables 2 and 3, respectively. The comparison of the unigenes with seven functional annotation databases yielded values of 94,646 (90.70%) for NR, 92,967 (89.09%) for NT, 75,069 (71.94%) for SwissProt, 77,746 (74.50%) for KOG, 76,181 (73.00%) for KEGG, 77,940 (74.69%) for GO, and 76,334 (73.15%) for Pfam (Table 4). In total, 81,056 CDSs were detected using Transdecoder and 4618 unigenes encoding transcription factors were predicted.

Table 1. Results of Trinity assembly.

Unigene	Total Length	Average Length	N50	GC%
104,353	166,491,728 bp	1595 bp	2183 bp	40.45%

Table 2. Quality statistics of filtered reads.

Sample	Total Raw Reads (M)	Total Clean Reads (M)	Total Clean Bases (Gb)	Clean Reads Q20 (%)	Clean Reads Q30 (%)	Clean Reads Ratio (%)
A1	45.57	42.35	6.35	97.42	93.19	92.93
A2	47.33	43.43	6.52	96.75	92.16	91.78
A3	45.57	42.64	6.4	96.75	92.12	93.56
B1	45.57	42.72	6.41	97.39	93.12	93.74
B2	47.33	42.27	6.34	96.72	92.07	89.31
B3	45.57	42.68	6.4	96.53	91.65	93.64
C1	45.57	42.68	6.4	97.42	93.2	93.65
C2	45.57	42.3	6.34	96.84	92.32	92.81
C3	45.57	42.43	6.36	96.67	91.98	93.09
D1	45.57	42.04	6.31	96.75	92.11	92.24
D2	45.57	42.36	6.35	96.58	91.77	92.95
D3	47.33	42.93	6.44	96.66	91.97	90.7
E1	47.33	43.24	6.49	96.66	91.96	91.37
E2	45.57	42.15	6.32	96.73	92	92.48
E3	47.33	42.28	6.34	96.81	92.29	89.34

Table 3. Result of reference genome alignment.

Sample	Total Clean Reads (M)	Total Mapping (%)	Uniquely Mapping (%)
A1	42.35	86.73	17.58
A2	43.43	87.39	17.72
A3	42.64	87.38	17.73
B1	42.72	86.51	17.88
B2	42.27	85.02	17.49
B3	42.68	87.02	18.1
C1	42.68	87.35	18.27
C2	42.3	87.03	18.42
C3	42.43	87.46	18.43
D1	42.04	84.99	17.81
D2	42.36	85.43	17.98
D3	42.93	84.19	17.7
E1	43.24	86.77	17.78
E2	42.15	86.38	17.71
E3	42.28	85.94	17.45

Table 4. Annotated number and ratio of the unigenes in various databases.

Databases	Number of Unigenes	Ratio (%)
All	104,353	100.00%
NR	94,646	90.07%
NT	92,967	89.09%
Swiss Prot	75,069	71.94%
KOG	77,746	74.50%
KEGG	76,181	73.00%
GO	77,940	74.69%
Pfram	76,334	73.15%

Compared with unigene group A, group B had 29,139 DEGs (14,669 upregulated and 14,470 down-regulated), group C had 32,366 DEGs (15,909 upregulated and 16,457 down-regulated), group D had 36,706 DEGs (21,108 upregulated and 15,598 down-regulated), and group E had 31,228 DEGs (15,505 upregulated and 15,723 down-regulated; Figure 1A). The number of upregulated genes relative to the control was largest after 3 h alkali treatment (Figure 1B). In total, 75,171 DEGs were common to the five groups; 630 and 635 genes were significantly expressed only after 3 h and 6 h acid treatment, respectively, and 4200 and 605 genes were significantly expressed only after 3 h and 6 h alkali treatment, respectively. Among DEGs, 10,281 genes were differentially expressed in all comparison groups; 3187 and 3943 genes were differentially expressed after 3 h and 6 h acid treatment, respectively, and 10,120 and 3744 genes were differentially expressed after 3 h and 6 h alkali treatment, respectively (Figure 1C,D). Volcano maps of the relative differential gene expressions are provided in Figure 2D. All upregulated and down-regulated DEGs are listed in Supplementary Files S1–S8.

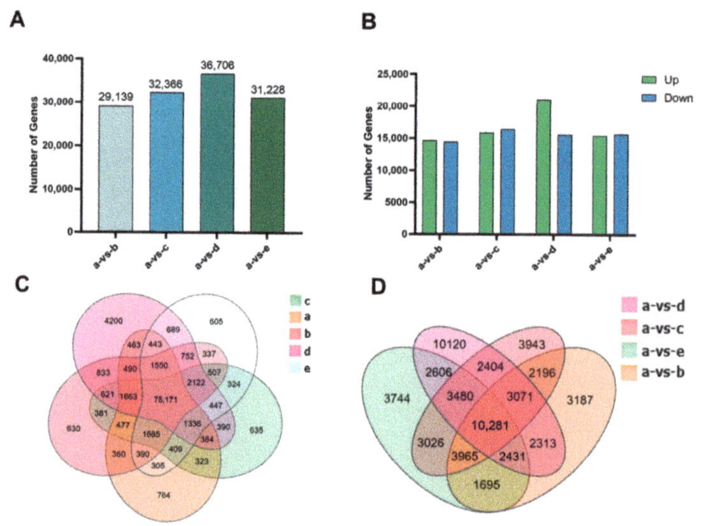

Figure 1. Differential expressed genes analysis after transcriptome sequencing. (**A**) The number of DEGs in each group compared with the control group; (**B**) the number of upregulated (green columns) and down-regulated (blue columns) genes in all DEGs in each group compared with the control group. (**C**) Venn map of DEGs numbers in each group. (**D**) Venn map of DEG numbers in each group compared with the control group. Group 'a' represents the control group, 'b' represents the acid treatment for 3 h, 'c' represents the acid treatment for 6 h, 'd' represents the alkali treatment for 3 h, and 'e' represents the alkali treatment for 6 h.

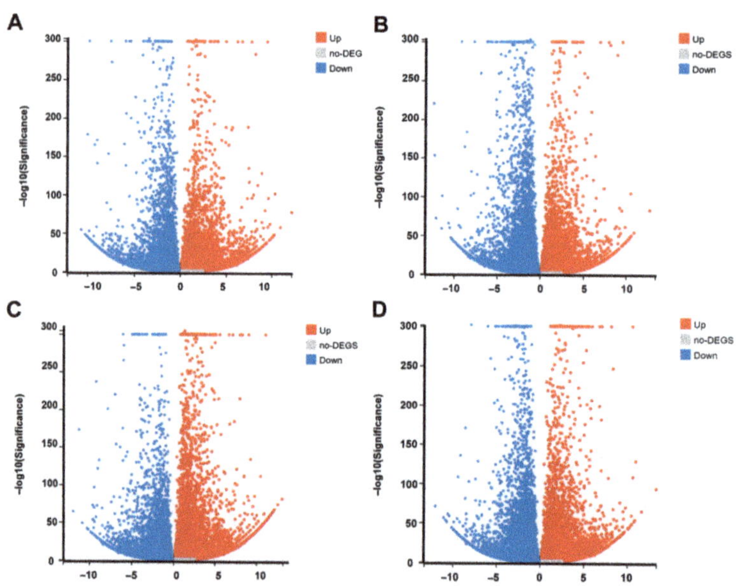

Figure 2. DEGs volcano map of each experimental group compared with the control group (pH 5.5). (**A**) pH 4.0 for 3 h and untreated. (**B**) pH 4.0 for 6 h and untreated. (**C**) pH 8.5 for 3 h and untreated. (**D**) pH 8.5 for 6 h and untreated.

2.2. Differential Gene Expression Under Acid and Alkali Treatments

Gene annotation using the GO database revealed little difference in DEG classification after acid or alkali treatment for 3 h or 6 h. Under acid and alkaline treatments, DEGs participated in processes related to cellular anatomical entities, binding, and catalytic activity, as well as cellular and metabolic processes. The number of DEGs in each of these categories differed between treatment types. Under 3 h and 6 h acid treatment, most DEGs in the biological process category were involved in cell and metabolic processes (Figure 3A,B). Most of those in the cellular component category were related to cellular anatomical entities, and most in the molecular function category were involved in binding and catalytic activity (Figure 3C,D). KEGG terms for corresponding pathways are shown in Figure 4A–D. Under 3 h acid or alkaline stress, most upregulated genes participated in ribosome pathways; under 6 h acid or alkaline treatment, genes involved in plant–pathogen interactions and the MAPK signaling pathway were upregulated. The expression of lupinus genes related to the synthesis of organic matter changed first at 3 h, and those related to the plant hormone and signal transduction response changed at 6 h. The largest number of transcription factors was from the MYB family. We found that there are no HD-Zip family genes in lupinus genomes, but in HB, MYB, C3H, AP2, bHLH, and other transcription factor genes (Supplementary File S9 and Figure 5A). There are a few HB genes in lupinus, and we chose *LaSBH1* as an example to show the conserved domains in them. Both *LaSBH1* and *GsHZ4* have the same HOX domain, but *GsHZ4* from the HD-ZIP family has the special 'Zip' domain at the end of HOX domain (Figure 5B). So, it was predicted that the mission of the 'Zip' domain in key genes may cause a tolerance reduction in lupinus.

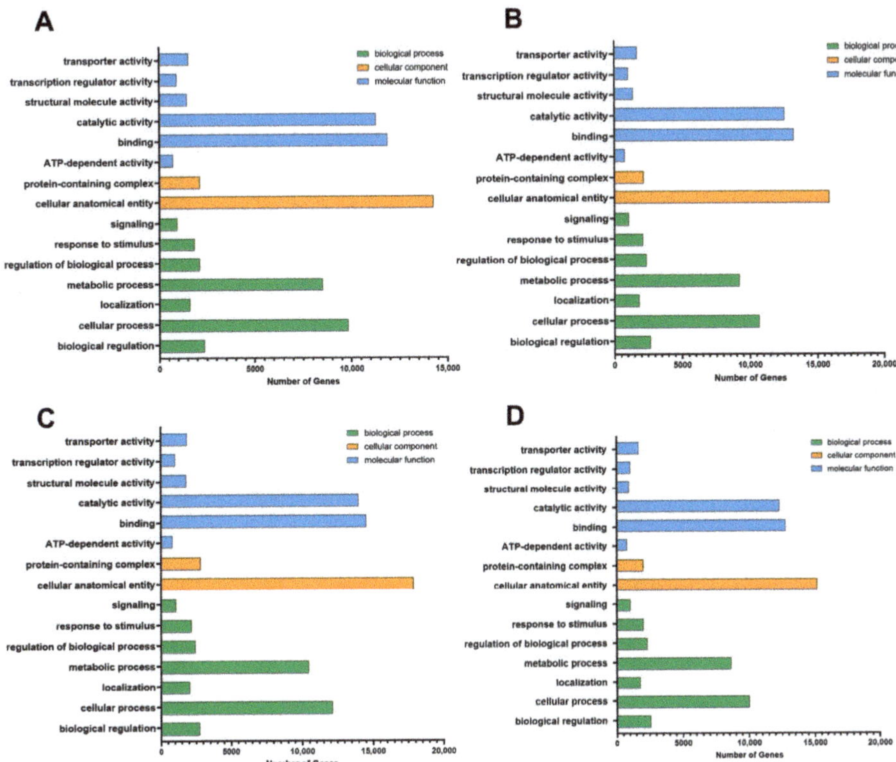

Figure 3. GO annotation diagram of DEGs in each experimental group compared with the control group. (**A**) pH 4.0 for 3 h versus untreated. (**B**) pH 4.0 for 6 h versus untreated. (**C**) pH 8.5 for 3 h versus untreated. (**D**) pH 8.5 for 6 h versus untreated.

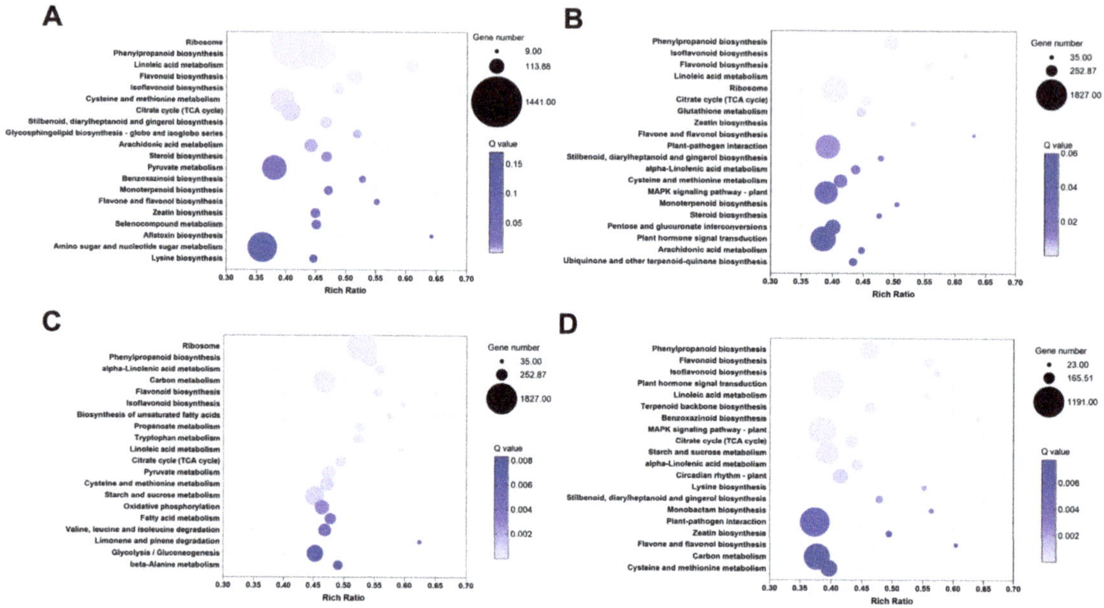

Figure 4. KEGG pathways of DEGs in each experimental group compared with the control group. (**A**) pH 4.0 treated 3 h versus untreated. (**B**) pH 4.0 treated 6 h versus untreated. (**C**) pH 8.5 for 3 h versus untreated. (**D**) pH 8.5 for 6 h versus untreated.

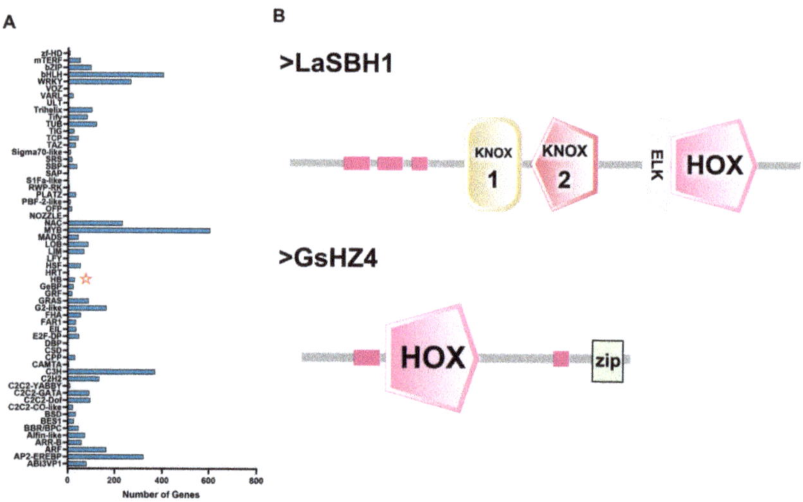

Figure 5. Transcription factor families and domains analysis in lupinus. (**A**) Transcription factors annotation and classification in all DEGs in lupinus. The HB family is marked with a star. (**B**) Conservative domain comparison between *LaSBH1* and *GsHZ4*. The pink squares represent the conserved motifs.

2.3. *GsHZ4 Overexpression in Lupinus Roots Improved Alkali Stress Tolerance*

We separately transformed the GFP empty vector and *GsHZ4*-GFP into the root tips of mung bean, this signal was distributed evenly in multiple areas of the root tips, indicating that *GsHZ4*-GFP could be stably expressed in the root tips (Figure 6A). The relative expression of *GsHZ4* increased with 0 mM, 50 mM, and 100 mM $NaHCO_3$ treatment

(Figure 6B). The experiments proved that lupinus could generate transgenic hairy roots, and that wild soybean genes could be expressed in response to alkali treatment. Relative to those in WT roots, the CAT and POD activity levels were significantly higher in overexpressed plants before treatment; CAT activity levels were higher under 20 mM, 40 mM, and 60 mM $NaHCO_3$ treatment (Figure 6C), and POD activity levels were higher at all concentrations less than 100 mM (Figure 6D). The SOD content of overexpressed plants was significantly greater than that of the WT under 40 mM, 60 mM, 80 mM, and 100 mM $NaHCO_3$ treatment (Figure 6E). The MDA content was significantly less, indicating less membrane damage in the overexpressed plants than in the WT under all treatments (Figure 6F). These results indicate that *GsHZ4* improved antioxidant enzyme activity and maintained cell membrane permeability in lupinus. The highest alkali tolerance threshold for transgenic plants was about 100 mM.

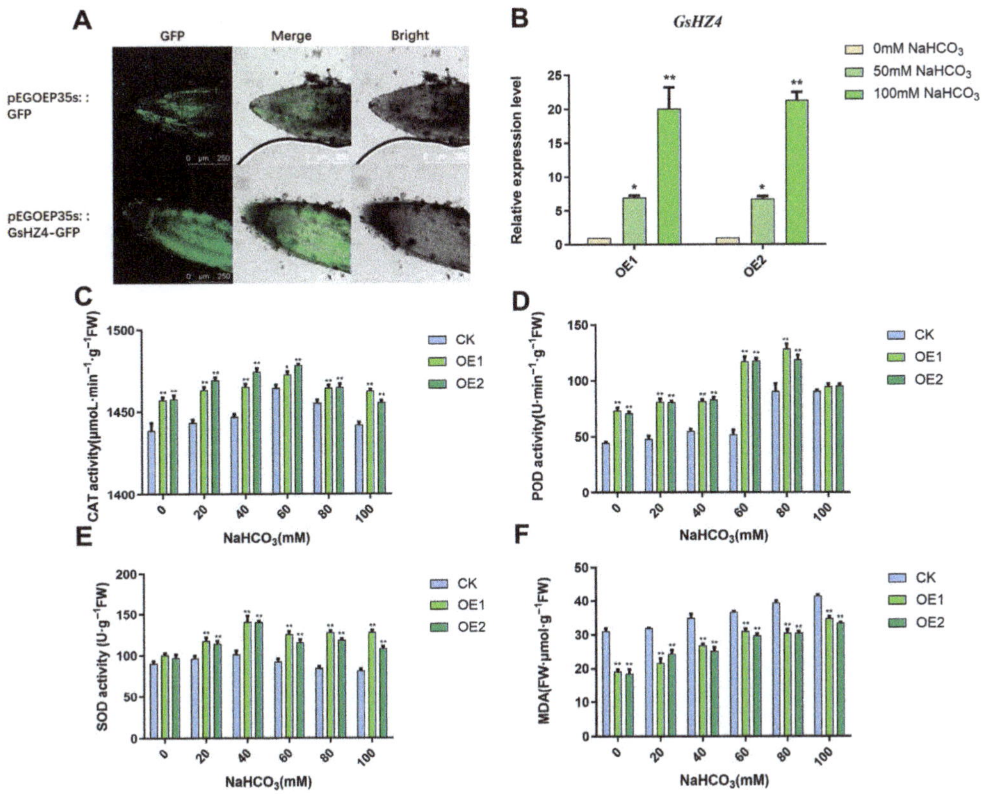

Figure 6. Overexpression of *GsHZ4* in lupinus hairy roots can improve alkaline resistance. (**A**) Laser confocal microscopy was used to detect the GFP signal in the root tips of new lupinus roots and the signal containing *GsHZ4*-GFP was more strongly expressed in the root tips. (**B**) The expression level of the *GsHZ4* gene in transgenic roots treated with 0 mM, 50 mM, and 100 mM $NaHCO_3$. (**C**) CAT enzyme activity. (**D**) POD enzyme activity. (**E**) SOD enzyme activity. (**F**) MDA content. For (**B–F**), a significant difference was calculated by Student's *t*-test (* $p < 0.05$, ** $p < 0.01$).

2.4. GsHZ4 Overexpression Increased Plant Growth Under Alkali Stress

Under control conditions, both the overexpression and control group plants exhibited comparable, normal growth phenotypes. The WT strains showed leaf wilting, yellowing, curling, and shrinking after 48 h treatment with 50 mM $NaHCO_3$, whereas the overexpressed plants showed no obvious change. After 48 h treatment with 100 mM $NaHCO_3$, the WT plants wilted completely, with thin stems and curled, wrinkled leaves; the overexpressed plants showed some leaf yellowing but grew normally (Figure 7A). The weight and root length of the overexpressed plants were significantly greater than those of the WT plants (Figure 7B–D). DAB and NBT staining were similar in untreated overexpressed and WT plant leaves, but weaker in overexpressed than in WT leaves, reflecting lesser H_2O_2 and O_2^- contents, under 100 mM $NaHCO_3$ (Figure 7E,F). Thus, *GsHZ4* overexpression improved the alkali tolerance of lupinus roots, leading to improved oxidation resistance and tolerance of the plants' aboveground parts.

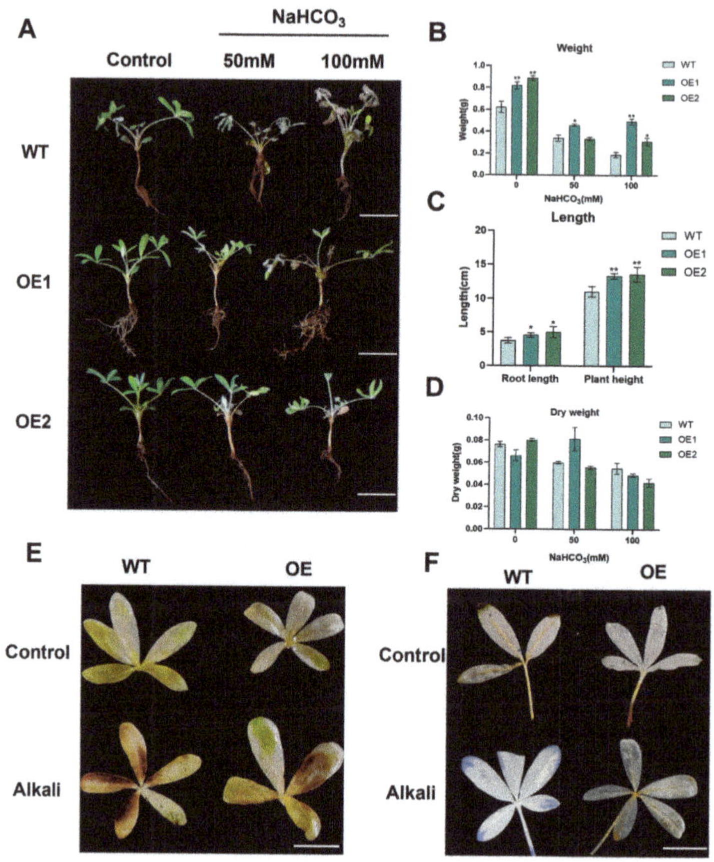

Figure 7. Phenotype observation of lupinus seedlings. (**A**) Plants (one control group and two groups of overexpressed plants) treated with 0 mM, 50 mM, and 100 mM $NaHCO_3$. Bar = 2 cm. (**B**) Determination of fresh plant weight in each group. (**C**) The dry weight of each group was determined. (**D**) The root length of each group was measured. (**E**) DAB staining experiments on the leaves of the control group and the experimental group under the treatment of 0 mM and 100 mM $NaHCO_3$. The brown part represents oxidative damage. Bar = 1 cm. (**F**) NBT staining of leaves in the control group and the experimental group under the treatment of 0 mM and 100 mM $NaHCO_3$. The blue part represents oxidative damage. Bar = 1 cm. For (**B–D**), a significant difference was calculated by Student's *t*-test (* $p < 0.05$, ** $p < 0.01$).

2.5. GsHZ4 Improved Plant Alkaline Tolerance by Regulating Alkali Stress-Related Genes

Alkali stress-related genes were detected in overexpressed plants to predict the alkali resistance regulatory pathway of *GsHZ4* [20]. In untreated plants, the relative expression of all genes was significantly lower than in the WT. *LaKIN* is a gene responsible for sustained responses to alkali stress. *LaMYB34* is a transcription factor gene that responds to alkali stress [9]. Under 50 mM $NaHCO_3$ treatment, the relative expression of *LaKIN* and *LaMYB34* in *GsHZ4*-overexpressing lines was greater than in WT plants (Figure 8A,B). The relative expression of *LaKIN* and *LaMYB34* was significantly decreased in overexpressed lines relative to that in WT plants under 100 mM $NaHCO_3$ treatment. The expression level of *LaDnaJA6* (a downstream functional gene) decreased with increasing alkaline concentration in WT, while it exhibited an upward trend in overexpression lines (Figure 8C). The expression level of *LaNFYC* (responsible for sustained responses to alkali stress) generally exhibited a downward trend (Figure 8D). These findings suggest that *GsHZ4* regulates the *LaKIN*-based response to alkali stress. In addition, *LaMYB34* may be located in the same pathway as *GsHZ4*.

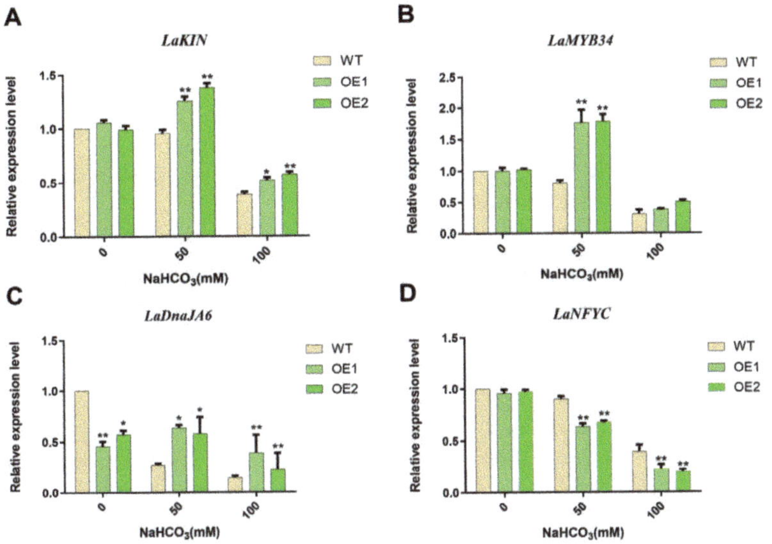

Figure 8. Relative expression levels of alkali stress related marker genes in WT and two groups of overexpressed plants treated with 0 mM, 50 mM, and 100 mM $NaHCO_3$. (**A**) *LaKIN* (**B**) *LaMYB34* (**C**) *LaDanJA6* (**D**) *LaNFYC*. For (**A–D**), a significant difference was calculated by Student's *t*-test (* $p < 0.05$, ** $p < 0.01$).

The alkaline response-related genes were screened out from the transcriptome sequencing results (Figure 9A). The expression level of *LaDNAJ1* was relatively low before any processing; its expression was significantly greater in overexpressed plants than in the WT under 50 and 100 mM $NaHCO_3$ treatment, and *LaDnaJ20* expression was significantly greater in overexpressed plants than in the WT under 0 and 50 mM $NaHCO_3$ treatment (Figure 9B,C). However, the expression levels of *LaDnaJ20*, *LaNAC22*, and *LaNAC35* were significantly lower in *GsHZ4*-overexpressing lines under 100 mM $NaHCO_3$ treatment. The relative expression of *LaNAC21* was significantly lower in overexpressed than in WT plants under all treatments (Figure 9D). The relative expression of *LaNAC22* was greater in overexpressed than in WT plants under 50 mM $NaHCO_3$ treatment (Figure 9E). The relative expression of *LaNAC35* was higher in untreated overexpressed plants than in the WT, but significantly lower than in the WT under all treatments (Figure 9F). These results suggest that *GsHZ4* transfer improved plant alkaline tolerance by regulating the expression of related genes (*LaDnaJ1*, *LaDnaJ20*, and *LaNAC22*) under alkali stress.

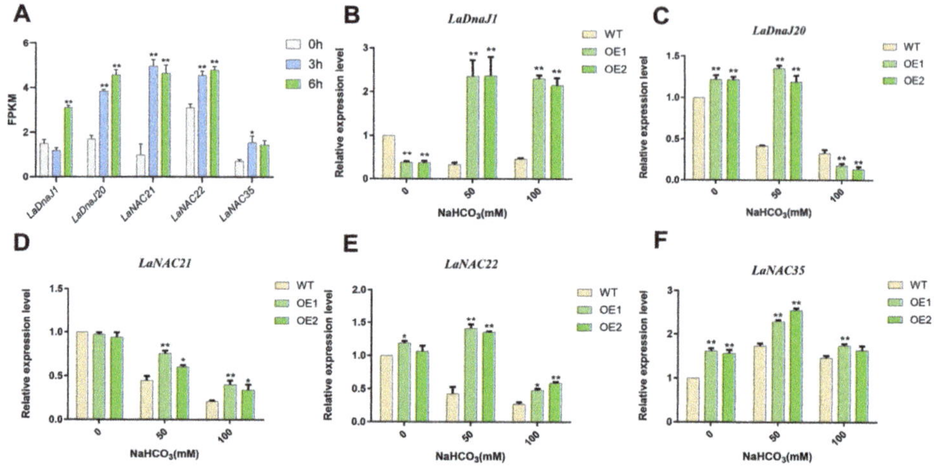

Figure 9. Relative expression levels of alkali stress-related genes from the RNA-seq data. WT and two groups of overexpressed plants were treated with 0 mM, 50 mM, and 100 mM $NaHCO_3$. (**A**) FPKM values of genes related to alkali stress (**B**) *LaDnaJ1*, (**C**) *LaDnaJ20*, (**D**) *LaNAC21*, (**E**) *LaNAC22*, and (**F**) *LaNAC35*. For (**A–F**) a significant difference was calculated by Student's *t*-test (* $p < 0.05$, ** $p < 0.01$).

2.6. GsHZ4 Was Induced by IAA in Lupinus Roots and Was Slightly Responsive to ABA, JA, and GA

GsHZ4 was significantly upregulated by IAA (Figure 10A–D). Under ABA (abscisic acid) treatment, its expression was down-regulated at 6 h and upregulated at 12 h. Under JA (jasmonic acid) treatment, *GsHZ4* expression was significantly upregulated at 3 h. *GsHZ4* expression was significantly down-regulated after 12 h of GA (gibberellin) treatment. These results indicate that *GsHZ4* in lupinus was induced mainly by IAA and regulated by ABA and was involved mainly in the IAA and ABA signaling pathways. In hormone treatment experiments, *GsHZ4* is initiated by its own promoter in roots of lupinus plants.

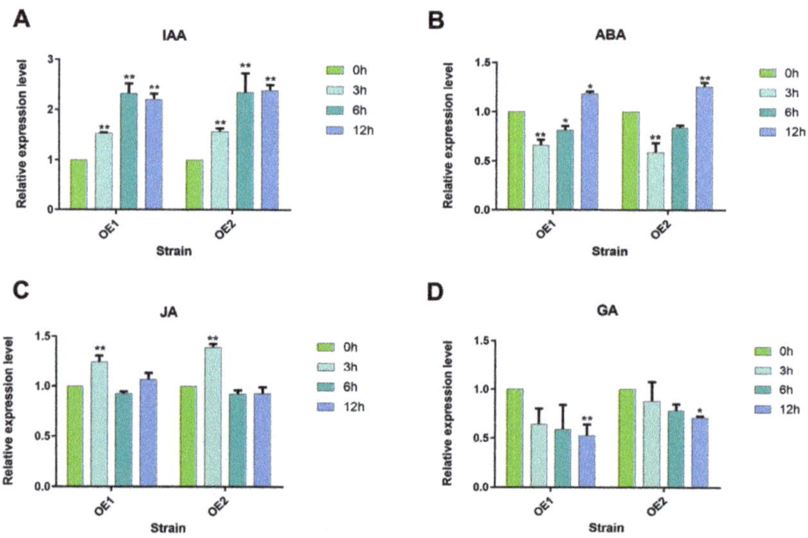

Figure 10. The relative expression of *GsHZ4* in the roots of lupinus seedlings (WT and *pGsHZ4::GsHZ4 overexpression lines*) treated with different hormones at 0 h, 3 h, 6 h, and 12 h was measured in OE groups. (**A**) 4 mg/L ABA (**B**) 5 μM IAA (**C**) 10 μM JA (**D**) 0.5 μM GA. For (**A–D**), a significant difference was calculated by Student's *t*-test (* $p < 0.05$, ** $p < 0.01$).

3. Discussion

3.1. Transcriptome Sequencing Analysis of Acid/Alkali Response Genes

Transcriptome sequencing is a high-throughput sequencing method applied in research conducted under defined environmental conditions. It has been used to examine plants' salt stress tolerance, a vital step in efforts to breed salt-resistant crops [21]. Different pH treatments can reveal differentially expressed genes in plants. For example, in tomato roots exposed to varying pH conditions, transpiration rates and stomatal conductance show distinct alterations [22]. Given the limited research progress reported on lupinus, we performed transcriptome sequencing on lupinus under different pH conditions to facilitate the identification of biologically significant insights. In this study, transcriptome sequencing revealed that lupinus lacks HD-Zip transcription factors, a finding that may explain its sensitivity to alkaline stress. The use of the *Agrobacterium rhizogenes*-mediated transformation and *GsHZ4* to obtain chimeric lupinus plants with improved alkaline tolerance is an innovative aspect of this research. *GsHZ4* was successfully transferred into lupinus hairy roots via *Agrobacterium rhizogenes* and expressed. The qPCR result analysis indicated that *GsHZ4* increased the expression levels of alkali stress-related genes (*LaKIN*, *LaMYB34*, *LaDnaJ1*, *LaDnaJ20*, and *LaNAC21*) which differed under alkali stress (Figures 8 and 9). These results indicate the feasibility of using various stress resistance regulation models from wild soybean to improve the ornamental characteristics of lupinus.

3.2. The Role of HD-ZIP Family Genes in Alkali Stress Response

The HD-ZIP (homeodomain-leucine zipper) family of transcription factors plays a pivotal role in regulating plant growth, developmental processes, and stress responses, particularly under alkali stress conditions. Alkali stress, characterized by elevated pH levels and excessive sodium carbonate/bicarbonate concentrations, induces significant osmotic and ionic toxicity in plants, resulting in growth retardation and cellular damage [23]. Emerging evidence indicates that HD-ZIP genes participate in plant adaptation to alkaline environments through the regulation of stress-responsive gene networks and maintenance of ion homeostasis. In *Arabidopsis thaliana*, for example, *AtHB7* (a member of the HD-ZIP I subfamily) has been shown to improve alkali tolerance by modulating ABA-mediated signaling cascades and osmotic adjustment mechanisms. Similarly, Medicago truncatula exhibits upregulation of *MtHB2* under alkali stress conditions, implying its functional involvement in stress adaptation [24]. Notably, recent studies have revealed that the HD-ZIP family contributes to the regulation of plant height in *Lagerstroemia* species and mediates stress response mechanisms. The evolutionary conservation of structural domains among HD-ZIP members, especially in their DNA-binding homeodomains and leucine zipper motifs, suggests the existence of conserved regulatory mechanisms across diverse plant species [25]. In light of increasing global soil alkalization problems, elucidating the molecular mechanisms underlying HD-ZIP-mediated alkali stress responses could provide valuable insights for developing stress-tolerant lupinus cultivars, thereby facilitating their cultivation in marginal agricultural lands. This study, based on the results of transcriptome sequencing, found that the HD-ZIP family genes were absent in the transcriptome of lupinus. However, as both belong to the legume family, there are significant differences in their alkaline tolerance between lupinus, wild soybean, and soybean. Since previous experiments demonstrated that *GsHZ4* enhances alkali tolerance in *Arabidopsis thaliana* [9], we initiated this study to introduce this exogenous gene into lupinus.

3.3. Mechanism of Alkali Stress Regulation by the Transcription Factor GsHZ4

Among the DEGs identified in this study, we found HB transcription factors that lack a zip domain (Figure 5). Gene Ontology analysis indicated that the products of these HB family genes are involved in cellular component organization and biological processes. *LaKIN1* expression was found to be increased in *GsHZ4*-overexpressing *Arabidopsis thaliana*, suggesting that *GsHZ4* positively regulates this expression and participates in the response to bicarbonate stress [9]. MYB is involved in the regulation of a variety of biological processes. *GmMYB14* may be involved in plant responses and resistance to environmental stresses via the regulation of related gene expression [26]. DnaJA proteins act as molecular chaperones related to environmental stresses in legumes. For example, *GmDnaJA6* is a candidate regulator of salt-alkali stress resistance that confers increased tolerance of this stress in soybeans overexpressing it [2]. The expression levels of *GmDnaJA9* and *GmDnaJA13* are higher under salt and alkali stresses, especially after 6 and 12 h, compared with the control conditions. These findings suggest that wild soybean genes can regulate lupinus genes and improve the stress tolerance of lupinus. However, we found that *GsHZ4* overexpression in lupinus inhibited the expression of genes such as *LaDnaJA6*, *LaNFYC*, *LaNAC21*, and *LaNAC35* (Figures 8 and 9). These genes may be located in different pathways. *GmNF-YC14*, which encodes an NF-YC member in soybean, forms heterotrimers with *GmNF-YA16* and *GmNF-YB2* to activate the *GMPYR1*-mediated ABA signaling pathway, thereby regulating soybean tolerance to drought and salt stresses [27]. These results suggest that *GsHZ4* may function through specific regulatory pathways and protein interactions, exerting its influence via coordinated action within the broader regulatory network. The diverse functions of *GsHZ4* in lupinus need to be studied further.

3.4. Hormone-Mediated Alkali Stress Tolerance

Hormones play important regulatory roles in plant growth. IAA induces crop root development [28]. Auxin regulates the expression of HD-Zip genes via the elegantly short TIR1/AFB signal transduction pathway [29]. Several auxin response elements (AuxREs) have been recognized in the promoters of wheat HD-Zip subfamily genes, with the upregulation of the transcript levels of *PIN1*, *PIN3*, and *PIN4*. Under saline conditions, auxin affects the direction and depth of root growth, helping plants to better absorb water and nutrients and adapt [30]. *AtHB2* encodes an HD-Zip transcription factor and functions as an auxin responsive gene in Arabidopsis [31]. Some HD-Zip genes are involved in the ABA-mediated pathway [32]. While no prior studies have reported the response of *GsHZ4* to IAA induction, its functional homolog, *GhHB12*, is known to be auxin-inducible. *GhHB12* has been implicated in regulating plant height in cotton by modulating auxin signal transduction and cell wall expansion [33]. In this study, the roots of untreated plants grew longer than those of the control, likely due to the participation of *GsHZ4* promoter in the IAA regulatory pathway, resulting in $NaHCO_3$ and alkaline stress tolerance. $NaHCO_3$ resistance processes may share the same pathway with IAA and ABA hormone regulation. In this project, we found that these response mechanisms could not be activated in the absence of HD-Zip genes, which may be the root cause of the difference in stress tolerance between wild soybean and lupinus. Further study of the molecular mechanisms of *GsHZ4*'s activity in lupinus is needed.

4. Materials and Methods

4.1. Plant Material and Culture Conditions

The experiment was conducted at the Horticulture Experimental Center of Northeast Agricultural University, situated at 45°44′28″ N and 126°43′16″ E. The lupinus seeds were purchased from the commercial supplier 'Hua You Xiu'. Seedlings were initially cultured in nutrient pots filled with a soil mixture (nutrient soil:vermiculite = 1:1). After the development of the third pair of leaves, they were transferred to a hydroponic system with pH 5.5 Hoagland solution. The bowls were maintained at a controlled temperature of

22–25 °C under a 16 h light/8 h dark photoperiod until the third pair of palmate compound leaves had fully developed. The seedlings were transferred to Hoagland solution for $NaHCO_3$ stress and hormone treatment. Wild soybean seeds were surface-sterilized by briefly soaking in concentrated sulfuric acid for 5 min, followed by rinsing with distilled water 5 times. The sample was placed in an incubator at 28 °C for 48 h and cultured in 1/4 Hoagland medium (pH 5.8) for 21 days. The temperature was 26-30 °C with 16 h light/8 h dark conditions [34].

4.2. Transcriptome Sequencing Sampling and Detection Methods

Lupinus plants were cultured in pH 5.5 Hoagland solution as a control (group A/a). Other lupinus plants were treated in pH 4.0 and pH 8.2 Hoagland solution for 3 h (group B/b and D/d) and 6 h (group C/c and E/e). Root tips were frozen with liquid nitrogen and stored in the -80 °C refrigerator. Each group of samples was taken 3 times. A total of 15 samples were sent to the BGI for transcriptome sequencing. The data were uploaded to the NCBI SRA database (SUB14514279).

4.3. Differential Expression Genes Analysis

SOAPnuke (v1.6.5), developed by the BGI (Beijing Genomics Institution), was used as a filter software to remove reads containing joints; reads with unknown base N content greater than 1% were removed. Low-quality reads were removed. The filtered 'Clean Reads' are saved in FASTQ format. Bowtie2 (https://bowtie-bio.sourceforge.net/bowtie2/index.shtml, accessed on 2 November 2025), a memory-efficient short-read aligner utilizing the Burrows–Wheeler transform and FM-indexing algorithm, was employed to perform sequence alignment and quantify gene/transcript expression levels (in FPKM/RPKM/TPM) by mapping the processed sequencing reads against the reference genome/transcriptome. This alignment-based quantification enabled comparative analysis of sequence utilization efficiency across different assembly strategies. Trinity (v2.13.2) was used for de novo assembly of clean reads (removing PCR duplicates to improve assembly efficiency). CD-HIT (CD-HIT: 4.6) was used for clustering to remove redundancy to obtain the unigenes. Unigenes were compared to the SwissProt database and Pfam database. TransDecoder.Predict (v5.5.0) was used to predict the CDS structures, such as 3′ UTR, 5′ UTR, CDS, and mRNA, etc. Venn diagrams were drawn using Venn (version 1.7). The volcano map of the DEGs was created using Sangerbox [35].

4.4. GO Annotation and KEGG Analysis

All unigene sequences were aligned to the NR (http://ftp.ncbi.nlm.nih.gov/blast/db, accessed on 2 November 2025), NT (http://ftp.ncbi.nlm.nih.gov/blast/db, accessed on 2 November 2025), Uniprot (https://www.uniprot.org, accessed on 2 November 2025), KEGG (https://www.kegg.jp, accessed on 2 November 2025), GO (http://geneontology.org, accessed on 2 November 2025), and COG (clusters of orthologous groups) (https://www.ncbi.nlm.nih.gov/COG/, accessed on 2 November 2025) databases. The GO (Gene Ontology) database consists of three branches: molecular functions, biological processes, and cellular components which were used to describe the function of gene products. According to KEGG annotation, the differential expressed genes were classified into different biological pathways. Generally, the Q value ≤ 0.05 was considered as significant enrichment. Trinity assembly of quality statistics of filtered reads is shown in Table 1. Results of the reference genome alignment are shown in Table 2. The annotated number and ratio of the unigenes in various databases is shown in Table 3. The annotated number and ratio of the unigenes in various databases is shown in Table 4.

4.5. Determination of Positive Transgenic Hairy Roots and Physiological Signs

Gene cloning primers are listed in Table 5. The empty vector pEGOEP35S (p35S)-GFP (Biorun, Wuhan, China) carries the green fluorescent protein (GFP) gene. The p35S-GFP construct was generated by replacing the GUS coding sequence in the pBI121 backbone with a functional GFP gene, thereby placing GFP expression under the control of the CaMV 35S promoter. Protoplasts were observed under incident-light fluorescence microscopy 24 h after electroporation, with 20–60% of protoplasts exhibiting intense green fluorescence upon excitation with blue light (450–490 nm). Agrobacterium rhizogenes strain K599 harboring either pEGOEP35S::*GsHZ4*-GFP or the empty pEGOEP35S::GFP plasmid was cultured at 28 °C. A 5 mL aliquot of LB liquid medium supplemented with kanamycin (50 μg/mL) was dispensed into a sterile 15 mL polypropylene tube. Single colonies were inoculated into the medium and incubated at 28 °C with shaking at 180 rpm for 48 h. The bacterial culture was then collected and centrifuged. After discarding the supernatant, the pellet was resuspended to a final volume of 400 μL, and the suspension was spread onto LB solid medium followed by incubation at 28 °C for 2 days until dense Agrobacterium growth was observed. After lupinus seed germination, roots were excised at a 45° angle approximately 1 cm below the cotyledons. The cut surfaces of both control and experimental seedlings were thoroughly coated with the corresponding Agrobacterium suspension and placed on moist filter paper in Petri dishes. All seedlings were maintained in an incubator at 28 °C for 48 h, after which they were transferred and planted in pre-soaked vermiculite. The vector was transferred into lupinus to determine the phenotype, activity of stress-related enzymes, and marker genes. RT-PCR identification is shown in Figure S1. In RT-qPCR experiments, the 35S promoter on pBI121-35S::*GsHZ4* was replaced by a promoter of *GsHZ4* in transgenic plants to detect the expression level of *GsHZ4* under hormone and alkaline treatments.

Table 5. Primer sequences for gene cloning and qPCR analysis.

Primer Name	Primer Sequences (5′-3′)
GsHZ4-F	CCACCTTTCCAAGACCAC
GsHZ4-R	CTCCATTGCCAGCCTATC
Ubiquitin-F	GGCAAGACCATCACTCTCGA
Ubiquitin-R	ACCTCAAGGGTGATGGTCT
LaKIN-F	AAGCTTCAACCTCACGAGCC
LaKIN-R	CAATTCCATAACTCCTGCGTGG
LaMYB34-F	GCAAAGTTTGGCAACAGATGG
LaMVB34-R	GCTTCTTGAGACAAGCGTGC
LaNFYC-F	ACATTGTTAATCGTTACAGCAAAGC
LaNFYC-R	TCCCATAGCCAGGTGATCCA
LaDnaJA6-F	GCAGTGGCTAAGCTAGCAGG
LaDnaJA6-R	TTTCCCAACCCAGTCGATCC
LaDnaJ1-F	GGAAATCCATTTGGTGGCGG
LaDnaJ1-R	CCAAGCTGACCTTGAGAGGG
LaDnaJ20-F	ATTACCATCATCATCAGGTGAGT
LaDnaJ20-R	GTTTTGACCCATCCTAGCAGC
LaNAC21-F	AGCCATGGGATATTCCTGAAACAG
LaNAC21-R	TCTCCCTTTTGGTGCTCTTCC
LaNAC22-F	CCCAAAGGAAATAAAACTGAGTGGG
LaNAC22-R	ACACAAAACCCAATCTTCCTTGG
LaNAC35-F	CCATGGGAATTGCCTGATTTGT
LaNAC35-R	AGGCCTTCCTCCACTACCAT

After the seedlings had developed three palmate compound leaves, root tips from transgenic (positive) and control plants were excised. These samples were then observed and imaged using a confocal laser scanning microscope (Leica, Wetzlar, Hessen, Germany) and photographed. Transgenic overexpression lines were identified and confirmed by quantifying *GsHZ4* expression levels using qPCR. The expression of *GsHZ4*, being an exogenous gene, was undetectable in wild-type (WT) plants. For qPCR analysis, the expression level of *GsHZ4* in each treated transgenic sample was normalized to an internal reference gene (*Ubiquitin*) and was calculated relative to its expression in the untreated transgenic control (set as 1.0) using the $2^{-\Delta\Delta Ct}$ method. Transgenic and non-transgenic (WT) plants were treated with $NaHCO_3$ solutions at concentrations of 0 mM, 20 mM, 40 mM, 60 mM, 80 mM, and 100 mM. The activities of POD, CAT, and SOD, as well as the MDA content, were measured using commercial assay kits according to the manufacturers' protocols. Kits (POD: G0107W; CAT: G0105W, SOD: G0101W; MDA: G0109W) were from Geruisi Biotechnology Co., LTD., Suzhou, China.

4.6. Root Phenotype Analysis

To generate transgenic hairy roots, *Agrobacterium rhizogenes* strain K599 harboring the pEGOEP35S::*GsHZ4*-GFP plasmid was used for infectionHairy roots induced by the empty vector (K599 without the *GsHZ4* insert) served as the control group. Plants were treated with 0 mM, 50 mM or 100 mM $NaHCO_3$ solution to observe phenotypic responses [20].

4.7. Diaminobenzidine (DAB) Staining

The DAB staining experiment can indicate the H_2O_2 content in plant tissues, and the experiment was conducted by using DAB powder (Biotopped, Beijing, China) to prepare a staining solution. We dissolved 0.1 g DAB powder in a brown bottle with 45 mL distilled water (prepared away from light). Then, we adjusted the pH to 5.7, and added distilled water to 50 mL. The leaves were placed in DAB dyeing solution for vacuum treatment and stained for 18 h. The dyed leaves were soaked in anhydrous ethanol, and anhydrous ethanol was added to them appropriately during the boiling process to prevent drying. They were removed thoroughly after decolorization and placed on a glass slide dripping with glycerin. They were then observed and recorded by a digital camera [36].

4.8. Nitrotetrazolium Chloride Blue Solution (NBT) Staining

NBT staining was used to detect the content of superoxide anion in plant tissue. Experiments were performed by using NBT powder (Biotopped, Beijing, China) to prepare a staining solution. NaH_2PO_4 buffer and Na_2HPO_4 buffer were configured. A total of 1.6 mL NaH_2PO_4 buffer and 8.4 mL Na_2HPO_4 buffer were mixed and distilled water was added to 200 mL. The pH was adjusted to 7.5. Then, 0.1 g of NBT powder was dissolved in 50 mL of the phosphate buffer in a brown bottle. Leaves were immersed in the NBT solution and incubated in the dark for 18 h. The subsequent steps of decolorization (in anhydrous ethanol), mounting (in glycerol), and image capture were performed as described for DAB staining. They were then observed and recorded by a digital camera [36].

4.9. Determination of Relative Expression Levels of Key Genes

Total RNA was extracted from root tips of both wild-type (WT) and transgenic lupinus seedlings treated with 0 mM, 50 mM or 100 mM $NaHCO_3$, using a TransZol Up kit (ET111-01-V2; TransGen Biotech, Beijing, China). Using the extracted RNA as template, cDNA was synthesized by reverse transcription with a SPARKscript RT Plus kit (with gDNA eraser; AG0304-B, Sparkjade, Shandong, China). The relative expression levels of *GsHZ4* and key alkali stress-related genes—*LaKIN*, *LaDnaJA6*, *LaNFYC*, *LaMYB34*, *LaDnaJ1*, *LaDnaJ20*, *LaNAC21*, *LaNAC22*, and *LaNAC35*—were determined by quantitative PCR (qPCR; AH0104-B, Sparkjade, Shandong, China). The relative expression of *GsHZ4* was also measured in roots of WT and transgenic lupinus seedlings treated with 4 mg/L ABA, 5 μM IAA, 0.5 μM GA, or 10 μM JA for 0 h, 3 h, 6 h, and 12 h, respectively [2]. Because *GsHZ4* is an exogenous gene, its expression was undetectable in wild-type lupinus. Therefore, in the results, the expression level of *GsHZ4* was compared to that in the corresponding untreated overexpression plants (0 mM $NaHCO_3$ or 0 h hormone treatment) as the control. All data were calculated using the $2^{-\Delta\Delta CT}$ method. The sequences of all primers used are provided in Table 5.

4.10. Statistical Analysis

Means and standard deviations of all study variables were calculated and compared between samples using Student's t test. For multiple comparisons, two-factor analysis of variance and post hoc Tukey testing were performed using GraphPad Prism software 10.4.1. In this project, '*' represented $p < 0.05$; '**' represented $p < 0.01$.

5. Conclusions

Lupinus is a legume that requires an acidic environment and is highly sensitive to alkaline stress. Wild soybean, also part of the legume family, has strong alkaline resistance. In this study, transcriptome sequencing was used to obtain different acid-base response genes in lupinus and GO annotation; KEGG pathway analysis of the genes showed that the acid and alkaline response pathways were completely different in lupinus. Comparative analysis revealed that lupinus lacks homologs of key alkali-tolerance genes present in wild soybean, such as those encoding HD-Zip transcription factors. *GsHZ4*, an alkali tolerance-related HD-Zip gene in wild soybean that is absent in lupinus, was transferred into hairy lupinus roots and was expressed in the root tips. *GsHZ4* expression in lupinus was induced by IAA via its own promoter, and lupinus plant tolerance of alkali stress was improved by the regulation of lupinus stress-related gene expression. This study demonstrates that the abundant resistance-gene resources in wild soybean can be utilized to enhance abiotic stress tolerance in lupinus, offering a viable strategy for expanding its cultivation.

Supplementary Materials: The following supporting information can be downloaded at: https://www.mdpi.com/article/10.3390/plants14223408/s1, Figure S1: RT-PCR identification of the GsHZ4 overexpression strain; Table S1: The upregulated expression genes compared between Group A and Group B; Table S2: The down-regulated expression genes compared between Group A and Group B; Table S3: The upregulated expression genes compared between Group A and Group C; Table S4: The down-regulated expression genes compared between Group A and Group C; Table S5: The upregulated expression genes compared between Group A and Group D; Table S6: The down-regulated expression genes compared between Group A and Group D; Table S7: The upregulated expression genes compared between Group A and Group E; Table S8: The down-regulated expression genes compared between Group A and Group E; Table S9: The gene ID of the transcription factor.

Author Contributions: Conceptualization, L.C.; Formal Analysis, J.T.; Investigation, J.T., M.L., Y.L., X.W., X.D., X.L., M.Z., H.M., Y.F. and Q.G.; Methodology, J.T.; Resources, L.C.; Supervision, J.Z. and L.C.; Writing-Original Draft, J.T.; Writing-Review and Editing, L.C. All authors have read and agreed to the published version of the manuscript.

Funding: National Natural Science Foundation of China (32001505, 32472750) to L.C. Natural Science Foundation of Heilongjiang Province (LH2023C006) to L.C. Research Foundation Project (HUDF2019104) to J.Z. Student Innovation Practical Training (X202410224259) to L.C.

Data Availability Statement: The data were uploaded to the NCBI SRA database (SUB14514279).

Acknowledgments: We are grateful to the College of Horticulture and Landscape Architecture at Northeast Agricultural University for providing the experimental platform. We also extend our sincere thanks to Haizhen Zhang and Tao Yang for their technical support.

Conflicts of Interest: The authors declare no competing interests.

References

1. Rao, Y.; Peng, T.; Xue, S. Mechanisms of plant saline-alkaline tolerance. *J. Plant Physiol.* **2023**, *281*, 153916. [CrossRef]
2. Zhang, B.; Liu, Z.; Zhou, R.; Cheng, P.; Li, H.; Wang, Z.; Liu, Y.; Li, M.; Zhao, Z.; Hu, Z.; et al. Genome-wide analysis of soybean DnaJA-family genes and functional characterization of *GmDnaJA6* responses to saline and alkaline stress. *Crop J.* **2023**, *11*, 1230–1241. [CrossRef]
3. Kuźniacka, J.; Biesek, J.; Banaszak, M.; Grabowicz, M.; Adamski, M. The Quality of Eggs from Rosa 1 Hens Fed Diets Containing Seeds of Legume Plants (*Lupinus luteus* L., *Lupinus angustifolius*, and *Pisum sativum*) in Two Laying Phases. *Animals* **2020**, *10*, 1942. [CrossRef] [PubMed]
4. Ding, W.; Clode, P.L.; Lambers, H. Is pH the key reason why some Lupinus species are sensitive to calcareous soil? *Plant Soil* **2019**, *434*, 185–201. [CrossRef]
5. Waller, Z.; Rejili, M.; Mars, M.; Brachmann, A.; Marín, M. Draft Genome Sequence of *Phyllobacterium endophythicum* mTS5, Isolated from *Lupinus micranthus* in Tunisia. *Microbiol. Resour. Announc.* **2019**, *8*, e00968-19. [CrossRef]
6. Hufnagel, B.; Marques, A.; Soriano, A.; Marquès, L.; Divol, F.; Doumas, P.; Sallet, E.; Mancinotti, D.; Carrere, S.; Marande, W.; et al. High-quality genome sequence of white lupin provides insight into soil exploration and seed quality. *Nat. Commun.* **2020**, *11*, 492. [CrossRef]
7. Aslam, M.M.; Karanja, J.K.; Yuan, W.; Zhang, Q.; Zhang, J.; Xu, W. Phosphorus uptake is associated with the rhizosheath formation of mature cluster roots in white lupin under soil drying and phosphorus deficiency. *Plant Physiol. Biochem.* **2021**, *166*, 531–539. [CrossRef] [PubMed]
8. Shaaban, A.; Al-Elwany, O.A.; Abdou, N.; Hemida, K.A.; El-Sherif, A.M.A.; Abdel-Razek, M.A.; Semida, W.M.; Mohamed, G.F.; El-Mageed, T.A.A. Filter Mud Enhanced Yield and Soil Properties of Water-Stressed *Lupinus termis* L. in Saline Calcareous Soil. *J. Soil Sci. Plant Nutr.* **2022**, *22*, 1572–1588. [CrossRef]
9. Cao, L.; Yu, Y.; DuanMu, H.; Chen, C.; Duan, X.; Zhu, P.; Chen, R.; Li, Q.; Zhu, Y.; Ding, X. A novel Glycine soja homeodomain-leucine zipper (HD-Zip) I gene, *Gshdz4*, positively regulates bicarbonate tolerance and responds to osmotic stress in Arabidopsis. *BMC Plant Biol.* **2016**, *16*, 184. [CrossRef]
10. Jin, T.; Sun, Y.; Zhao, R.; Shan, Z.; Gai, J.; Li, Y. Overexpression of Peroxidase Gene *GsPRX9* Confers Salt Tolerance in Soybean. *Int. J. Mol. Sci.* **2019**, *20*, 3745. [CrossRef]
11. Feng, X.; Feng, P.; Yu, H.; Yu, X.; Sun, Q.; Liu, S.; Minh, T.N.; Chen, J.; Wang, D.; Zhang, Q.; et al. GsSnRK1 interplays with transcription factor *GsERF7* from wild soybean to regulate soybean stress resistance. *Plant Cell Environ.* **2020**, *43*, 1192–1211. [CrossRef]
12. Ma, Q.; Xia, Z.; Cai, Z.; Li, L.; Cheng, Y.; Liu, J.; Nian, H. *GmWRKY16* Enhances Drought and Salt Tolerance Through an ABA-Mediated Pathway in *Arabidopsis thaliana*. *Front. Plant Sci.* **2019**, *9*, 1979. [CrossRef] [PubMed]
13. Zhou, G.; Chang, R.; Qiu, L. Overexpression of soybean ubiquitin-conjugating enzyme gene *GmUBC2* confers enhanced drought and salt tolerance through modulating abiotic stress-responsive gene expression in *Arabidopsis*. *Plant Mol. Biol.* **2010**, *72*, 357–367. [CrossRef]
14. Liu, G.; Li, X.; Jin, S.; Liu, X.; Zhu, L.; Nie, Y.; Zhang, X. Overexpression of rice NAC gene *SNAC1* improves drought and salt tolerance by enhancing root development and reducing transpiration rate in transgenic cotton. *PLoS ONE* **2014**, *9*, e86895. [CrossRef]
15. Kang, T.; Yu, C.; Liu, Y.; Song, W.; Bao, Y.; Guo, X.; Li, B.; Zhang, H. Subtly Manipulated Expression of *ZmmiR156* in Tobacco Improves Drought and Salt Tolerance Without Changing the Architecture of Transgenic Plants. *Front. Plant Sci.* **2020**, *10*, 1664. [CrossRef]
16. Zhang, Y.; Wan, S.; Xing, B.; Peng, C.; Zhu, J.; Shao, Q.; Lv, A. An HD-zip transcription factor *ArHDZ22* regulates plant height and decreases salt tolerance in *Anoectochilus roxburghii*. *Ind. Crops Prod.* **2025**, *223*, 120251. [CrossRef]
17. Zhao, S.; Gao, H.; Jia, X.; Zhou, K.; Wang, H.; Mao, K.; Ma, F. MdHB7-like confers drought tolerance and improves water-use efficiency through modulating stomatal density in apple (*Malus domestica*). *Sci. Hortic.* **2022**, *294*, 110758. [CrossRef]
18. Oh, S.K.; Yoon, J.; Choi, G.J.; Jang, H.A.; Kwon, S.Y.; Choi, D. *Capsicum annuum* homeobox 1 (CaHB1) is a nuclear factor that has roles in plant development, salt tolerance, and pathogen defense. *Biochem. Biophys. Res. Commun.* **2013**, *442*, 116–121. [CrossRef]
19. Cao, L.; Yu, Y.; Ding, X.; Zhu, D.; Yang, F.; Liu, B.; Sun, X.; Duan, X.; Yin, K.; Zhu, Y. The *Glycine soja* NAC transcription factor *GsNAC019* mediates the regulation of plant alkaline tolerance and ABA sensitivity. *Plant Mol. Biol.* **2017**, *95*, 253–268. [CrossRef]
20. Liu, M.; Tang, J.; Ma, S.; Liu, Y.; Wang, X.; Du, X.; Sun, X.; Zeng, Y.; Zeng, Y.; Ding, X.; et al. *GsEXPA8* Enhances Soybean Tolerance of $NaHCO_3$ Stress by Regulating Root Morphology. *Agronomy* **2025**, *15*, 16. [CrossRef]
21. Ren, H.; Zhang, B.; Zhang, F.; Liu, X.; Wang, X.; Zhang, C.; Zhao, K.; Yuan, R.; Lamlom, S.F.; Abdelghany, A.M.; et al. Integration of Physiological and Transcriptomic Approaches in Investigating Salt-Alkali Stress Resilience in Soybean. *Plant Stress* **2024**, *11*, 100375. [CrossRef]
22. Kang, Y.I.; Park, J.M.; Kim, S.H.; Kang, N.J.; Park, K.S.; Lee, S.Y.; Jeong, B.R. Effects of root zone PH and nutrient concentration on the growth and nutrient uptake of tomato seedlings. *J. Plant Nutr.* **2011**, *34*, 640–652. [CrossRef]

23. Yang, C.; Shi, D.; Wang, D. Comparative effects of salt and alkali stresses on growth, osmotic adjustment and ionic balance of an alkali-resistant halophyte *Suaeda glauca* (Bge.). *Plant Growth Regul.* **2008**, *56*, 179–190. [CrossRef]
24. Song, S.; Chen, Y.; Zhao, M.; Zhang, W. A novel Medicago truncatula HD-Zip gene, *MtHB2*, is involved in abiotic stress responses. *Environ. Exp. Bot.* **2012**, *80*, 1–9. [CrossRef]
25. Ariel, F.D.; Manavella, P.A.; Dezar, C.A.; Chan, R.L. The true story of the HD-ZIP family. *Trends Plant Sci.* **2007**, *12*, 419–426. [CrossRef]
26. Chen, L.; Yang, H.; Fang, Y.; Guo, W.; Chen, H.; Zhang, X.; Dai, W.; Chen, S.; Hao, Q.; Yuan, S.; et al. Overexpression of *GmMYB14* improves high-density yield and drought tolerance of soybean through regulating plant architecture mediated by the brassinosteroid pathway. *Plant Biotechnol. J.* **2021**, *19*, 702–716. [CrossRef] [PubMed]
27. Yu, T.; Liu, Y.; Fu, J.; Ma, J.; Fang, Z.; Chen, J.; Zheng, L.; Lu, Z.; Zhou, Y.; Chen, M.; et al. The NF-Y-PYR module integrates the abscisic acid signal pathway to regulate plant stress tolerance. *Plant Biotechnol. J.* **2021**, *19*, 2589–2605. [CrossRef]
28. Xue, L.; Wang, Y.; Fan, Y.; Jiang, Z.; Wei, Z.; Zhai, H.; He, S.; Zhang, H.; Yang, Y.; Zhao, N.; et al. IbNF-YA1 is a key factor in the storage root development of sweet potato. *Plant J.* **2024**, *118*, 1991–2002. [CrossRef]
29. He, G.; Liu, P.; Zhao, H.; Sun, J. The HD-ZIP II Transcription Factors Regulate Plant Architecture through the Auxin Pathway. *Int. J. Mol. Sci.* **2020**, *21*, 3250. [CrossRef]
30. Qi, B.; Zhang, X.; Mao, Z.; Qin, S.; Lv, D. Integration of root architecture, root nitrogen metabolism, and photosynthesis of 'Hanfu' apple trees under the cross-talk between glucose and IAA. *Hortic. Plant J.* **2023**, *9*, 631–644. [CrossRef]
31. Liu, H.; Zhang, W.; Guo, Z.; Yu, Z.; Chen, Z.; Kong, J.; Zheng, Q.; Li, W.; Su, Y.; Wang, G.; et al. Suppressing an AUX/IAA gene *GhIAA43* expression activates the SA-mediated immune pathway and enhances Verticillium wilt resistance in cotton. *Plant Sci.* **2025**, *362*, 112797. [CrossRef] [PubMed]
32. Wang, Y.; Fan, J.; Wu, X.; Guan, L.; Li, C.; Gu, T.; Li, Y.; Ding, J. Genome-Wide Characterization and Expression Profiling of HD-Zip Genes in ABA-Mediated Processes in *Fragaria vesca*. *Plants* **2022**, *11*, 3367. [CrossRef] [PubMed]
33. Liu, Y.; Wang, W.; Zhang, L.; Zhu, L.; Zhang, X.; He, X. The HD-Zip transcription factor *GhHB12* represses plant height by regulating the auxin signaling in cotton. *J. Integr. Agric.* **2023**, *22*, 2015–2024. [CrossRef]
34. Chen, C.; Xu, L.; Zhang, X.; Wang, H.; Nisa, Z.; Jin, X.; Yu, L.; Jing, L.; Chen, C. Exogenous strigolactones enhance tolerance in soybean seedlings in response to alkaline stress. *Physiol. Plant.* **2022**, *174*, e13784. [CrossRef]
35. Zhang, X.; Xue, Y.; Wang, H.; Nisa, Z.; Jin, X.; Yu, L.; Liu, X.; Yu, Y.; Chen, C. Genome-wide identification and characterization of *NHL* gene family in response to alkaline stress, ABA and MEJA treatments in wild soybean (*Glycine soja*). *PeerJ* **2022**, *10*, e14451. [CrossRef]
36. Yu, Z.; Chen, X.; Chen, Z.; Wang, H.; Shah, S.H.A.; Bai, A.; Liu, T.; Xiao, D.; Hou, X.; Li, Y. BcSRC2 interacts with BcAPX4 to increase ascorbic acid content for responding ABA signaling and drought stress in pak choi. *Hortic. Res.* **2024**, *11*, uhae165. [CrossRef] [PubMed]

plants

MDPI

Article

Genome-Wide Characterization, Expression, and Functional Analysis of Acyl-CoA-Binding Protein (ACBP) Gene Family in *Puccinellia tenuiflora*

Hongxia Zheng [1,2,†], Ying Li [1,†], Sizhu Wang [1], Xin Ni [1], Xiaofeng Xu [2,*] and Shaojun Dai [2,*]

[1] Key Laboratory of Saline-Alkali Vegetation Ecology Restoration, Ministry of Education, College of Life Sciences, Northeast Forestry University, Harbin 150040, China; zhenghongxia2020@foxmail.com (H.Z.); ly7966@nefu.edu.cn (Y.L.); wangsizhu202510@163.com (S.W.); 13136869041@139.com (X.N.)

[2] Development Center of Plant Germplasm Resources, College of Life Sciences, Shanghai Normal University, Shanghai 200234, China

* Correspondence: xiaofengxu@shnu.edu.cn (X.X.); daishaojun@shnu.edu.cn (S.D.)

† These authors contributed equally to this work.

Abstract

Acyl-CoA-binding proteins (ACBPs) possess a conserved acyl-CoA-binding (ACB) domain that facilitates binding to acyl-CoA esters. In addition to their typical role in lipid metabolism, plant ACBPs have been shown to participate in various physiological processes, such as membrane biogenesis, stress response pathways and plant immunity mechanisms. Here, we identified five *PutACBP* members in alkaligrass (*Puccinellia tenuiflora*), which were divided into four distinct classes based on a phylogenetic tree constructed from 86 *ACBP* genes from 12 plant species. Promoter analysis identified numerous *cis*-acting elements linked to abiotic stresses (e.g., light, drought, heat, and cold) and hormone responses. Expression profile analyses revealed that *PutACBPs* exhibit broad expression patterns across many organs and respond to salinity-alkali, cold, H_2O_2, and $CdCl_2$ stresses. Transient expression of five PutACBP-GFPs in tobacco (*Nicotiana tabacum*) revealed PutACBP1 and PutACBP2 localized to the plasma membrane, cytoplasm, and cell nucleus, while PutACBP3, PutACBP4, and PutACBP5 localized around the plasma membrane and cytoplasm. Furthermore, heterologous constitutive expression of *PutACBP3* in *Arabidopsis* (*Arabidopsis thaliana*) enhanced the resistance of transgenic plants to salinity stress, possibly through alterations in the levels of lipid metabolism-related and stress-responsive genes. The *ACBP* gene family is highly conserved across different plant species. This study provides the first comprehensive genomic and functional characterization of the PutACBP family in alkaligrass, elucidating its evolutionary conservation, phylogenetic classification, and stress-response roles. Notably, overexpression of *PutACBP3* in *Arabidopsis* significantly enhanced salt tolerance, suggesting its critical function in salt-stress adaptation in alkaligrass.

Keywords: acyl-CoA-binding proteins (ACBPs); gene family analysis; salt response; *Puccinellia tenuiflora*

1. Introduction

Lipids are essential for maintaining membrane stability, energy supply, and stress adaptation in plants [1]. Acyl-CoA-binding proteins (ACBPs), characterized by conserved acyl-CoA-binding (ACB) domains, function as lipid carrier proteins that mediate lipid binding, transport, and metabolism across cellular compartments [2]. ACBPs were initially identified in mammals as Diazepam Binding Inhibitor (DBI) [3,4], while plant ACBPs were first reported in oilseed rape (*Brassica napus*) [5]. Subsequently, ACBP families have been identified at least in 27 plant species, ranging phylogenetically from algae, moss, ferns, and gymnosperms to angiosperms [6,7]. Based on their size and domain architecture, plant ACBPs are classified into four classes: Class I (single ACB domain), Class II (ACB domain with C-terminal ankyrin repeats), Class III (extended proteins with a C-terminal ACB domain), and Class IV (multi-domain proteins with C-terminal kelch motifs) [8–11].

ACBPs are evolutionarily conserved across eukaryotes and play diverse roles in lipid metabolism, membrane biosynthesis, and stress adaptation [12–14]. For instance, yeast ACBPs are critical for plasma membrane assembly, and their mutations disrupt key pathways such as fatty acid biosynthesis, glycerol metabolism, and stress responses [15,16]. In plants, *ACBPs* exhibit evolutionarily conserved isoform specialization in lipid binding [17], as evidenced by systematic characterization of *Arabidopsis* (*Arabidopsis thaliana*) isoforms (AtACBP1–6). AtACBP1 and AtACBP2 proteins preferentially recognize C18:2-CoA, C18:3-CoA esters, phosphatidylcholine (PC), and phosphatidic acid (PA) [18–21], while AtACBP3 exhibits broader specificity for arachidonyl-CoA (C20:4), PC, and phosphatidylethanolamine (PE) [21,22]. AtACBP4 and AtACBP5 show a distinct preference for C18:1-CoA and PC [23–26], whereas AtACBP6 selectively binds 16:0-CoA and 18:2-CoA [27]. These isoform-specific binding profiles establish plant ACBPs as modular transporters that orchestrate compartmentalized lipid trafficking via precision-tuned molecular recognition.

Soil salinity impairs plant growth by inducing osmotic, ionic, and oxidative stress [28–33]. Salt stress signaling involves phospholipids such as PA and the salt overly sensitive (SOS) pathway, in which PA activates mitogen-activated protein kinase 6 (MPK6) to phosphorylate SOS1, thereby regulating Na^+ homeostasis [34–39]. *ACBPs* play critical roles in salt stress adaptation. For example, *Chlorella* sp. *ChACBP1* enhances salinity tolerance, and its overexpression in yeast and *Arabidopsis* likely improves stress resilience via phospholipid metabolism [40]. In *Arabidopsis*, *AtACBP1* overexpression increased sensitivity to NaCl during seed germination and seedling establishment, whereas the *atacbp1* mutant showed reduced sensitivity during germination [41]. Conversely, overexpression of *OsACBP4* in transgenic rice (*Oryza sativa*) and *AtACBP2* in transgenic *Arabidopsis* conferred enhanced salt tolerance in both species [42]. In maize (*Zea mays*), NaCl induces expression of *ZmACBP1* and *ZmACBP3*, and overexpression of *ZmACBP3* (*ZmACBP3-OEs*) in *Arabidopsis* promoted better growth and longer roots under salt stress. *ZmACBP3-OEs* also exhibited significantly upregulated expression of lipid metabolism genes (*FAD2*, *DGAT*, *PLA2*, *PLC3*, and *ACX*) and stress-responsive genes (*COR47*, *AREB1*, *RAB*, *ABI1*, *RD29A*, and *RD29B*) compared to WT (wild-type) plants, suggesting *ZmACBP3* enhances tolerance by modulating lipid metabolism and activating stress-responsive gene expression [7]. Similarly, soybean (*Glycine max*) *GmACBPs* exhibit differential expression under salt stress, suggesting functional diversification among *ACBP* family members [43].

Beyond salinity, *ACBPs* also contribute to diverse stress responses. *AtACBP1* and *AtACBP2* in *Arabidopsis* facilitate membrane repair by binding to phospholipid precursors under heavy metal stress [20,44]. Additionally, *AtACBP2*-overexpressing plants exhibit increased sensitivity to abscisic acid (ABA) during seed germination and development, where ABA-mediated reactive oxygen species (ROS) accumulation in guard cells enhances drought tolerance by promoting stomatal closure [45]. Furthermore, overexpression of *AtACBP3* in *Arabidopsis* enhances disease resistance by upregulating pathogen-associated genes, increasing cell mortality, and accumulating hydrogen peroxide [46]. These findings underscore the importance of *ACBPs* in plant stress responses, particularly through their roles in lipid signaling, membrane repair, and ABA-mediated stress adaptation.

Alkaligrass (*Puccinellia tenuiflora*) is a halophytic monocot forage grass that thrives in saline-alkali soils, making it an ideal model for studying salt adaptation mechanisms [47]. Comparative genomic and transcriptomic analyses have identified thousands of salinity-responsive genes in alkaligrass [48]. Quantitative proteomic studies have systematically mapped stress-induced changes in protein abundance and post-translational modifications under diverse stressors, including salts (NaCl, Na_2CO_3, and $NaHCO_3$), low temperatures, and oxidative stress [49–56]. Integrated multi-omics data reveal that alkaligrass employs unique adaptive strategies for salinity tolerance, including the proteins involved in fatty acid metabolism particularly through dynamic regulation of its robust antioxidant systems for ROS scavenging [49]. Notably, four root *PutACBPs* (*PutACBP1-PutACBP4*) and two leaf *PutACBPs* (*PutACBP1* and *PutACBP3*) are differentially expressed under NaCl, Na_2CO_3, and $NaHCO_3$ stresses [48], suggesting their involvement in lipid-mediated stress responses. However, a comprehensive analysis of the *ACBP* gene family in alkaligrass, including its phylogenetic evolution, expression dynamics and functional validation, has not yet been conducted.

In this study, we identified five *PutACBP* genes through whole-genome screening. We systematically analyzed the phylogenetic evolution, organ-specific and stress-responsive expression profiles, and association with agronomic traits of *PutACBPs*. Our findings demonstrate that *PutACBP3* enhances salinity resistance through phospholipid-mediated membrane remodeling and ROS homeostasis, providing novel insights into halophyte stress adaptation mechanisms.

2. Results

2.1. Identification and Phylogenetic Analysis of ACBP Genes in Alkaligrass

To identify ACBPs in alkaligrass, we performed a BLASTp search using known ACBP sequences from *Arabidopsis* and rice. This initial search identified 15 candidate genes predicted to encode ACBPs. To validate these candidates and confirm the presence of the defining acyl-CoA-binding domain, we screened all 15 sequences against the conserved ACB domain (PF00887). Furthermore, candidates were analyzed for the presence of Ankyrin and Kelch domains characteristic of specific ACBP classes. This combined domain analysis confirmed five bona fide *ACBP* genes in alkaligrass, designated PutACBP1 to PutACBP5 with locus identifiers Pt_Chr0101998, Pt_Chr0106094, Pt_Chr0300051, Pt_Chr0402341, and Pt_Chr0402344 (Table S1). The deducted PutACBPs range in size from 89 to 655 amino acids (AA) and exhibit isoelectric points (pI) between 4.45 and 5.08. All five proteins are hydrophilic, as indicated by negative Grand Average of Hydropathy (GRAVY) values. PutACBP1 and PutACBP2 show high sequence similarity. Detailed characteristics, including molecular weight (MW), coding sequence (CDS) length, AA number, pI, GRAVY, subfamily classification, and the predicted subcellular localization for each gene, are provided in Table S1.

To classify the identified PutACBPs and determine their evolutionary relationships, we constructed a phylogenetic tree using the neighbor-joining method. The tree was based on the alignment of full-length ACBP sequences from alkaligrass and *Arabidopsis* (Figure 1). Phylogenetic analysis revealed that the PutACBPs share high similarity with rice OsACBPs and cluster into the four established plant ACBP classes. Specifically, PutACBP1 and PutACBP2 group within Class I (characterized by a single ACB domain). PutACBP3, PutACBP4, and PutACBP5 were classified into Class II (ACB domain + ankyrin repeats), Class III (large protein with C-terminal ACB domain), and Class IV (multi-domain with kelch motifs), respectively. This provides a crucial foundation for subsequent functional characterization of these ACBPs in abiotic stress responses.

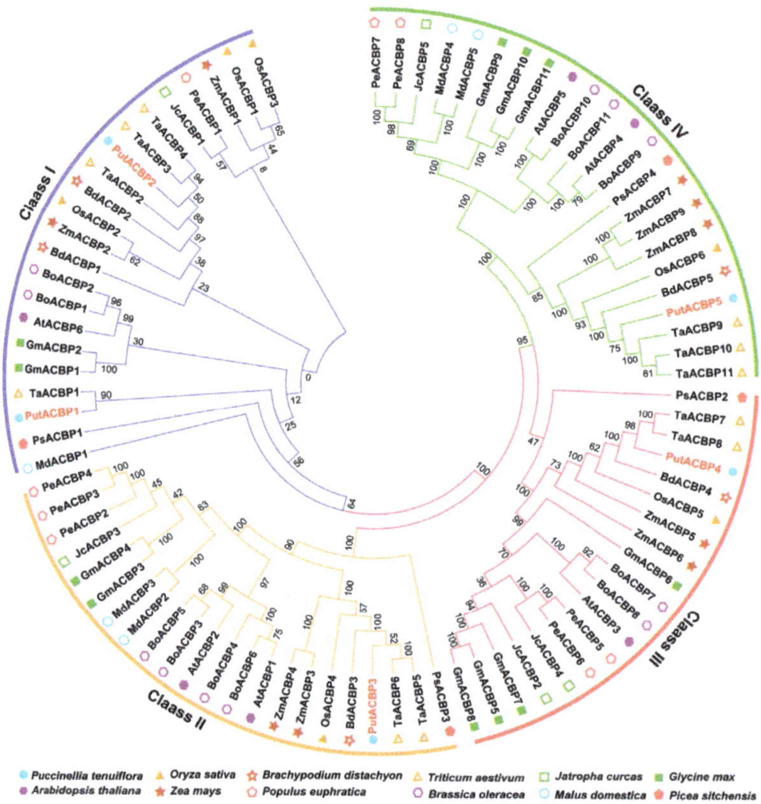

Figure 1. Phylogenetic tree of acyl-CoA-binding proteins (ACBPs) from alkaligrass (*Puccinellia tenuiflora*) and other representative plant species. Alignments of 86 ACBPs in alkaligrass, rice (*Oryza sativa*), *Brachypodium distachyon*, wheat (*Triticum aestivum*), leper tree (*Jatropha curcas*), soybean (*Glycine max*), *Arabidopsis*, maize (*Zea mays*), poplar (*Populus euphratica*), cabbage (*Brassica oleracea*), apple (*Malus domestica*), and spruce (*Picea sitchensis*) were performed using the default parameter of ClustalW and the phylogenetic tree was constructed using the Neighbor-Joining tree method with 1000 bootstrap replicates in MEGA 7.0 software. These ACBPs were divided into four classes. Different colored circles represent different plant species.

2.2. *Evolutionary Genomics Analysis Reveals Chromosomal Distribution and Conservation of ACBP Genes in Alkaligrass*

Analysis of chromosomal distribution revealed that the five *PutACBP* genes are located on three chromosomes (1, 3, and 4) of alkaligrass, displaying an uneven distribution pattern (Figure 2A). *PutACBP1* and *PutACBP2* are localized in close proximity on the same chromosome, suggesting that they may have arisen from a common ancestral gene and potentially share similar structures and functions. This non-random localization mirrors observations in *Arabidopsis* and rice, where *ACBP* genes are also unevenly distributed across their respective genomes [9,57]. These findings suggest that the chromosomal arrangement of *ACBP* genes may be a conserved feature across plant species.

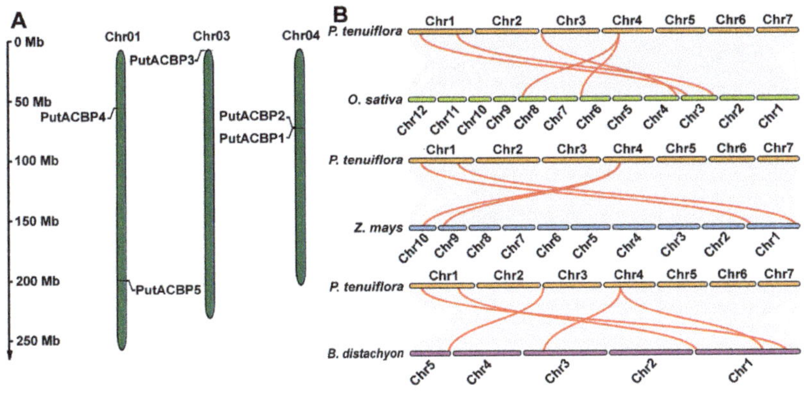

Figure 2. Distribution of *PutACBPs* on chromosomes and collinearity analysis of *ACBPs* between alkaligrass and other plant species. (**A**) Chromosomal localization of the *PutACBP* genes. The diagram was drawn using the MapGene2Chromorome V2 (v 2.1), and five *PutACBP* genes were located on three chromosomes. Chromosomal distances are given in Mb. (**B**) Collinearity analysis of *ACBPs* among alkaligrass, rice, maize, and *B. distachyon*. Gray lines in the background indicated the collinear blocks within alkaligrass and other plant genomes, while the red lines highlighted the syntenic *ACBP* gene pairs.

To elucidate evolutionary relationships within the ACBP family, we performed collinearity analysis using homologous *ACBP* genes from alkaligrass, rice, maize, and *B. distachyon*. The analysis identified five, four, and five homologous gene pairs between alkaligrass *ACBPs* and their counterparts in rice, maize, and *B. distachyon*, respectively (Figure 2B). Notably, no maize ortholog was identified for *PutACBP3*. This absence suggests lineage-specific evolutionary divergence, potentially linked to extensive gene fractionation following the ancestral tetraploidization event in maize, which may have led to the loss of this specific ancestral *ACBP* homolog. In contrast, *B. distachyon* and rice appear to have retained orthologs of *PutACBP3* under stronger purifying selection, implying functional conservation potentially critical for niche adaptation. The conservation of *PutACBP3* orthologs in *B. distachyon* and rice further suggests its involvement in fundamental cellular processes. Its loss in maize may have been evolutionarily permissible due to functional redundancy, potentially compensated by paralogous genes or alternative metabolic pathways. Collectively, the collinearity patterns indicate that *PutACBP* genes exhibit relatively greater evolutionary conservation with *B. distachyon* and rice than with maize.

2.3. Comparison in Architecture of ACBP Domain from Arabidopsis, Rice, and Alkaligrass

According to different domains and amino acid lengths, the five ACBPs were divided into four subfamilies (Figure 3). All ACBPs contain an ACB domain, and some of them also contain the Ankyrin domain or the Kelch domain. PutACBP1 and PutACBP2 belong to Small-ACBP in Class I and contain only one ACB domain. PutACBP3 contains an ACB domain and an Ankyrin domain and was classified as ankyrin-repeat ACBPs (ANK-ACBP) in Class II. PutACBP4 was categorized as Large-ACBP in Class III and only contains one ACB domain in the C-terminal. PutACBP5 belongs to Kelch-ACBPs in Class IV and contains one ACB domain in the N-terminal and three Kelch domains in the center (Figure 3).

When compared PutACBPs with those from *Arabidopsis* and rice, we found that the AA sequences of PutACBPs are more similar with those from rice. In *Arabidopsis*, Class II and Class IV contain two ACBPs, while rice and alkaligrass contain only one ACBP gene (Figure 3, Table S4). Collectively, PutACBPs show higher similarity to rice homologs.

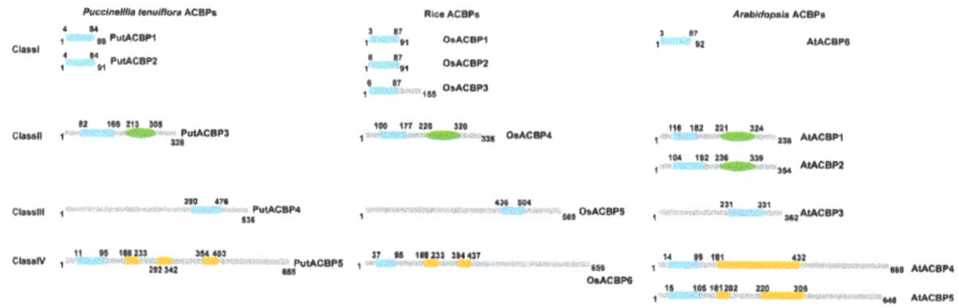

Figure 3. Architecture of ACBPs in alkaligrass. Comparation of the architecture of ACBPs among alkaligrass, rice, and *Arabidopsis*. Blue indicates the acyl-CoA-binding (ACB) domain, green indicates the ankyrin repeats (Ankyrin) domain, and yellow indicates the Kelch domain.

2.4. Analysis of Gene Structure, Cis-Acting Elements, and Conserved Motif Distribution

Gene structure analysis of the five *PutACBPs* was performed using the Gene Structure Display Server (http://gsds.cbi.pku.edu.cn, accessed on 27 June 2025). Based on genomic sequences, exon numbers were largely conserved within each *PutACBP* subfamily, while introns varied in both length and position (Figure 4A).

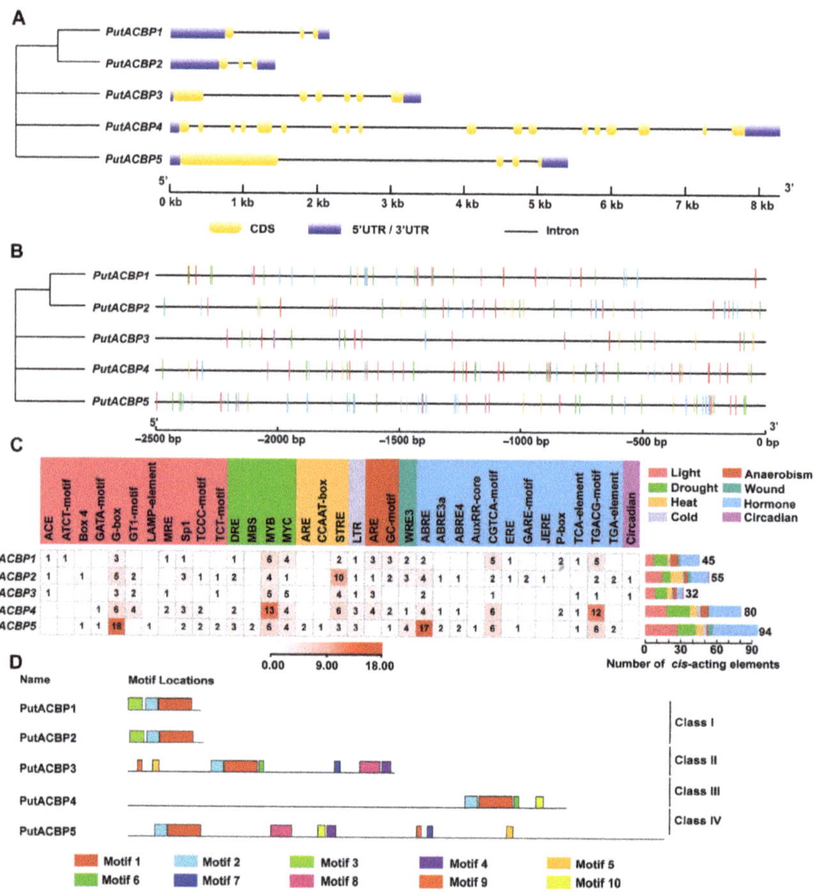

Figure 4. Structure, *cis*-acting elements, and conversed protein motifs of PutACBPs. (**A**) Phylogenetic tree and gene structure analyses of *PutACBPs*. The UTR, exon and intron were represented by blue

rectangle, yellow rectangle, and black line, respectively. (**B**) Predicted *cis*-acting elements in the promoter regions of *PutACBPs*. (**C**) The names and numbers of *cis*-acting elements in *PutACBP* promoters. The heatmap in grid and the color columns indicated the numbers of *cis*-acting elements. ABRE: ABA responsive element; ARE: Anaerobic-responsive element; DRE: Dehydration-responsive element; JERE: Jasmonate and/or elicitor responsive element; LTR: Low temperature-responsive element; MBS: MYB-binding site; MYB: V-myb avian myeloblastosis viral oncogene homolog; MYC: Myelocytomatosis; WRE3: Wound response element 3. (**D**) Conversed motif analysis in PutACBPs. The conserved motifs were predicted using MEME motif search analysis, and the maximum number parameter was set to ten. Various motifs were represented by different colors. The sequence information for each motif was provided in Table S2.

To investigate the potential functions of *PutACBPs*, a prediction of *cis*-acting elements was made in five *PutACBP* promoter regions and a schematic map was created of the prevalent distributions of stress and hormone responsive *cis*-acting elements (Figure 4B). A total of 306 *cis*-acting elements in *PutACBPs* were predicted, which were associated with various abiotic stresses, including light (74 elements), drought (61), heat (28), anaerobic conditions (19), wounding (10), cold (9), circadian (2), and hormones (102) (Figure 4C). Among these, certain elements responsive to drought (e.g., DRE, MBS, MYB, and MYC), light-responsive elements (e.g., ACE, ATCT-motif, Box4, GATA-motif, G-box, GT1-motif, LAMP-element, MRE, Sp1, TCCC-motif, and TCT-motif), heat-responsive elements (e.g., ARE, CCAAT-box, and STRE), and phytohormones together account for a significant portion of the total in each of the *PutACBPs*, some of which are conserved among the five *PutACBPs* (Figure 4C). These findings suggest that *PutACBPs* play vital roles in responding to drought, light, heat, and hormonal signals. It is noteworthy that a total of 102 hormone-responsive elements were detected, including 29 ABA (also known as ABRE), 40 methyl jasmonate (MeJA) (specifically TGACG and CGTCA motifs), and 4 salicylic acid (SA) (identified as TCA element), 11 gibberellin (GA) (specifically P-box and TATC-box), and 5 auxin (specifically TGA-element and AuxRR-core) (Figure 4C). The conserved elements found in the promoter region indicate that *PutACBPs* play a crucial role in responding to abiotic stresses and phytohormone treatments.

Motif analysis using the MEME Suite systematically identified ten conserved structural elements (motifs 1–10) across PutACBP isoforms (Figures 4D and S1). Regarding motif distribution per class (Figure 4D): Class I contained three motifs, while Class II, III, and IV contained eight, four, and eight motifs, respectively. Importantly, Motif 1 contains the signature ACB domain essential for protein function. Subfamily members exhibited similar motif composition and arrangement (Figure 4D). For instance, motif 1 and motif 2 were present in all *PutACBP* genes, while motif 3 was specific to Class I. Motifs 4, 5, 7, 8, and 9 were found in Class II and Class IV. Motif 6 was observed in Class II and Class III, and motif 10 was present in Class III and Class IV. The distinct distribution of these motifs across subfamilies suggest functional diversity.

2.5. Expression Pattern of PutACBP Genes in Different Organs

To investigate the organ-specific expression profiles of *PutACBP* genes, we performed RT-qPCR analysis across seven plant organs. The expression of *PutACBP* genes in different organs (i.e., roots, stems, leaves, sheaths, flowers, spikes, and seeds) were analyzed by RT-qPCR. The results showed that five *PutACBP* genes were detected in all the organs (Figure 5A, Table S6). *PutACBP1* and *PutACBP3* exhibited high expression levels in leaves. *PutACBP1* expression was lower in roots and stems compared to other organs, while *PutACBP2* showed reduced expression in roots, stems, and leaves (Figure 5A). *PutACBP4* expression was lowest in seeds relative to other organs (Figure 5A). These distinct expression patterns suggest functional diversification within the *PutACBP* family.

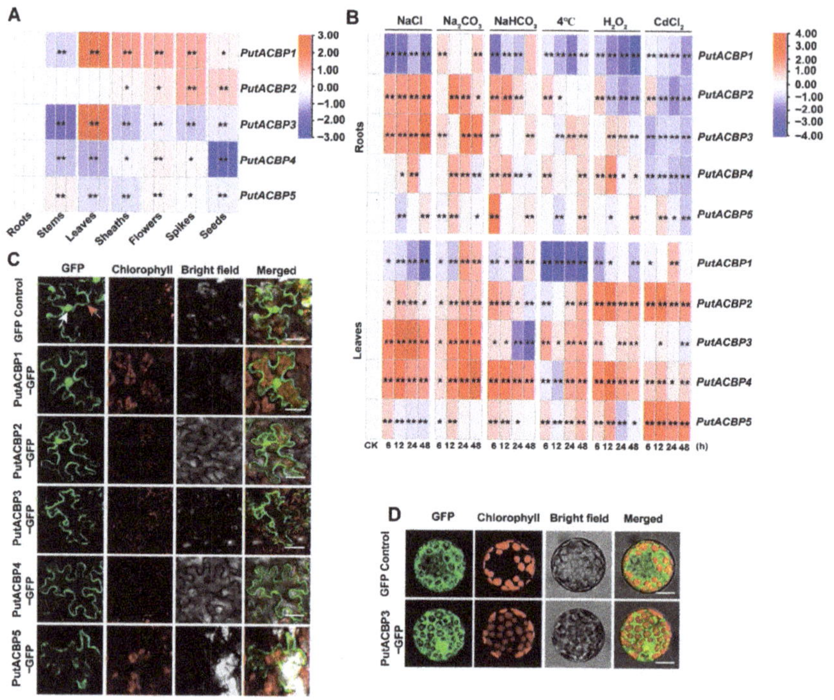

Figure 5. Organ specific expression, different stress expression, and subcellular localization of PutACBPs. (**A**) Organ specific expression *PutACBPs* were analyzed in roots, stems, leaves, sheaths, flowers, spikes, and seeds using RT-qPCR analysis. (**B**) Different stress expression patterns in roots and leaves of *PutACBPs*. The heatmap of *PutACBPs* expression based on the RT-qPCR data standardized by log2 conversion, blue indicates down regulated expression and red indicates upregulated expression. (**C**) Subcellular localization of PutACBPs in tobacco epidermal cells. The epidermal cells of tobacco leaves were examined under a confocal laser scanning microscope. Chloroplasts were visualized by chlorophyll autofluorescence. The red arrow indicates the cell membrane, while the white arrow indicates the cell nucleus. Scale bars = 25 μm. (**D**) Subcellular localization of PutACBP3 in *Arabidopsis* protoplasm cells. The protoplasts were extracted from *Arabidopsis* transgenic seedlings and examined under a confocal laser scanning microscope. Chloroplasts were visualized by chlorophyll autofluorescence. Scale bars = 10 μm. Significant differences compared to the control group were determined by Student's *t*-test, ** $p < 0.01$, and * $p < 0.05$.

2.6. Expression Profiling of PutACBP Genes in Response to Abiotic Stress

To investigate the potential role of *PutACBP* genes under abiotic stress, their expression was assessed via RT-qPCR following treatment with salts (200 mM NaCl, 75 mM Na_2CO_3, 100 mM $NaHCO_3$), low temperature (4 °C), oxidation (10 mM H_2O_2), and heavy metal (160 μM $CdCl_2$) stresses (Figure 5B, Table S7). Under NaCl stress, *PutACBP1* was down-regulated in leaves and roots, while *PutACBP2*, *PutACBP3*, and *PutACBP4* in leaves, as well as PutACBP2 and PutACBP3 in roots, were significantly up-regulated (Figure 5B). Similarly, Na_2CO_3 and $NaHCO_3$ stresses induced *PutACBP2*, *PutACBP3*, and *PutACBP4* expression in leaves or roots at certain timepoints (Figure 5B). Furthermore, low temperature (4 °C) induced *PutACBP2* to *PutACBP5* expression at certain timepoints, but significantly reduced *PutACBP1* in roots and leaves. Oxidative stress (H_2O_2) reduced *PutACBP1* in roots and leaves while inducing *PutACBP2* to *PutACBP5* in leaves. In addition, $CdCl_2$ induced *PutACBP2*, *PutACBP4*, and *PutACBP5* in leaves but reduced the expression of all *PutACBPs* in roots (Figure 5B). These results indicate that *PutACBPs* play important roles in alkaligrass tolerance to different abiotic stresses.

2.7. Subcellular Localization of ACBPs in Alkaligrass

To determine the subcellular localization of PutACBPs in alkaligrass, we performed transient transformation in tobacco leaves using *35S::GFP* and *35S::PutACBP-GFP* constructs, and stable transformation of *35S::PutACBP3-GFP* in *Arabidopsis* seedlings. In tobacco epidermal cells, PutACBP1-GFP and PutACBP2-GFP displayed fluorescence in the plasma membrane, cytoplasm, and nucleus. In contrast, fluorescence signals from the PutACBP3-GFP, PutACBP4-GFP, and PutACBP5-GFP proteins were primarily localized to the plasma membrane and cytoplasm (Figure 5C). This membrane-associated localization pattern of PutACBP3-GFP was further supported by experiments in *Arabidopsis* protoplasts (Figure 5D).

2.8. PutACBP3 Is Essential for Seedling Salt Tolerance

To investigate the function of *PutACBP3* in response to salt stress, we constructed *Arabidopsis PutACBP3*-overexprssing seedlings and compared their phenotypes with WT. Under salt stress (100 mM NaCl and 125 mM NaCl), *PutACBP3*-overexprssing seedlings exhibited significantly longer roots and higher fresh weight than those of WT (Figure 6A–C), which indicates *PutACBP3* enhanced the salt tolerance.

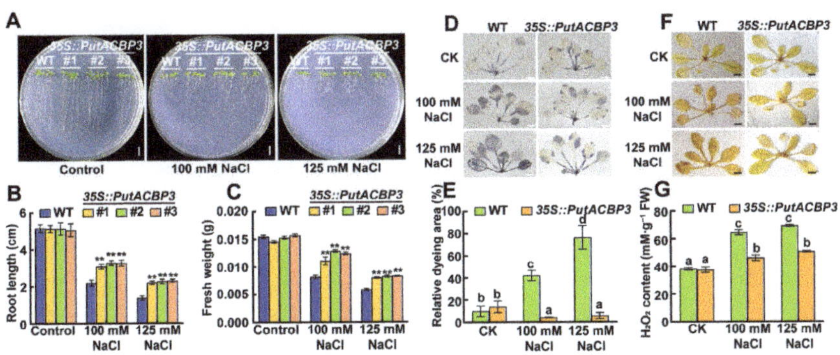

Figure 6. The phenotypes of WT (wild-type) and *PutACBP3*-overexpressing *Arabidopsis* plants. (**A**) Seedlings phenotypes of WT, *35S::PutACBP3* (#1, #2, and #3) under normal condition (1/2 MS medium), 100 mM NaCl, and 125 mM NaCl stresses for ten days. The scale bar is 2 cm. (**B**) Root growth WT and three *35S::PutACBP3* alleles. (**C**) Fresh weight of WT and three *35S::PutACBP3* alleles under normal conditions (1/2 MS medium) NaCl stress conditions. Each value represents the mean ± SD ($n = 15$). Error bars indicate SD. Student's *t*-test. ** $p < 0.01$. (**D**) Visualization of $O^{2\cdot-}$ by NBT staining, in leaves of three-week-old *Arabidopsis* WT and transgenic seedlings under normal condition (1/2 MS), 100 mM NaCl, and 125 mM NaCl stresses for 48 h. The blue colors show the accumulation of $O^{2\cdot-}$. The scale bar = 0.5 cm. (**E**) The statistics of dyeing area of leaves in (**D**). The values were presented as mean ± SE ($n = 3$). (**F**) Visualization of H_2O_2 by DAB staining, in leaves of three-week-old *Arabidopsis* WT and transgenic seedlings under normal condition (1/2 MS), 100 mM NaCl, and 125 mM NaCl stresses for 48 h. The brown colors show the accumulation of H_2O_2. The scale bar = 0.5 cm. (**G**) H_2O_2 contents in 3-week-old *Arabidopsis* WT and transgenic seedlings under normal condition (1/2 MS), 100 mM NaCl, and 125 mM NaCl stresses for 48 h. The values were presented as mean ± SD ($n = 3$). One-way ANOVA was used for statistical analyses in (**E**,**G**). Different letters indicate significant differences.

The NBT and DAB staining was applied to detect O^{2-} accumulation and H_2O_2 content in the leaves, respectively. Both O^{2-} accumulation and H_2O_2 contents were much lower in transgenic seedlings than the WT under the treatments of NaCl (Figure 6D–F). The H_2O_2

contents in transgenic seedlings were not different from that in WT seedlings in normal condition. However, the H_2O_2 contents were obviously lower in all the transgenic lines under NaCl treatment (Figure 6G).

2.9. Phosphatidylcholine-Specific Binding of PutACBP3 Associates with PLDδ-Mediated Membrane Remodeling for Enhanced Salinity Tolerance

To elucidate the biochemical properties and physiological function of PutACBP3, we conducted lipid-binding assays and stress response analyses. Recombinant His-PutACBP3 protein migrated at the predicted molecular mass (~54 kDa) on Coomassie-stained SDS-PAGE gels (Figure S2). In vitro lipid binding assays revealed a specific affinity of PutACBP3 for PC, including the variants 16:0-PC, 18:0-PC, and 18:1-PC. In contrast, no significant binding was detected to phosphatidylserine (PS), PA, PE, phosphatidylglycerol (PG), or dimyristoylphosphatidylcholine (DMPC) (Figure 7A,B).

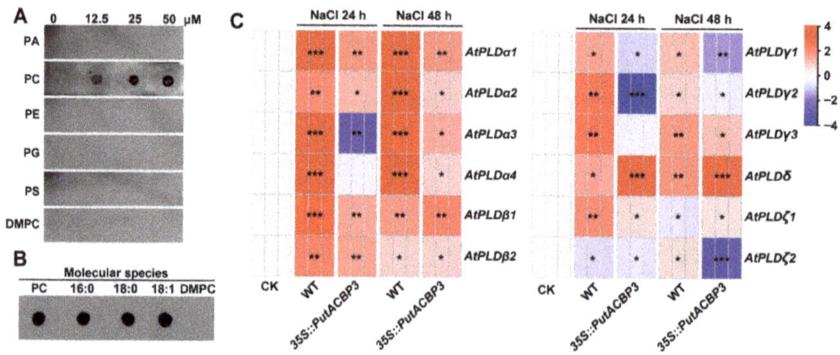

Figure 7. Recombinant protein His-PutACBP3 binding to phospholipids in vitro and expression profiles of *PLD* genes in *PutACBP3*-overexpressing *Arabidopsis* seedlings. (**A**) His-PutACBP3 binds to phosphatidylcholine (PC) on filters. Various concentrations (0, 12.5, 25, and 50 μM) of lipids including phosphatidic acid (PA), PC, phosphatidylethanolamine (PE), phosphatidyl glycerol (PG), phosphatidylserine (PS), and Dimyristoylphosphatidylcholine (DMPC) were spotted onto nitrocellulose and incubated with 1 μg of purified His-PutACBP3 proteins. The His-PutACBP3 binding was detected by immunoblotting with HRP-conjugated anti-penta-His antibodies. (**B**) Effect of PC acyl species on His-PutACBP3/PC binding. 50 μM lipids (PC, 16:0-PC, 18:0-PC, 18:1-PC or DMPC) were spotted onto nitrocellulose and then were incubated with 1 μg ml^{-1} of purified His-PutACBP3 protein. The His-PutACBP3 lipid binding was detected by immunoblotting with HRP-conjugated anti-penta-His antibodies. (**C**) Expression profiles of the 12 *AtPLD* genes in WT and *PutACBP3*-overexpressing *Arabidopsis* seedlings illustrated by a heat map under normal conditions (1/2 MS medium), and NaCl stress conditions. Significant differences compared to the control group were determined by Student's *t*-test, *** $p < 0.001$, ** $p < 0.01$, and * $p < 0.05$.

Given this specific PC-binding activity and the enhanced salinity tolerance observed in *PutACBP3*-overexpressing *Arabidopsis* (Figure 6), we investigated its potential link to phospholipid signaling pathways known to be involved in stress responses. Considering the established roles of phospholipase D (PLD) isoforms in stress signaling and membrane lipid remodeling, we analyzed the expression of *Arabidopsis PLD* genes in WT and *35S::PutACBP3* seedlings under 100 mM NaCl treatment. RT-qPCR revealed that the expression level of *AtPLDδ* was significantly higher in *35S::PutACBP3* seedlings compared to WT at both 24 h and 48 h after salt treatments (Figure 7C, Table S8). These results suggest that the enhanced salt tolerance conferred by *PutACBP3* overexpression may be associated with elevated *PLDδ* expression. This observation is consistent with a model where PutACBP3, through its specific interaction with PC, influences phospholipid metabolism and signaling, potentially facilitating PLDδ-mediated membrane remodeling processes critical for salinity adaptation.

3. Discussion

3.1. Evolutionary Conservation and Functional Diversification of ACBPs

ACBPs have been functionally annotated in several higher plants, including *Arabidopsis*, rice, and *B. napus* [57]. Phylogenetic analyses indicate that ACBPs are conserved across all plants, including green microalgae (e.g., *Chlorella* sp.), bryophytes (e.g., *Physcomitrella*), and vascular plants [7], suggesting an origin in early photosynthetic eukaryotes. Our phylogenetic analysis revealed that the five PutACBPs were grouped into four subfamilies (Figure 1), consistent with finding in *Arabidopsis* and rice [9]. Comparative analyses further reveal species-specific variations in the types and numbers of ACBP genes (Figure 1), potentially reflecting functional specialization across plant lineages. This systematic characterization sheds light on the evolutionary diversification and stress adaptation mechanisms of ACBPs in plants adapted to extreme environments. Sequence alignment analysis showed that PutACBP shares similar functional domains with ACBP homologs from rice (Figure 3). Gene structure analysis of ACBP genes indicated that genes within the same branch generally possessed similar exon-intron structures (Figure 4A). In addition, the motif analysis of ACBPs also found that the ACBPs that were clustered in the same subfamily shared a similar motif composition (Figure 4D). The high sequence similarity observed across species suggests that these genes are likely to share conserved functions.

3.2. Stress-Responsive Regulation of PutACBPs

Cis-acting elements are essential in many biological processes and stress responses [58]. The G-Box, ARE, P-Box, ABRE, and CGTCA-motif are *cis*-acting elements that respond to light, heat, and hormonal induction [3,59–65]. We found that G-Box, ARE, P-Box, ABRE, and CGTCA-motifs exit in the promoter of *PutACBP* genes (Figure 4B), implying that *PutACBPs* might respond to light, heat and hormone treatments. However, unlike Class II *PtACBP3*, which lacks the hypoxia-specific inducible GC-motif, members of Classes I, III, and IV appear to participate in plant responses to abiotic stress and hypoxia, a functional similarity shared with Class III *AtACBP3* [66]. In our study, both *PutACBP1* and *PutACBP2* were classified into Class I. Promoter analysis revealed that they contain similar *cis*-acting elements, suggesting functional overlap in certain biological processes. In contrast, previous studies have shown that overexpression of *OsACBP4* and *AtACBP2*, both Class II members, enhances salt tolerance in rice and *Arabidopsis*, respectively [42]. Consistent with these findings, our results demonstrate that overexpression of *PutACBP3*, another Class II protein, also improves salt tolerance in *Arabidopsis* (Figure 6). Notably, all three genes contain stress-related *cis*-elements in their promoters. Together, these observations support the hypothesis that Class II *ACBPs* play a conserved and important role in the plant salt stress response.

This study showed that the expressions of *PutACBP2* and *PutACBP3* were the highest in the leaves of alkaligrass (Figure 5A), while previous studies showed that the expression levels of *AtACBP1* and *AtACBP2* were higher in the leaves of *Arabidopsis* [26,44,67]. The expression level of *OsACBP4* in rice leaves was about eight times that in seeds [9]. We found that the expression of *PutACBP1* was the highest in the spike, *PutACBP4* was the highest in the flower, and *PutACBP5* was the highest in the stem (Figure 5A). In *Arabidopsis*, *AtACBP3* was highly expressed in flowers and vegetative tissues, suggesting that extracellular localization of *AtACBP3* plays a role in plant defense [68], and *OsACBP5* was highly expressed at reproductive stage [57].

PutACBPs genes were mostly upregulated in roots and leaves under different stresses (Figure 5B), suggesting that *PutACBPs* may be involved in responses to salt, alkaline, low temperature, oxidation, and heavy metal stresses. Previous studies have shown that *OsACBP5* plays an important role in salt stress, drought stress, pathogen stress and physical damage response [9]. We propose that *PutACBPs* could facilitate improving the repair ability of cell membrane and energy supply.

3.3. *PutACBP3 Enhances Salinity Tolerance via PC Binding and PLDδ Activation*

We found that SOD activity was enhanced but ROS accumulation was reduced in *PutACBP3*-overexpressing *Arabidopsis* lines (Figure 6D–G). Moreover, PutACBP3 has unique PC binding specificity (Figure 7A,B). This indicates its role in PC-mediated membrane stabilization and ROS scavenging. This mechanism integrates lipid trafficking with oxidative stress mitigation under salinity [36,56].

ACBP-mediated lipid modulation likely serves as a critical interface between stress perception and physiological adaptation. The observed variations in *ACBP* expression patterns under stress conditions could drive dynamic changes in lipid abundance and composition. For instance, *Arabidopsis AtACBP6* overexpression elevates *PLDδ* expression during freezing stress [27], mirroring our finding that *PutACBP3* overexpression upregulates *PLDδ* under salinity (Figure 7C). PLDδ catalyzes phospholipid hydrolysis to PA, a central lipid mediator in osmotic stress responses, ABA signaling, and pathogen defense [44]. In this study, *PutACBP3* transgenic *Arabidopsis* exhibited enhanced salt tolerance (Figure 6), likely through initiating lipid-dependent signaling Via PA production and/or modulating membrane architecture through PC binding. While PLDδ upregulation in *PutACBP3*-overexpressing lines suggests PC-derived PA signaling (Figure 7C), future lipidomic studies are needed to quantify PA flux and clarify its role in SOS pathway activation [37,39]. Notably, PutACBP3 and AtACBP1 exhibit distinct lipid-binding specificities: PutACBP3 preferentially binds to PC, whereas AtACBP1 has a higher affinity for PA [45], may determine pathway selectivity across species. This evolutionary divergence highlights how monocots like alkaligrass have uniquely optimized lipid-mediated signaling for extreme environments. This lipid-mediated signal amplification regulates stress-related gene expression, thereby positioning *PutACBP3* as a key node in halophyte salt adaptation.

4. Materials and Methods

4.1. *Identification of ACBP Family Genes in Alkaligrass*

To identify PutACBPs, the protein database of Alkaligrass was downloaded from http://www.xhhuanglab.cn/data/alkaligrass.html (accessed on 20 June 2025) [48], and Hidden Markov Model (HMM) profile file (PF00887) was downloaded from the Pfam database (http://pfam.xfam.org, accessed on 20 June 2025) [69]. The HMM file was exploited as a query to identify PutACBPs in the alkaligrass protein database using the hmmer search command of the HMMER (version 3.0) software. AtACBPs (AT5G53470, AT4G27780, AT4G24230, AT3G05420, AT5G27630, and AT1G31812) were used as the query sequence in a BlastP search against the alkaligrass genome database online. The sequences obtained by the above two methods were used for the subsequent selection, and the screening criteria were based on the ACB domain at the N-terminus, ankyrin (ANK) domain, and the Kelch domain at the C-terminus [9]. Peptide length, MW, and pI of each PutACBP were calculated using the online ExPASy program (https://www.expasy.org, accessed on 20 June 2025) [70]. Detailed information of PutACBPs can be found in Table S1. The CDS and protein sequences of PutACBPs were displayed in Tables S2 and S3.

4.2. *Phylogenetic Analysis of PutACBP Genes*

To investigate the phylogenetic relationships of the *ACBP* gene families among 12 plants. ACBPs were aligned using the BioEdit program. A neighbor-joining (NJ) phylogenetic tree was constructed for these proteins with MEGA 7.0 software (https://www.megasoftware.net, accessed on 22 June 2025) [71]. Bootstrapping was performed with 1000 repetitions. The protein sequences of ACBPs were displayed in Table S3.

4.3. Chromosomal Localization of the PutACBP Genes

The chromosomal localization of the *PutACBP* genes were obtained from http://www.xhhuanglab.cn/data/alkaligrass.html (accessed on 25 June 2025) [48], and the diagram was drawn using the MapInspect software (http://mg2c.iask.in/mg2c_v2.0, accessed on 25 June 2025).

4.4. Collinearity Analysis of ACBPs Between Alkaligrass and Other Plants

The collinearity analysis of *ACBPs* between the rice, maize, and *B. distachyon* was verified and visualized using One Stem MCScanX and Dual Systeny Plot for MCScanX in TBtools (v2.322) software, respectively [72].

4.5. Exon/Intron Structure, Conserved Motifs and Promoter Cis-Acting Element Analysis

The distribution patterns of exons and introns in *PutACBP* genes were predicted using the Gene Structure Display Server (GSDS2.0, http://gsds.cbi.pku.edu.cn, accessed on 27 June 2025) [73]. The *cis*-acting elements of *PutACBP* promoters were analyzed to further understand the *PutACBP* gene family. The 2500 bp upstream sequences of the *PutACBPs* promoter regions were downloaded FASTA format from the http://www.xhhuanglab.cn/data/alkaligrass.html (accessed on 27 June 2025) [48], and used to identify the putative *cis*-acting elements in PlantCARE (http://bioinformatics.psb.ugent.be/webtools/plantcare/html, accessed on 27 June 2025) [74]. The *cis*-acting elements were visualized on the GSDS 2.0 website. The conserved motifs in PutACBPs (with a maximum of 10 motifs) were predicted using the online MEME tools (http://meme-suite.org/tools/meme, accessed on 27 June 2025) [75].

4.6. Plant Materials, Growth Conditions, and Stress Treatments

The experimental materials for organ expression analysis were provided by Anda Experimental Base, Research Center for Biological Resources and Environment in Saline-Alkali Land, Northeast Forestry University. Roots, stems, leaves, sheaths, flowers, spikes, and seeds of wild alkaligrass from salt-alkali land were harvested, immediately frozen in liquid nitrogen, and then stored at −80 °C for further analysis.

For the specific expression experiment under various stresses, the cleaned seeds were soaked in darkness for 12 h, then evenly sown on the gauze of hydroponic culture and cultured in the greenhouse under long-day conditions (16 h light/8 h dark) at 25 °C, and supplied with water every two days. Seedlings that germinated after three weeks were subjected to the following conditions, 4 °C (cold), 200 mM NaCl, 100 mM $NaHCO_3$, 75 mM Na_2CO_3, 10 mM H_2O_2, and 160 uM $CdCl_2$, respectively. The control (untreated) and treated roots and leaves were harvested, respectively, at 0 h, 6 h, 12 h, 24 h and 48 h after treatments, respectively. All samples were frozen in liquid nitrogen and stored at −80 °C until further use.

The *35S::PutACBP3-GFP* was amplified with appropriate primers (Table S5) and cloned into the *pBI121* vector and then introduced into WT Via *Agrobacterium tumefaciens* (GV3101 strain) using the floral dip method [76]. Seeds were surface sterilized by fumigation with chlorine gas for 4 h, then plated on solidified half strength Murashige and Skoog (MS) medium. All the seeds were held for two days at 4 °C, and finally grown in a greenhouse delivering a 16 h photoperiod with a constant temperature of 22 °C.

4.7. RNA Extraction and Reverse Transcription Quantitative Polymerase Chain Reaction (RT-qPCR)

Total RNA was extracted from roots and leaves using Trizol reagent (9109, TaKaRa, Tokyo, Japan) following the manufacturer's protocol. The quality of the RNA was determined using the NanoDrop 2000 spectrophotometer (Thermo Scientific, Waltham, MA, USA). After removing genomic DNA contamination with DNaseI, cDNA was synthesized by using the PrimeScriptTMRT Reagent Kit (RR092A, Takara, Tokyo, Japan). For RT-qPCR analysis, the ChamQ SYBR Color qPCR Master Mix (Q411, Vazyme, Nanjing, China) was used with *Putactin* and *PutGADPH* serving as an internal control. Primers of *PutACBP* genes were designed using Primer3 online tools (https://bioinfo.ut.ee/primer3-0.4.0/, accessed on 7 July 2025). All the primers were listed in Table S5. RT-qPCR analyses were performed using three independent biological replicates, and the data were normalized using the $2^{-\Delta\Delta Ct}$ method. The heat map of gene expression pattern was visualized using TBtools software.

4.8. Subcellular Localization of PutACBP and Microscopy Analysis

For subcellular localization analysis, the full-length CDS of *PutACBPs* were cloned into the vector *pBI121-GFP*, transformed into *A. tumefaciens* strain GV3101. The *A. tumefaciens* strain GV3101 containing *35S::PutACBPs-GFP* and *35S::GFP* constructions were incubated shaking until the OD_{600} reached 0.6–0.8, respectively. Centrifuge the cells at 4000× *g* for 5 min, then discard the supernatant and resuspend the pellet with induction buffer (0.01 M MES, 0.01 M $MgCl_2$, 150 μM acetosyringone) to find OD_{600} = 0.6, inject the solution into three-week-old tobacco leaves. After 48 h incubation, the fluorescence signal of GFP was observed using a fluorescence microscope LSM 780 (Zeiss, Baden-Württemberg, Germany), with excitation at 488 nm.

4.9. Phenotype Analysis

The seedlings were grown on 1/2 MS for seven days then transferred to media supplemented with various concentration of NaCl. Position of root tips marked on the plate and grown for 10 days. The root lengths were determined from digital images using Image J software (v1.41). Fresh weight was also measured at the same time.

4.10. Nitroblue Tetrazolium (NBT) Staining

NBT staining was used to detect $O^{2\cdot-}$ using a method modified from that described by Dunand et al. [77]. Fifteen-day-old seedlings were treated with salt stress for two days. The untreated seedlings were used as experimental controls under the same conditions, then the seeding incubated in 2 mM NBT (Sigma, Darmstadt, Germany) in 20 mM K phosphate/0.1 M NaCl at pH 6.1 for 15 min, then transferred to distilled water.

4.11. Isolation of His-PutACBP3 Proteins from E. coli

The CDS of *PutACBP3* was amplified by PCR and cloned into *pET32a* expression vector. The construction was transformed into *Escherichia coli* strain BL21 (DE3) for protein expression. His-tagged proteins were purified using Ni-NTA resin according to instructions of the manufacturer.

4.12. Purification of Recombinant His-PutACBP3 for Filter-Binding Assays

The binding of His-PutACBP3 to the various lipids on the filters was performed as previously described [78] with minor modifications. Briefly, various concentrations of lipids were spotted onto nitrocellulose and incubated at room temperature for 1 h in dark. LysoPC, PC, PA, 18:0-PC, and 18:2-PC were purchased from Sigma, and 16:0-PC, 18:1-PC, and DMPC were purchased from Echelon Biosciences. The lipid-bound filter was blocked with Tris-buffered saline (TBS) with 1% nonfat milk for 1 h. After incubation with 1 μg·mL^{-1} purified His-ACBP3 protein in blocking buffer for 2 h, the filter was gently washed three times with TTBS (TBS plus 0.1% Tween 20), each for 10 min. Following incubation with the horseradish peroxidase (HRP)-conjugated anti-His antibodies (1:2000; Qiagen, Düsseldorf, Germany, Catalog No. 1014922) for 1 h at room temperature, the filter was again washed three times with TTBS, each for 10 min, and then detected with the ECL Western blotting Detection Kit (Amersham, Uppsala, Sweden) following the manufacturer's protocols.

5. Conclusions

We identified five alkaligrass *ACBP* genes, which could be classified into four subfamilies. We confirmed that *PutACBP* genes displayed a great evolutionary divergence in alkaligrass by gene structure, conserved domains and phylogenetic analysis. Furthermore, we identified that *ACBP* genes showed special organ expression profiles, indicating their potential divergent functions in alkaligrass growth and development. *Cis*-acting elements analysis and expression profiles under different stresses suggested that ACBP family genes might be related to the responses of abiotic stresses and hormone stimuli. Overexpressing *PutACBP* gene in *Arabidopsis* seedlings revealed that *PutACBP3* was predominantly expressed in roots and mediated ROS homeostasis primarily under salinity stress. Moreover, we focused on the expression of *PLDs* in *Arabidopsis* overexpressing *PutACBP3*, the RT-qPCR showed that after 100 mM NaCl stress, a higher expression level in *PLDδ* than WT, suggesting that salt stress tolerance of *PutACBP3* may be related to the enhancement of *PLDδ* expression. Taken together, we established a foundation for further functional characterization of *PutACBP* genes in response to salinity stress in the future.

Supplementary Materials: The following supporting information can be downloaded at https://www.mdpi.com/article/10.3390/plants14233551/s1, Figure S1: Predicted amino acid sequence for each motif. Motifs were predicted by MEME motif searching, and 10 were selected as the maximum number parameter; Figure S2: Induction and Purification of $(His)_6$-PutACBP3 recombinant protein; Table S1: Features of *PutACBP* genes in *Puccinellia tenuiflora;* Table S2: List of coding sequences of *PuACBPs* from *Puccinellia tenuiflora;* Table S3: List of 86 ACBPs from *Puccinellia tenuiflora, Oryza sativa, Arabidopsis thaliana* and other Plants; Table S4: List of structural domain protein sequence from *Puccinellia tenuiflora*, *Oryza sativa*, and *Arabidopsis thaliana;* Table S5: List of primers used in cloning and RT-qPCR analysis for *PutACBPs;* Table S6: Organ- specific expression analysis of *PutACBPs;* Table S7: Expression analysis of *PutACBPs* in roots and leaves under differernt stress; Table S8: Expression profiles of *PLD* genes in *PutACBP3*-overexpressing *Arabidopsis* seedlings.

Author Contributions: Conceptualization, S.D., X.X. and Y.L.; methodology, H.Z.; software, H.Z.; validation, H.Z., S.W. and X.N.; formal analysis, H.Z.; investigation, S.D., X.X., Y.L. and H.Z.; resources, H.Z.; data curation, H.Z.; writing—original draft preparation, S.D., X.X. and H.Z.; writing—review and editing, S.D., X.X. and H.Z.; visualization, H.Z.; supervision, S.D., X.X. and Y.L.; project administration, S.D., X.X. and Y.L.; funding acquisition, S.D. All authors have read and agreed to the published version of the manuscript.

Funding: This work was funded by the National Natural Science Foundation of China (No. 32441006 and 32070300); and the Fund of Shanghai Engineering Research Center of Plant Germplasm Resources, China (No. 17DZ2252700) to Shaojun Dai.

Data Availability Statement: Data is contained within the article and Supplementary Materials.

Conflicts of Interest: The authors declare no conflicts of interest.

References

1. Feddersen, S.; Neergaard, T.B.F.; Knudsen, J.; Færgeman, N.J. Transcriptional regulation of phospholipid biosynthesis is linked to fatty acid metabolism by an acyl-CoA-binding-protein-dependent mechanism in *Saccharomyces cerevisiae*. *Biochem. J.* **2007**, *407*, 219–230. [CrossRef]
2. Xiao, S.; Chye, M. New roles for acyl-CoA-binding proteins (ACBPs) in plant development, stress responses and lipid metabolism. *Prog. Lipid Res.* **2011**, *50*, 141–151. [CrossRef] [PubMed]
3. Guidotti, A.; Forchetti, C.M.; Corda, M.G.; Konkel, D.; Bennett, C.D.; Costa, E. Isolation, characterization, and purification to homogeneity of an endogenous polypeptide with agonistic action on benzodiazepine receptors. *Proc. Natl. Acad. Sci. USA* **1983**, *80*, 3531–3535. [CrossRef]
4. Knudsen, J.; Mandrup, S.; Rasmussen, J.T.; Andreasen, P.H.; Poulsen, F.; Kristiansen, K. The function of acyl-CoA-binding protein (ACBP) Diazepam binding inhibitor (DBI). *Mol. Cell. Biochem.* **1993**, *123*, 129–138. [CrossRef]
5. Hills, M.J.; Dann, R.; Lydiate, D.; Sharpe, A. Molecular cloning of a cDNA from *Brassica napus* L. for a homologue of acyl-COA binding protein. *Plant Mol. Biol.* **1994**, *25*, 917–920. [CrossRef]
6. Lung, S.; Chye, M. Deciphering the roles of acyl-CoA-binding proteins in plant cells. *Protoplasma* **2016**, *253*, 1177–1195. [CrossRef] [PubMed]
7. Zhu, J.; Li, W.; Zhou, Y.; Pei, L.; Liu, J.; Xia, X.; Che, R.; Li, H. Molecular characterization, expression and functional analysis of acyl-CoA-binding protein gene family in maize (*Zea mays*). *BMC Plant Biol.* **2021**, *21*, 94. [CrossRef]
8. Leung, K.; Li, H.; Mishra, G.; Chye, M. ACBP4 and ACBP5, novel *Arabidopsis* acyl-CoA-binding proteins with kelch motifs that bind oleoyl-CoA. *Plant Mol. Biol.* **2005**, *55*, 297–309. [CrossRef]
9. Meng, W.; Su, Y.C.F.; Saunders, R.M.K.; Chye, M. The rice acyl-CoA-binding protein gene family: Phylogeny, expression and functional analysis. *New Phytol.* **2011**, *189*, 1170–1184. [CrossRef]
10. Nie, Z.; Wang, Y.; Wu, C.; Yu, L.; Kang, G.; Qin, H.; Zeng, R. Acyl-CoA-binding protein family members in laticifers are possibly involved in lipid and latex metabolism of *Hevea brasiliensis* (the Para rubber tree). *BMC Genom.* **2018**, *19*, 5. [CrossRef] [PubMed]
11. Raboanatahiry, N.; Wang, B.; Yu, L.; Li, M. Functional and structural diversity of acyl-CoA binding proteins in oil crops. *Front. Genet.* **2018**, *9*, 182. [CrossRef]
12. Fan, J.; Liu, J.; Culty, M.; Papadopoulos, V. Acyl-coenzyme A binding domain containing 3 (ACBD3; PAP7; GCP60): An emerging signaling molecule. *Prog. Lipid Res.* **2010**, *49*, 218–234. [CrossRef] [PubMed]
13. Elle, I.C.; Simonsen, K.T.; Olsen, L.C.B.; Birck, P.K.; Ehmsen, S.; Tuck, S.; Le, T.T.; Færgeman, N.J. Tissue-and paralogue-specific functions of acyl-CoA-binding proteins in lipid metabolism in Caenorhabditis elegans. *Biochem. J.* **2011**, *437*, 231–241. [CrossRef]
14. Neess, D.; Bek, S.; Engelsby, H.; Gallego, S.F.; Færgeman, N.J. Long-chain acyl-CoA esters in metabolism and signaling: Role of acyl-CoA binding proteins. *Prog. Lipid Res.* **2015**, *59*, 1–25. [CrossRef]
15. Faergeman, N.J.; Feddersen, S.; Christiansen, J.K.; Larsen, M.K.; Schneiter, R.; Ungermann, C.; Mutenda, K.; Roepstorff, P.; Knudsen, J. Acyl-CoA-binding protein, Acb1p, is required for normal vacuole function and ceramide synthesis in *Saccharomyces cerevisiae*. *Biochem. J.* **2004**, *380*, 907–918. [CrossRef]
16. Zhang, J.; Diaz, A.; Mao, L.; Ahlquist, P.; Wang, X. Host acyl coenzyme a binding protein regulates replication complex assembly and activity of a positive-strand RNA virus. *J. Virol.* **2012**, *86*, 5110–5121. [CrossRef] [PubMed]
17. Xiao, S.; Chye, M. An *Arabidopsis* family of six acyl-CoA-binding proteins has three cytosolic members. *Plant Physiol. Biochem.* **2009**, *47*, 479–484. [CrossRef] [PubMed]
18. Chye, M.; Li, H.; Yung, M. Single amino acid substitutions at the acyl-CoA-binding domain interrupt ^{14}[C]palmitoyl-CoA binding of ACBP2, an *Arabidopsis* acyl-CoA binding protein with ankyrin repeats. *Plant Mol. Biol.* **2000**, *44*, 711–721. [CrossRef]
19. Leung, K.; Li, H.; Xiao, S.; Tse, M.; Chye, M. *Arabidopsis* ACBP3 is an extracellularly targeted acyl-CoA-binding protein. *Planta* **2006**, *223*, 871–881. [CrossRef]
20. Gao, W.; Xiao, S.; Li, H.; Tsao, S.; Chye, M. *Arabidopsis thaliana* acyl-CoA-binding protein ACBP2 interacts with heavy-metal-binding farnesylated protein AtFP6. *New Phytol.* **2009**, *181*, 89–102. [CrossRef]
21. Xiao, S.; Gao, W.; Chen, Q.; Chan, S.; Zheng, S.; Ma, J.; Wang, M.; Welti, R.; Chye, M. Overexpression of *Arabidopsis* acyl-CoA binding protein ACBP3 promotes starvation-induced and age-dependent leaf senescence. *Plant Cell* **2010**, *22*, 1463–1482. [CrossRef]
22. Wang, X. Lipid signaling. *Curr. Opin. Plant Biol.* **2004**, *7*, 329–336. [CrossRef]
23. Walz, C.; Giavalisco, P.; Schad, M.; Juenger, M.; Klose, J.; Kehr, J. Proteomics of curcurbit phloem exudate reveals a network of defence proteins. *Phytochemistry* **2004**, *65*, 1795–1804. [CrossRef] [PubMed]
24. Zhou, Y.; Atkins, J.B.; Rompani, S.B.; Bancescu, D.L.; Petersen, P.H.; Tang, H.; Zou, K.; Stewart, S.B.; Zhong, W. The mammalian Golgi regulates numb signaling in asymmetric cell division by releasing ACBD3 during mitosis. *Cell* **2007**, *129*, 163–178. [CrossRef] [PubMed]

25. Yoshimoto, K.; Jikumaru, Y.; Kamiya, Y.; Kusano, M.; Consonni, C.; Panstruga, R.; Ohsumi, Y.; Shirasu, K. Autophagy negatively regulates cell death by controlling NPR1-dependent salicylic acid signaling during senescence and the innate immune response in *Arabidopsis*. *Plant Cell* **2009**, *21*, 2914–2927. [CrossRef]
26. Chen, Q.; Xiao, S.; Qi, W.; Mishra, G.; Ma, J.; Wang, M.; Chye, M. The *Arabidopsis acbp1acbp2* double mutant lacking acyl-CoA-binding proteins ACBP1 and ACBP2 is embryo lethal. *New Phytol.* **2010**, *186*, 843–855. [CrossRef] [PubMed]
27. Chen, Q.; Xiao, S.; Chye, M. Overexpression of the *Arabidopsis* 10-kilodalton acyl-coenzyme A-binding protein ACBP6 enhances freezing tolerance. *Plant Physiol.* **2008**, *148*, 304–315. [CrossRef]
28. Tsugane, K.; Kobayashi, K.; Niwa, Y.; Ohba, Y.; Wada, K.; Kobayashi, H. A recessive *Arabidopsis* mutant that grows photoautotrophically under salt stress shows enhanced active oxygen detoxification. *Plant Cell* **1999**, *11*, 1195–1206. [CrossRef]
29. Hong, Z.; Lakkineni, K.; Zhang, Z.; Verma, D.P.S. Removal of feedback inhibition of delta(1)-pyrroline-5-carboxylate synthetase results in increased proline accumulation and protection of plants from osmotic stress. *Plant Physiol.* **2000**, *122*, 1129–1136. [CrossRef]
30. Ashraf, M. Some important physiological selection criteria for salt tolerance in plants. *Flora* **2004**, *199*, 361–376. [CrossRef]
31. Zhu, J.; Fu, X.; Koo, Y.D.; Zhu, J.; Jenney, F.E.; Adams, M.W.W.; Zhu, Y.; Shi, H.; Yun, D.-J.; Hasegawa, P.M.; et al. An enhancer mutant of *Arabidopsis salt overly sensitive 3* mediates both ion homeostasis and the oxidative stress response. *Mol. Cell. Biol.* **2007**, *27*, 5214–5224. [CrossRef]
32. Li, J.; Liu, J.; Wang, G.; Cha, J.-Y.; Li, G.; Chen, S.; Li, Z.; Guo, J.; Zhang, C.; Yang, Y.; et al. A chaperone function of no catalase activity1 is required to maintain catalase activity and for multiple stress responses in *Arabidopsis*. *Plant Cell* **2015**, *27*, 908–925. [CrossRef]
33. Yang, Y.; Guo, Y. Elucidating the molecular mechanisms mediating plant salt-stress responses. *New Phytol.* **2017**, *217*, 523–539. [CrossRef] [PubMed]
34. Shi, H.; Quintero, F.J.; Pardo, J.M.; Zhu, J. The putative plasma membrane Na^+/H^+ antiporter SOS1 controls long-distance Na^+ transport in plants. *Plant Cell* **2002**, *14*, 465–477. [CrossRef]
35. Shabala, L.; Cuin, T.A.; Newman, I.A.; Shabala, S. Salinity-induced ion flux patterns from the excised roots of *Arabidopsis sos* mutants. *Planta* **2005**, *222*, 1041–1050. [CrossRef] [PubMed]
36. Testerink, C.; Munnik, T. Phosphatidic acid: A multifunctional stress signaling lipid in plants. *Trends Plant Sci.* **2005**, *10*, 368–375. [CrossRef] [PubMed]
37. Yu, L.; Nie, J.; Cao, C.; Jin, Y.; Yan, M.; Wang, F.; Liu, J.; Xiao, Y.; Liang, Y.; Zhang, W. Phosphatidic acid mediates salt stress response by regulation of MPK6 in *Arabidopsis thaliana*. *New Phytol.* **2010**, *188*, 762–773. [CrossRef]
38. Testerink, C.; Munnik, T. Molecular, cellular, and physiological responses to phosphatidic acid formation in plants. *J. Exp. Bot.* **2011**, *62*, 2349–2361. [CrossRef]
39. Qiu, Q.; Guo, Y.; Dietrich, M.A.; Schumaker, K.S.; Zhu, J. Regulation of SOS1 a plasma membrane Na^+/H^+ exchanger in *arabidopsis thaliana* by SOS2 and SOS3. *Proc. Natl. Acad. Sci. USA* **2022**, *99*, 8436–8441. [CrossRef]
40. Qiao, K.; Wang, M.; Takano, T.; Liu, S. Overexpression of *acyl-CoA-binding protein 1* (*ChACBP1*) from saline-alkali-tolerant *chlorella* sp. enhances stress tolerance in *Arabidopsis*. *Front. Plant Sci.* **2018**, *9*, 1772. [CrossRef]
41. Chen, M.; Hu, T.; Xue, Y.; Zhu, F.; Du, Z.; Lo, C.; Chye, M. *Arabidopsis* Acyl-Coenzyme-A-Binding Protein ACBP1 interacts with AREB1 and mediates salt and osmotic signaling in seed germination and seedling growth. *Environ. Exp. Bot.* **2018**, *156*, 130–140. [CrossRef]
42. Guo, Z.; Pogancev, G.; Meng, W.; Du, Z.; Liao, P.; Zhang, R.; Chye, M. The overexpression of rice acyl-CoA-binding protein 4 improves salinity tolerance in transgenic rice. *Environ. Exp. Bot.* **2021**, *183*, 104349. [CrossRef]
43. Azlan, N.S.; Guo, Z.; Yung, W.; Wang, Z.; Lam, H.; Lung, S.; Chye, M. *In silico* analysis of acyl-CoA-binding protein expression in Soybean. *Front. Plant Sci.* **2021**, *12*, 646938. [CrossRef]
44. Xiao, S.; Gao, W.; Chen, Q.; Ramalingam, S.; Chye, M. Overexpression of membrane-associated acyl-CoA-binding protein ACBP1 enhances lead tolerance in *Arabidopsis*. *Plant J.* **2008**, *54*, 141–151. [CrossRef] [PubMed]
45. Du, Z.; Chen, M.; Chen, Q.; Xiao, S.; Chye, M. *Arabidopsis* acyl-CoA-binding protein ACBP1 participates in the regulation of seed germination and seedling development. *Plant J.* **2013**, *74*, 294–309. [CrossRef]
46. Xiao, S.; Chye, M. Overexpression of *Arabidopsis* ACBP3 enhances NPR1-dependent plant resistance to *Pseudomonas syringe* pv *tomato* DC3000. *Plant Physiol.* **2011**, *156*, 2069–2081. [CrossRef]
47. Zhang, X.; Wei, L.; Wang, Z.; Wang, T. Physiological and molecular features of *Puccinellia tenuiflora* tolerating salt and alkaline-salt stress. *J. Integr. Plant Biol.* **2013**, *55*, 262–276. [CrossRef]
48. Zhang, W.; Liu, J.; Zhang, Y.; Qiu, J.; Li, Y.; Zheng, B.; Hu, F.; Dai, S.; Huang, X. A high-quality genome sequence of alkaligrass provides insights into halophyte stress tolerance. *Sci. China Life Sci.* **2020**, *63*, 1269–1282. [CrossRef]
49. Yu, J.; Chen, S.; Zhao, Q.; Wang, T.; Yang, C.; Diaz, C.; Sun, G.; Dai, S. Physiological and Proteomic Analysis of Salinity Tolerance in *Puccinellia tenuiflora*. *J. Proteome Res.* **2011**, *10*, 3852–3870. [CrossRef]

50. Yu, J.; Chen, S.; Wang, T.; Sun, G.; Dai, S. Comparative proteomic analysis of *Puccinellia tenuiflora* leaves under Na_2CO_3 stress. *Int. J. Mol. Sci.* **2013**, *14*, 1740–1762. [CrossRef] [PubMed]
51. Meng, X.; Zhao, Q.; Jin, Y.; Yu, J.; Yin, Z.; Chen, S.; Dai, S. Chilling-responsive mechanisms in halophyte *Puccinellia tenuiflora* seedlings revealed from proteomics analysis. *J. Proteom.* **2016**, *143*, 365–381. [CrossRef]
52. Zhao, Q.; Suo, J.; Chen, S.; Jin, Y.; Ma, X.; Yin, Z.; Zhang, Y.; Wang, T.; Luo, J.; Jin, W.; et al. Na_2CO_3-responsive mechanisms in halophyte *Puccinellia tenuiflora* roots revealed by physiological and proteomic analyses. *Sci. Rep.* **2016**, *6*, 32717. [CrossRef] [PubMed]
53. Yu, J.; Zhang, Y.; Liu, J.; Wang, L.; Liu, P.; Yin, Z.; Guo, S.; Ma, J.; Lu, Z.; Wang, T.; et al. Proteomic discovery of H_2O_2 response in roots and functional characterization of *PutGLP* gene from alkaligrass. *Planta* **2018**, *248*, 1079–1099. [CrossRef] [PubMed]
54. Yin, Z.; Zhang, H.; Zhao, Q.; Yoo, M.; Zhu, N.; Yu, J.; Yu, J.; Guo, S.; Miao, Y.; Chen, S.; et al. Physiological and comparative proteomic analyses of saline-alkali $NaHCO_3$-responses in leaves of halophyte *Puccinellia tenuiflora*. *Plant Soil* **2019**, *437*, 137–158. [CrossRef]
55. Zhang, Y.; Zhang, Y.; Yu, J.; Zhang, H.; Wang, L.; Wang, S.; Guo, S.; Miao, Y.; Chen, S.; Li, Y.; et al. NaCl-responsive ROS scavenging and energy supply in alkaligrass callus revealed from proteomic analysis. *BMC Genom.* **2019**, *20*, 990. [CrossRef]
56. Suo, J.; Zhang, H.; Zhao, Q.; Zhang, N.; Zhang, Y.; Li, Y.; Song, B.; Yu, J.; Cao, J.; Wang, T.; et al. Na_2CO_3-responsive photosynthetic and ROS scavenging mechanisms in chloroplasts of alkaligrass revealed by phosphoproteomics. *Genom. Proteom. Bioinf.* **2020**, *18*, 271–288. [CrossRef]
57. Du, Z.; Arias, T.; Meng, W.; Chye, M. Plant acyl-CoA-binding proteins: An emerging family involved in plant development and stress responses. *Prog. Lipid Res.* **2016**, *63*, 165–181. [CrossRef]
58. Ibraheem, O.; Botha, C.E.J.; Bradley, G. In silico analysis of *cis*-acting regulatory elements in 5′ regulatory regions of sucrose transporter gene families in rice (*Oryza sativa* Japonica) and *Arabidopsis thaliana*. *Comput. Biol. Chem.* **2010**, *34*, 268–283. [CrossRef] [PubMed]
59. Busk, P.K.; Pagès, M. Regulation of abscisic acid-induced transcription. *Plant Mol. Biol.* **1998**, *37*, 425–435. [CrossRef]
60. Mattick, J.S.; Gagen, M.J. The evolution of controlled multitasked gene networks: The role of introns and other noncoding RNAs in the development of complex organisms. *Mol. Biol. Evol.* **2001**, *18*, 1611–1630. [CrossRef]
61. Hattori, T.; Totsuka, M.; Hobo, T.; Kagaya, Y.; Yamamoto-Toyoda, A. Experimentally determined sequence requirement of ACGT-containing abscisic acid response element. *Plant Cell Physiol.* **2002**, *43*, 136–140. [CrossRef]
62. Turner, J.G.; Ellis, C.; Devoto, A. The jasmonate signal pathway. *Plant Cell* **2002**, *14*, S153–S164. [CrossRef]
63. Gómez-Porras, J.L.; Riaño-Pachón, D.M.; Dreyer, I.; Mayer, J.E.; Mueller-Roeber, B. Genome-wide analysis of ABA-responsive elements ABRE and CE3 reveals divergent patterns in *Arabidopsis* and rice. *BMC Genom.* **2007**, *8*, 260. [CrossRef]
64. Staswick, P.E. JAZing up jasmonate signaling. *Trends Plant Sci.* **2008**, *13*, 66–71. [CrossRef]
65. Yang, W.; Zhang, W.; Wang, X. Post-translational control of ABA signalling: The roles of protein phosphorylation and ubiquitination. *Plant Biotechnol. J.* **2017**, *15*, 4–14. [CrossRef]
66. Xie, L.; Yu, L.; Chen, Q.; Wang, F.; Huang, L.; Xia, F.; Zhu, T.; Wu, J.; Yin, J.; Liao, B.; et al. *Arabidopsis* acyl-CoA-binding protein ACBP3 participates in plant response to hypoxia by modulating very-long-chain fatty acid metabolism. *Plant J.* **2015**, *81*, 53–67. [CrossRef]
67. Manjithaya, R.; Anjard, C.; Loomis, W.F.; Subramani, S. Unconventional secretion of *Pichia pastoris* Acb1 is dependent on GRASP protein, peroxisomal functions, and autophagosome formation. *J. Cell Biol.* **2010**, *188*, 537–546. [CrossRef] [PubMed]
68. Zheng, S.; Xiao, S.; Chye, M. The gene encoding *Arabidopsis* acyl-CoA-binding protein 3 is pathogen inducible and subject to circadian regulation. *J. Exp. Bot.* **2012**, *63*, 2985–3000. [CrossRef] [PubMed]
69. Bateman, A.; Coin, L.; Durbin, R.; Finn, R.D.; Hollich, V.; Sam Griffiths, J.; Khanna, A.; Marshall, M.; Moxon, S.; Sonnhammer, E.L.L.; et al. The Pfam protein families database. *Nucleic Acids Res.* **2004**, *32*, D138–D141. [CrossRef] [PubMed]
70. Wilkins, M.R.; Gasteiger, E.; Bairoch, A.; Sanchez, J.; Williams, K.L.; Appel, R.D.; Hochstrasser, D.F. Protein identification and analysis tools in the ExPASy server. *Methods Mol. Biol.* **1999**, *112*, 531–552.
71. Larkin, M.A.; Blackshields, G.; Brown, N.P.; Chenna, R.; McGettigan, P.A.; McWilliam, H.; Valentin, F.; Wallace, I.M.; Wilm, A.; Lopez, R.; et al. Clustal W and Clustal X version 2.0. *Bioinformatics* **2007**, *23*, 2947–2948. [CrossRef] [PubMed]
72. Chen, C.; Chen, H.; Zhang, Y.; Thomas, H.R.; Frank, M.H.; He, Y.; Xia, R. TBtools: An integrative toolkit developed for interactive analyses of big biological data. *Mol. Plant* **2020**, *13*, 1194–1202. [CrossRef]
73. Hu, B.; Jin, J.; Guo, A.; Zhang, H.; Luo, J.; Gao, G. GSDS 2.0: An upgraded gene feature visualization server. *Bioinformatics* **2015**, *31*, 1296–1297. [CrossRef]
74. Rombauts, S.; Déhais, P.; Montagu, M.V.; Rouzé, P. PlantCARE, a plant *cis*-acting regulatory element database. *Nucleic Acids Res.* **1999**, *27*, 295–296. [CrossRef]
75. Bailey, T.L.; Boden, M.; Buske, F.A.; Frith, M.; Grant, C.E.; Clementi, L.; Ren, J.; Li, W.W.; Noble, W.S. MEME SUITE: Tools for motif discovery and searching. *Nucleic Acids Res.* **2009**, *37*, W202–W208. [CrossRef]

76. Zhang, X.; Henriques, R.; Lin, S.; Niu, Q.; Chua, N. *Agrobacterium*-mediated transformation of *Arabidopsis thaliana* using the floral dip method. *Nat. Protoc.* **2006**, *1*, 641–646. [CrossRef] [PubMed]
77. Dunand, C.; Crèvecoeur, M.; Penel, C. Distribution of superoxide and hydrogen peroxide in *Arabidopsis* root and their influence on root development: Possible interaction with peroxidases. *New Phytol.* **2007**, *174*, 332–341. [CrossRef] [PubMed]
78. Zhang, W.; Qin, C.; Zhao, J.; Wang, X. Phospholipase Dα1-derived phosphatidic acid interacts wirh ABI1 phosphatase 2C and regulates abscisic acid signaling. *Proc. Natl. Acad. Sci. USA* **2004**, *101*, 9508–9513. [CrossRef]

plants

MDPI

Article

Rice Cinnamoyl CoA Reductase-like Gene *OsCCR14* Involved in Heat Stress via Regulation Lignin and Flavonoid Accumulation

Hongwei Wang [1,2], Wei Tian [3], Yulu Teng [3], Yuxin Xue [3], Simin Qin [3], Jiaxin Liu [3] and Shuangcheng Ding [1,2,*]

1 MARA Key Laboratory of Sustainable Crop Production in the Middle Reaches of the Yangtze River (Co-Construction by Ministry and Province), Yangtze University, Jingzhou 434025, China; wanghw@yangtzeu.edu.cn
2 Hubei Key Laboratory of Food Crop Germplasm and Genetic Improvement, Wuhan 430064, China
3 Agricultural College, Yangtze University, Jingzhou 434000, China
* Correspondence: shchding@yangtzeu.edu.cn

Abstract

Cinnamoyl-CoA reductases (CCRs) and their homologs, CCR-like proteins, play key roles in plant secondary metabolism and defense against environmental stresses; however, their functions in heat stress responses remain unclear. In this study, phylogenetic and sequence alignment analyses revealed that OsCCR14 encodes a CCR-like protein. qRT-PCR assays showed that OsCCR14 is predominantly expressed in roots and is induced by heat stress. Notably, overexpression of OsCCR14 increased lignin content, and transgenic plants with enhanced OsCCR14 expression exhibited higher flavonoid accumulation in roots. Conversely, knockout of OsCCR14 reduced flavonoid content and impaired seedling heat tolerance. Furthermore, OsCCR14 overexpression improved heat tolerance, accompanied by increased root lignification and flavonoid accumulation. These results indicate that OsCCR14 acts as a critical regulator of lignin and flavonoid metabolism, thereby providing a potential target for enhancing crop heat tolerance.

Keywords: CCR-like protein; thermotolerance; lignin level; flavonoid metabolism

1. Introduction

Rice (*Oryza sativa* L.), which serves as a staple food for nearly half of the global population, frequently suffers from various environmental stress induced by climate warming. High-temperature (HT) stress, which has emerged as a major threat to rice production, occurs when temperatures rise 5 °C above rice's optimal growth range and trigger a series of cellular and metabolic responses in the plant to sustain its survival and growth [1]. HT stress adversely affects rice growth across all developmental stages, spanning from germination and seedling establishment to anthesis, grain filling, and grain maturation [2,3]. Notably, it has been documented that each 1 °C increase in global temperature leads to a 3.2% reduction in rice yield [4]. Thus, enhancing stress resilience in rice has become an imperative strategy to protect global agricultural productivity.

HT stress primarily damages plants by disrupting their metabolic homeostasis, which leads to the excessive accumulation of reactive oxygen species (ROS) free radicals [5,6]. While at physiological levels ROS function as signaling molecules that regulate plant adaptation to various stresses, excessive ROS production often inflicts severe harm on plant cells [7]. Thus, maintaining strict control over ROS homeostasis is essential for

https://doi.org/10.3390/plants14233626

plant survival. Numerous studies have demonstrated that enhanced antioxidant enzyme activity contributes to improved plant heat tolerance [8]. Beyond antioxidant systems, secondary metabolites also play a crucial role in enabling plants to adapt to adverse environmental conditions, including temperature stress [9,10]. Lignin is an important polymeric phenylpropanoid derivative that is mainly deposited in the secondary cell wall to maintain cell wall integrity during plant heat stress responses [11]. Formation of the lignin polymer requires three core monolignols, p-coumaryl alcohol, coniferyl alcohol, and sinapyl alcohol. The synthesis of these monolignols originates from phenylalanine and was catalyzed by some enzymes, including phenylalanine ammonia lyase (PAL), cinnamic acid 4-hydroxylase (C4H), and 4-coumarate-CoA ligase (4CL), driving early reactions to produce common phenylpropanoid intermediates [12–14]. These intermediates are directed to the monolignol-specific pathway for lignin production, they can also diverge into flavonoid-specific pathways (e.g., via chalcone synthase, CHS) to synthesize flavonoids. Flavonoids act as antioxidants, with the ability to reduce oxidative damage that arises from ROS accumulation; this accumulation is induced by extreme temperatures [15–17]. Therefore, changes in the activity and gene expression levels of phenylpropanoid pathway affect the synthesis of plant secondary metabolites, which exerts a crucial role in enhancing the capacity of plants to deal with high-temperature stress [18,19]. Although the flavonoid biosynthetic pathway and lignin biosynthetic pathway have been extensively investigated, the relationship between flavonoid biosynthesis and lignin biosynthesis within the plant phenylpropanoid pathway, along with how they are regulated and how they respond to high-temperature stress, remains unclear.

CCRs, the first enzyme in the monolignol-specific branch of the phenylpropanoid pathway, catalyze the conversion of cinnamoyl-CoA into their corresponding cinnamaldehydes [20,21]. To date, *CCR* genes have been characterized in a wide range of plant species [22–27]. They are highly diverse, encompassing members classified as genuine CCRs and CCR-like genes. In rice, it has been identified as 33 CCRs and CCR-like genes. For instance, OsCCR1, as an OsRac1 effector and lignin biosynthetic enzyme, participates in defense signaling [28]. *OsCCR10* overexpression improves drought tolerance during vegetative growth stages, whereas *OsCCR10* knock-out mutants mediated by CRISPR/Cas9 displayed decreased root lignin accumulation and substantially decreased drought tolerance compared to WT controls [29]. A NWYCY motif is recognized as a signature motif of CCR, distinguishing genuine CCRs from CCR-like proteins due to its significant impact on CCR catalytic activity. Another more reliable classification feature is the novel H202(X)2K205 motif (H-X-X-K, CCR-SBM or substrate binding motif), reported by Chao et al. [30]. Like CCRs, CCR-like genes belong to the short-chain dehydrogenase/reductases (SDR) superfamily, with encoded proteins containing common motifs (e.g., catalytic triads, NADPH-binding motifs), but they lack activity in converting cinnamoyl-CoA esters to hydroxy-cinnamaldehydes [25], and their functions remain unclear.

Previous transcriptome analysis revealed a complex network of gene expression changes under heat stress [31]. Subsequently, it was observed that Os09g31514, previously identified as a CCR-like gene and designated as *OsCCR14* [25], exhibited upregulated expression patterns in heat-tolerant genotyping cultivator rice plants, suggesting an important role in heat stress response. However, the specific role of this gene in regulating phenylpropanoid metabolism and its impact on heat tolerance remained unclear. Notably, Borah et al. proposed that this specific gene might encode a dihydroflavonol-4-reductase, a function distinct from the canonical CCR activity [32]. Therefore, in this study, *OsCCR14* was cloned and analyzed to determine its phylogenetic relationship. Additionally, the pattern of *OsCCR14* gene expression in response to heat stress and various hormone treatments, and its protein subcellular localization were analyzed. Importantly, ectopic expression of

OsCCR14 in Arabidopsis significantly increases lignin and flavonoid content, enhancing plant heat tolerance. The heat tolerance of *osccr14* mutants and *OsCCR14* overexpression lines at seedling stage was also evaluated in rice, revealing that *OsCCR14* positively regulated heat tolerance. The findings from this study would provide valuable insights into the molecular mechanisms underlying heat tolerance in rice and help to identify potential targets for crop improvement.

2. Results

2.1. Sequence and Phylogenetic Analysis of OsCCR14

OsCCR14 was previously identified as a gene encoding cinnamoyl CoA reductases; it uses NAD(P)(H) as a co-factor for catalysis and belongs to SDR superfamily [25]. Given that DFR and CCR are proposed to share a common ancestral origin, a phylogenetic tree was constructed with a focus on rice-derived sequences. This tree included two Arabidopsis *CCR* genes (*AtCCRs*), thirty-two rice *CCR* genes (*OsCCRs*), and nine putative rice and Arabidopsis *DFR* genes (eight *OsDFRs* and one *AtDFR1*) [33]. Phylogenetic analysis revealed two distinct monophyletic groups: Group I and Group II (Figure 1A). Group I encompasses the majority of *CCR* genes, including the Arabidopsis homologs *AtCCR1* and *AtCCR2*. Group II is further subdivided into two well-supported clades, each containing a mixture of *OsDFR* and *OsCCR* genes. Notably, Clade II comprises 11 *OsCCR* members (including *OsCCR14*) and two *OsDFR* genes (Figure 1A).

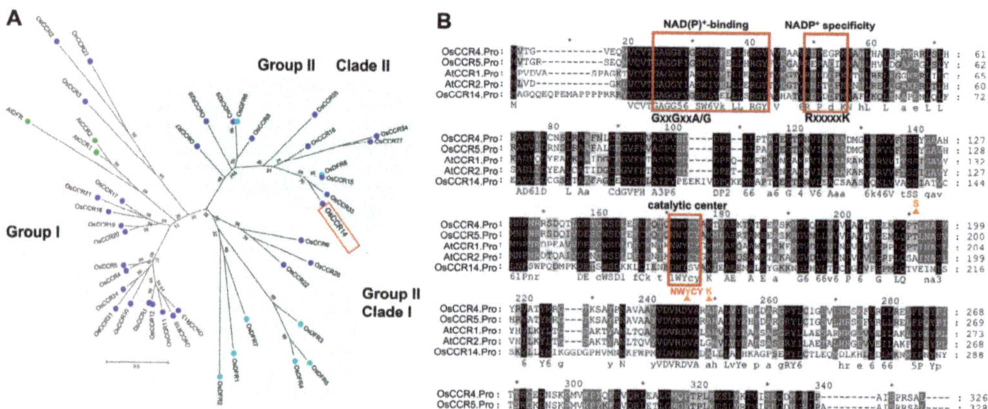

Figure 1. Phylogenetic and sequence analyses of OsCCR14. (**A**) Phylogenetic relationships of CCRs and DFRs from *Oryza sativa* and *Arabidopsis thaliana*. The evolutionary history was inferred by using the Maximum Likelihood method based on the JTT matrix-based model. Branch support was evaluated via 1000 bootstrap replicates, and bootstrap values are indicated at each branch node. The species name and the accession number of sequences used to construct the phylogenetic tree are listed in Table S1. (**B**) Alignment of OsCCR14 and other CCR proteins. The conserved motifs are indicated with a red box and the catalytic triad Ser–Tyr–Lys is indicated with an orange triangle. In sequence alignment, asterisks (*) serve as a visual annotation to indicate the length ranges of the aligned sequences.

Then, the protein sequence of OsCCR14 was aligned with those of functionally characterized CCR proteins. Sequence alignment uncovered that the most prominent homology across all analyzed peptide sequences resides in the N-terminal NAD(P)-binding motif. This region includes three key conserved elements: an NADP$^+$-binding motif harboring the signature G-X-X-G-X-X-A/G segment, an NADP$^+$-specific motif containing the R-X-

X-X-X-X-K sequence, and the Ser-Tyr-Lys catalytic triad—all of which are critical for SDR superfamily enzyme function (Figure 1B). However, a notable divergence was observed in the NWYCY motif: this signature sequence, which is indispensable for CCR catalytic activity, is replaced by DWYSV in OsCCR14, with amino acid mismatches at the 1st (asparagine [N] → aspartic acid [D]), 4th (cysteine [C] → serine [S]), and 5th (tyrosine [Y] → valine [V]) positions. Additionally, another CCR-specific signature motif, containing the H-X-X-K sequence that mediates binding to the highly electronegative region of CoA, exhibits complete sequence variation in OsCCR4, OsCCR5, and OsCCR14 (Figure 1B). Collectively, these bioinformatic analyses confirm that OsCCR14 might encode a CCR-like protein.

2.2. Expression Analysis of OsCCR14

To gain insights into the potential biological function of OsCCR14, we first examined its spatiotemporal expression patterns in 3-week-old wild-type (Nipponbare) seedlings, as well as in plants at the flowering stage. Quantitative real-time PCR (qRT-PCR) results showed that at the seedling stage, *OsCCR14* transcripts accumulated relatively abundantly in roots, whereas expression levels in above-ground tissues were significantly lower (Figure 2A). At the flowering stage, *OsCCR14* also exhibited notably high expression in roots, followed by floral organs, with the lowest expression detected in stems (Figure 2A).

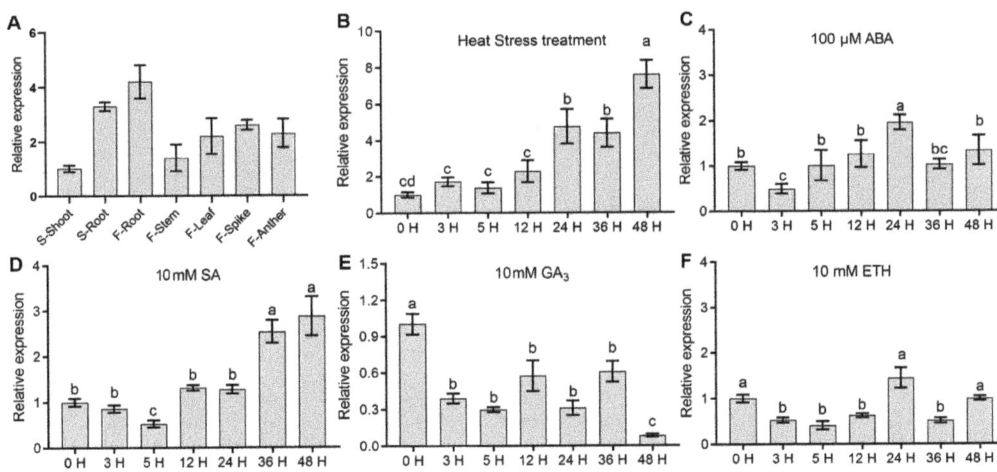

Figure 2. Expression profiles analysis of *OsCCR14*. (**A**) qRT-PCR analysis of *OsCCR14* in different tissues during different developmental stages. S and F stand for the seedling stage and flowering stage, respectively. (**B**) Expression of *OsCCR14* in response to 45 °C heat stress. (**C**) Expression of *OsCCR14* in response to 100 μM ABA. (**D**) Expression of *OsCCR14* in response to 10 mM SA. (**E**) Expression of *OsCCR14* in response to 10 mM GA_3. (**F**) Expression of *OsCCR14* in response to 10 mM ETH (ethephon). *Tubulins* gene transcript level was used as an internal control for data normalization. Data were presented as mean ± SD, based on three independent biological replicates. Bars with different letters indicate statistically significant differences ($p \leq 0.05$, ANOVA with Duncan's test).

We further investigated *OsCCR14* expression in response to heat stress and treatments with various exogenous hormones. As shown in Figure 2B, high-temperature stress induced a dramatic upregulation of *OsCCR14* expression. With prolonged treatment duration, the expression fold change gradually increased, reaching ~4-fold and ~8-fold at 24 h and 48 h post-treatment, respectively. In contrast, *OsCCR14* showed only mild responsiveness to exogenous abscisic acid (ABA) and salicylic acid (SA) treatments. Relative to untreated control plants, its expression was elevated by approximately 2.0-fold under these two hormone treatments (Figure 2C,D). Notably, *OsCCR14* expression displayed a significant downregulation trend in response to gibberellic acid (GA_3) treatment: expression levels decreased progressively, reaching a minimum at 48 h post-treatment with an approximate 11-fold downregulation compared to the control (Figure 2E). By comparison, *OsCCR14* exhibited only weak responsiveness to ethephon (Figure 2F). Collectively, these data suggests that *OsCCR14* genes may play important roles in rice development and abiotic stress responses, especially heat stress responses.

2.3. Subcellular Localization of OsCCR14

To determine the subcellular localization of the OsCCR14 protein, we transiently expressed two constructs in rice protoplasts: a control vector with green fluorescent protein (GFP) driven by the enhanced *CaMV 35S* promoter, and a fusion construct (*35S:OsCCR14-GFP*) encoding OsCCR14 linked to GFP. Nuclei were stained with 4′,6-diamidino-2-phenylindole (DAPI) for visualization. The fluorescence signal of OsCCR14-GFP was detected in both the cytoplasm and nucleus (Figure 3A). Additionally, we analyzed stable transgenic *Arabidopsis thaliana* plants expressing *35S:OsCCR14-GFP*. In these plants, the OsCCR14-GFP fusion protein was predominantly observed in roots and guard cells (Figure 3B). Specifically, in the root tip and maturation zone of transgenic Arabidopsis, 35S:OsCCR14-GFP localized to both the cytoplasm and nucleus—consistent with the pattern observed in rice protoplasts.

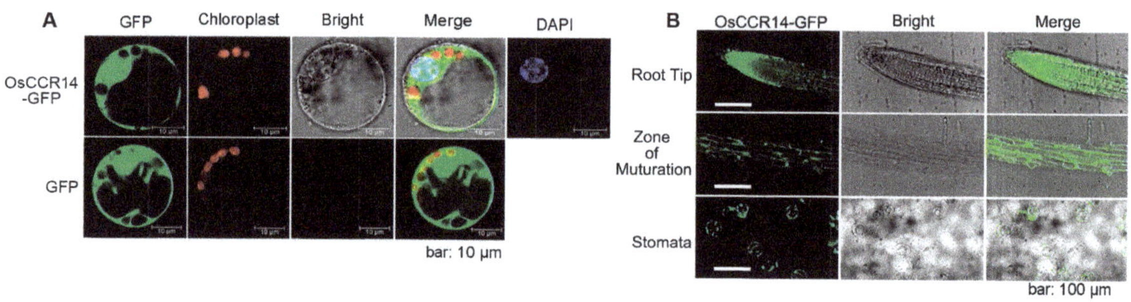

Figure 3. Subcellular and tissues localization of OsCCR14. (**A**) Subcellular localization of the OsCCR14-GFP fusion protein in rice cells. The *35S:GFP* and *35S:OsCCR14-GFP* constructs were transformed into rice leaf protoplasts. DAPI was used as a nuclear localization signal marker. (**B**) Root and stomatal localization of OsCCR14-GFP in *35S:OsCCR14-GFP* transgenic plants. Ten-day-old seedlings of *35S:OsCCR14-GFP* transgenic plants were observed under a fluorescence microscope.

2.4. Impact of OsCCR14-Overexpression on Lignin Content, Flavonol Fluorescence, and Heat Tolerance in Arabidopsis

To further clarify the effect of *OsCCR14* expression level on lignin content, we analyzed the lignin level in *OsCCR14*-transgenic Arabidopsis lines. A construct containing the *OsCCR14* coding sequence driven by the *35S* promoter was generated and subsequently introduced into wild-type Arabidopsis thaliana (Col-0) via genetic transformation. RT-PCR analysis confirmed that OE1, OE2, and OE3 were representative overexpression lines, which were then used for subsequent experiments (Figure 4A). The impact of *OsCCR14* overexpression on lignin content was evaluated first. A statistically significant increase in lignin content was observed in OE1, OE2, and OE3 compared with the wild-type control, with an average of approximately 20% (Figure 4B). Considering the "metabolic flux competition" between lignin and flavonoids, we examined the effect of *OsCCR14* overexpression on root flavonol content by visualizing in vivo flavonol fluorescence. Seven-day-old seedlings were soaked in a solution containing diphenylboric acid-2-aminoethyl ester (DPBA), and flavonols were observed under a confocal microscope. Interestingly, a stronger fluorescence signal was detected in *OsCCR14* overexpression lines, with OE1 showing the most intense signal (Figure 4C).

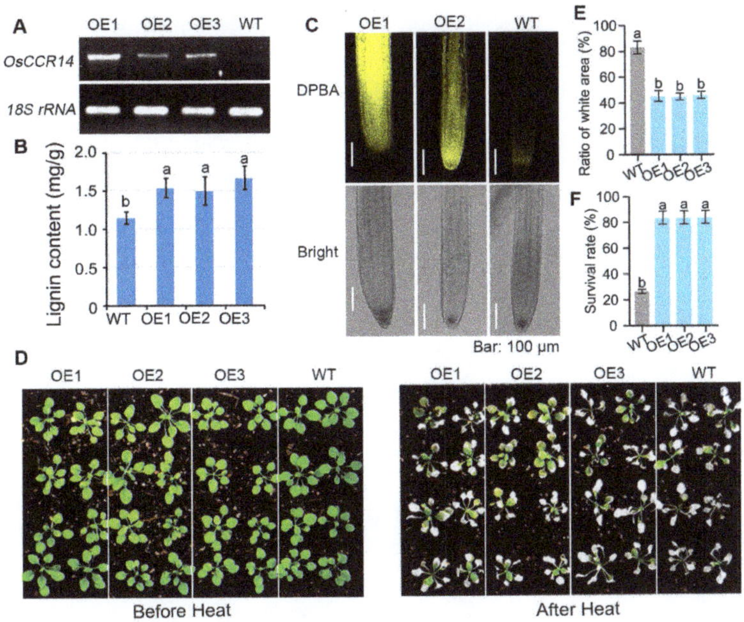

Figure 4. Analysis of lignin, flavonoid, and heat tolerance phenotypes in *OsCCR14*-transgenic Arabidopsis lines. (**A**) RT-PCR analysis of *OsCCR14* expression. (**B**) Lignin content quantified with the thioacidolysis method from 2-week-old transgenic plants. (**C**) DPBA staining showing flavonoid accumulation patterns in roots of 7-day-old seedlings. (**D**) Phenotype of *OsCCR14* overexpressing Arabidopsis thaliana lines under high temperature stress. (**E**) Proportion of white leaf area of *OsCCR14* overexpressing Arabidopsis lines under HT stress. (**F**) Survival rate of *OsCCR14* overexpressing Arabidopsis lines under HT stress. Values represented mean ± standard deviation of three independent biological replicates. Different letters above bars denoted significant differences ($p \leq 0.05$, ANOVA with Duncan's test). Colors are used solely to distinguish between different materials.

Given that *OsCCR14* expression is induced by SA and high temperature (HT), and that *OsCCR14* promotes flavonoid accumulation, we hypothesized that *OsCCR14* contributes to plant responses to heat stress. To test this hypothesis, we compared the heat tolerance of *OsCCR14* overexpression lines (OE1, OE2, OE3) and wild-type Arabidopsis plants using 2-week-old seedlings grown under normal conditions. Seedlings were subjected to 42 °C heat treatment for 48 h, followed by a 7-day recovery period at 22 °C. As shown in Figure 4D, the survival rates of *OsCCR14* overexpression lines were consistently higher (ranging from 83.4% to 93.6%) compared to wild-type plants (~26.5%) (Figure 4F). Furthermore, we examined the effect of *OsCCR14* overexpression on leaf bleaching—a typical symptom of heat-induced damage (Figure 4E). A marked reduction in leaf bleaching area was observed in OE1, OE2, and OE3 (45.3%, 44.9%, and 46.1%, respectively), whereas the wild type exhibited a much higher bleaching area (~83.2%) (Figure 4E). Collectively, these results suggest that *OsCCR14* overexpression affects level of lignin and flavonols, thereby enhancing plant heat tolerance.

2.5. *osccr14* Mutants Decrease Heat Tolerance at Seedling Stage

To further elucidate the biological function of OsCCR14, two *OsCCR14* knockout mutants were first identified and characterized. The first mutant, designated *osccr14-1* (accession number: 04Z11IV01), is a T-DNA insertion line isolated from the Rice Mutant Database (RMD). Genomic DNA amplification confirmed that the T-DNA was inserted into the untranslated region (UTR) 444 bp upstream of the ATG start codon (Figure 5A). RT-PCR analysis revealed a significant downregulation of *OsCCR14* transcript levels in OsCCR14-1 compared to the wild type (Figure 5B). The second knockout mutant, *osccr14-2*, was generated via CRISPR/Cas9-mediated genome editing. This mutant carries a 2-bp ("TA") deletion at positions 117–118 of the first exon, which induces a severe frameshift mutation in the *OsCCR14* coding sequence, likely resulting in the loss of functional protein production (Figure 5A).

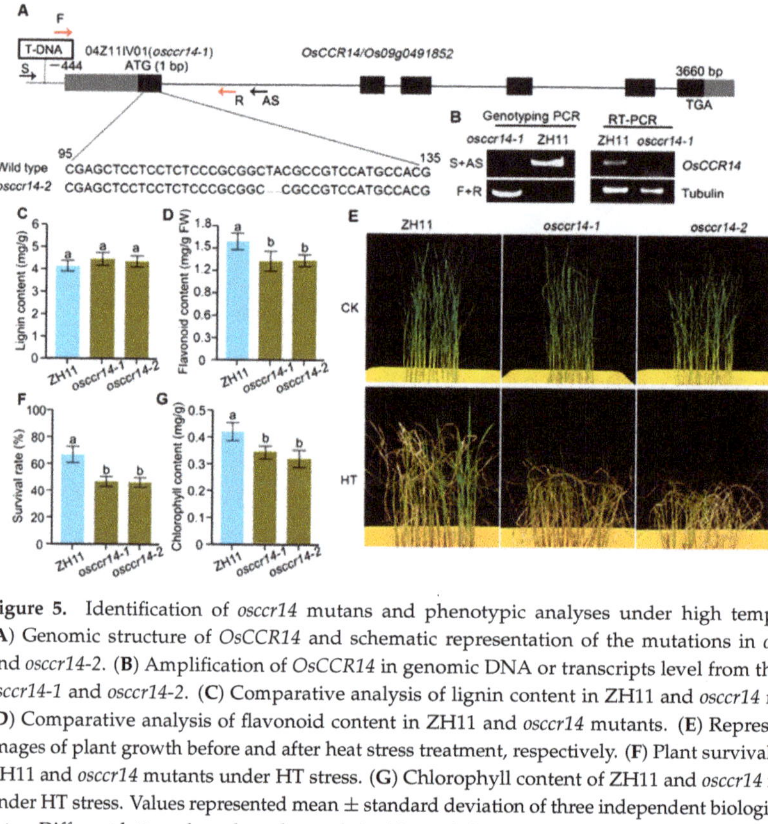

Figure 5. Identification of *osccr14* mutans and phenotypic analyses under high temperature. (**A**) Genomic structure of *OsCCR14* and schematic representation of the mutations in *osccr14-1* and *osccr14-2*. (**B**) Amplification of *OsCCR14* in genomic DNA or transcripts level from the ZH11, *osccr14-1* and *osccr14-2*. (**C**) Comparative analysis of lignin content in ZH11 and *osccr14* mutants. (**D**) Comparative analysis of flavonoid content in ZH11 and *osccr14* mutants. (**E**) Representative images of plant growth before and after heat stress treatment, respectively. (**F**) Plant survival rates of ZH11 and *osccr14* mutants under HT stress. (**G**) Chlorophyll content of ZH11 and *osccr14* mutants under HT stress. Values represented mean ± standard deviation of three independent biological replicates. Different letters above bars denoted significant differences ($p \leq 0.05$, ANOVA with Duncan's test). Colors are used solely to distinguish between different materials.

We next analyzed lignin and flavonoid contents in the two *osccr14* mutants and their wild-type background (ZH11) at the seedling stage (Figure 5C). Total lignin levels in *osccr14-1* and *osccr14-2* were comparable to those in ZH11. In contrast, a slight but detectable decrease in flavonoid content was observed in both mutants relative to the wild type (Figure 5D). To assess the role of *OsCCR14* in heat stress responses, we compared the heat tolerance of 16-day-old hydroponically grown *osccr14-1*, *osccr14-2* and ZH11 seedlings. Seedlings were exposed to 45 °C for 30 h to induce heat stress, followed by a 3-day recovery period at ambient temperature. Survival rates were then scored, and chlorophyll content was measured. As shown in Figure 5F, the survival rates of the mutants were significantly lower than that of ZH11: ~43.2% for *osccr14-1* and ~44.6% for *osccr14-2*, compared to ~68% for the wild type. Consistent with the survival phenotype, chlorophyll content was also reduced in both mutants after heat stress (Figure 5G). Collectively, these results indicate that OsCCR14 is involved in the regulation of heat tolerance in plants.

2.6. *Overexpression OsCCR14-Enhanced Plant Thermotolerance*

To further validate the role of *OsCCR14* in heat tolerance, we generated transgenic rice plants overexpressing *OsCCR14*, using the enhanced CaMV *35S* promoter to drive transgene expression. RT-PCR analysis confirmed that *OsCCR14* transcript levels were significantly upregulated in all transgenic overexpression lines (Figure 6A). Consistent with the phenotype of *OsCCR14*-overexpressing Arabidopsis, overexpression of *OsCCR14* also increased flavonoid levels in rice roots. The DPBA-stained fluorescent signal (indicative of flavonol accumulation) in lines OE17 and OE20 was significantly stronger than that in the wild type (Figure 6E).

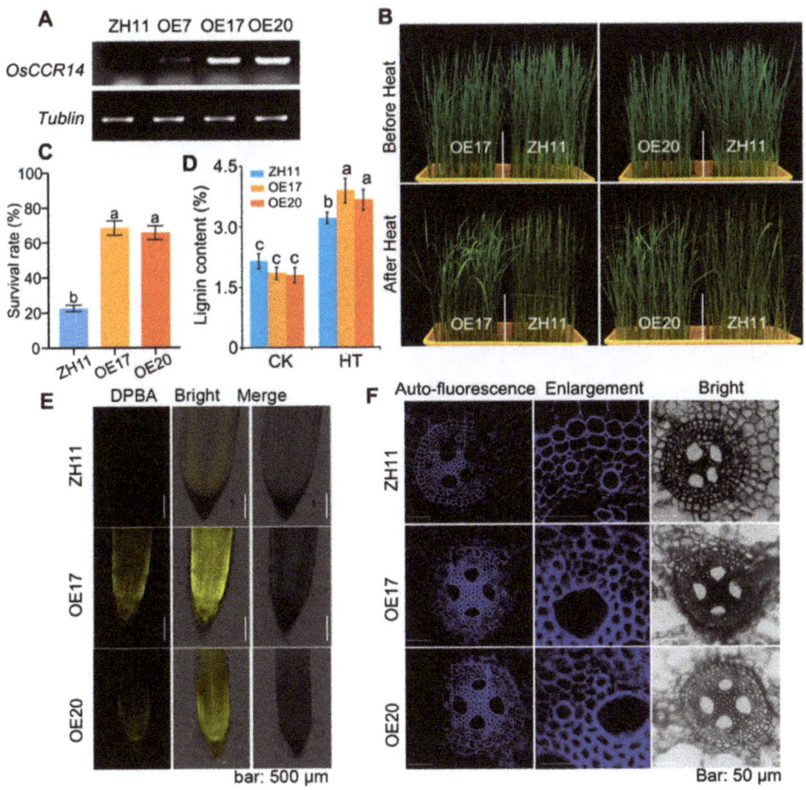

Figure 6. Phenotypic identification of *OsCCR14* overexpressing plants under high temperature. (**A**) RT-PCR analysis of OsCCR14 overexpressing rice lines. (**B**) Phenotype of OsCCR14 overexpressing rice lines under high temperature stress. (**C**) Survival rate of *OsCCR14* overexpressing rice lines under HT stress. (**D**) Analysis of total lignin content in OsCCR14 overexpressing rice plants before and after high temperature treatment. (**E**) DPBA staining of roots in OsCCR14 overexpressing rice lines. Values represented mean $\pm$ standard deviation of three independent biological replicates. (**F**) Lignin autofluorescence in root of OsCCR14 overexpressing rice lines under heat stress. Different letters above bars denoted significant differences ($p \leq 0.05$, ANOVA with Duncan's test).

Next, we compared the heat tolerance of the transgenic rice overexpression lines and their wild-type background (ZH11) at the seedling stage. Under normal growth conditions, transgenic lines OE17 and OE20 exhibited growth comparable to that of ZH11. After exposure to heat stress, however, the transgenic lines displayed markedly improved leaf vitality, whereas the wild type showed more severe heat-induced damage (Figure 6B). The survival rates of OE17 and OE20 were ~42.5% and ~45.4%, respectively, significantly higher than that of the wild type (~8.6%) (Figure 6C). In addition, we determined the total lignin content of 2-week-old seedlings using the acetyl bromide method. Under normal growth conditions, there was no significant difference in lignin content between the overexpression lines and the wild type. However, high temperature treatment significantly increased lignin levels in all plants; notably, the overexpression lines exhibited a significantly higher lignin content than the wild type after heat stress (Figure 6D). Additionally, observations of lignin autofluorescence in root cross-sections revealed that secondary wall thickening was correspondingly reduced in root of OE17 and OE20 after heat stress (Figure 6F). Taken together, these results demonstrate that overexpression of *OsCCR14* enhances plant thermotolerance, and this enhancement is potentially mediated by increased lignin and flavonoid accumulation.

3. Discussion

The increasing frequency and severity of climate-related stresses pose a significant threat to global food security, driving the urgent need for strategies to enhance crop resilience [34]. As CCR catalyzes the first committed step of the lignin branch pathway, it plays a crucial role in secondary metabolism [35,36]. According to analysis of gene expression pattern, a significant increase in *OsCCR14* expression was detected following heat exposure. This observation aligns with previous findings demonstrating the involvement of CCRs in stress responses in various plant species. For instance, *zmccr1* mutants in maize exhibited altered lignin structure and changes in sclerenchyma cell wall development [37], highlighting the link between CCR activity and cell wall modifications under stress. In addition, the contrasting expression patterns elicited by ABA and SA, and the inhibitory effect of GA_3 on *OsCCR14* expression, implied the intricate interplay of hormonal signaling pathways in regulating gene expression and stress tolerance. Exogenous SA has emerged as an effective strategy to mitigate heat stress (HT)-induced damage by upregulating the expression of genes in the phenylpropanoid pathway, leading to lignin accumulation [38]. In this study, we demonstrated that SA induces *OsCCR14* expression, implying that SA fine-tunes *OsCCR14* transcription to support lignin accumulation under heat stress, which aligns with previously findings. By contrast, the mild responsiveness of *OsCCR14* to ABA suggests this hormone does not act as a core regulator of *OsCCR14*-mediated lignin synthesis in rice. Notably, GA signaling is closely associated with the biosynthesis of secondary cell wall components: for example, HCT-RNAi alfalfa plants exhibit a dwarf phenotype accompanied by reduced GA levels, impaired GA perception, and decreased lignin content [39]. The observed repression of OsCCR14 by GA_3 in our study may therefore be attributed to GA's role in inhibiting *OsCCR14* to prevent excessive lignification—an outcome that could otherwise impair normal rice growth by restricting tissue flexibility and resource allocation to development.

The NWYCY motif is a signature of CCRs, distinguishing genuine CCRs from CCR-like proteins by its key role in CCR catalytic activity. However, OsCCR5 contains the NWYCY sequence but is classified as a CCR-like protein with no detectable CCR activity [25]. While *Selaginella moellendorffii* SmCCR1 and SmCCR2-1, with NWYCY substituted by NGYCL and EWYCL, respectively, have been verified as genuine CCRs in vitro [40]. Several researches also showed that the expression level of *CCR-like* gene in rice is consistent with the variation trend of lignin accumulation. A CCR-like gene *Snl6*, is required for pathogenesis-related gene expression and resistance to the bacterial pathogen, *Xanthomonas oryzae* pv. *oryzae* [41]. After *snl6* is mutated, the lignin content in the plant decreases significantly [41]. Another CCR-like gene, *OsCCRL1* involved in tapetum and pollen development. The *osccrl1* mutant exhibited reduced CCR enzyme activity, decreased lignin accumulation, delayed tapetum degradation, and disrupted phenylpropanoid metabolism [42]. In this study, we found that overexpression of *OsCCR14* led to increased lignin content and flavonoid levels in roots of transgenic Arabidopsis at normal growth condition (Figure 4). While the *osccr14* mutant exhibited a lignin level similar to that of the wild type at normal growth condition, but a clear increase after heat stress treatments (Figure 5). Notably, the unaltered lignin content in osccr14 mutants under normal conditions may stem from metabolic redundancy in lignin biosynthesis pathways, yet they showed reduced flavonoid levels and impaired heat tolerance. Similarly, a significantly enhanced lignin autofluorescence, indicative of secondary wall thickening, was observed in the roots of *OsCCR14*-overexpressing transgenic rice under heat stress (Figure 6). This lignin accumulation is consistently linked to higher survival rates in *OsCCR14*-overexpressing lines of both Arabidopsis and rice, and it underscores lignin might as the key driver of enhanced thermotolerance. Collectively, these results showed that OsCCR14 involved in lignification processes, especially in heat stress conditions.

Flavonoids, a group encompassing anthocyanins, flavonols, and proanthocyanidins (PAs), function as antioxidants to scavenge reactive oxygen species (ROS) and regulate ROS accumulation [43,44]. The flavonoid biosynthesis pathway belongs to one branch of the phenylpropanoid metabolic pathway, sharing the intermediate product coumaroyl-CoA with lignin biosynthesis pathway. Several key enzymes involved in the production of major flavonoid classes have been well characterized, including chalcone isomerase (CHI), flavanone 3-hydroxylase (F3H), dihydroflavonol 4-reductase (DFR), and anthocyanidin synthase (ANS) [45,46]. It was noticed that Borah et al. previously proposed that OsCCR14 might encode a dihydroflavonol-4-reductase, a function distinct from the canonical CCR activity [32]. In this study, we found that *OsCCR14* overexpression helps flavonoid levels accumulation in roots of both transgenic Arabidopsis seedlings and transgenic rice seedlings at normal growth conditions. We have not yet determined whether the OsCCR14 protein exhibits DFR activity. From the evolutionary and sequence analysis, OsCCR14 might present a high similarity to DFR protein (Figure 1). Given that *OsCCR14* is a *CCR-like* gene, it shares similarities with CCR proteins and, like DFR proteins, belongs to the short-chain dehydrogenase/reductase (SDR) superfamily; notably, both the CCR and DFR families are deeply rooted in the phylogeny of land plants [47]. One possible reason why overexpression of *OsCCR14* leads to an increase in both lignin and flavonoid levels, rather than a decrease in flavonoids. The possible explanation can be primarily attributed to the divergent substrate specificity of CCR-like protein. We hypothesize that OsCCR14 may catalyze certain hydroxycinnamoyl-CoA derivatives, which may alleviate feedback inhibition or competitive inhibition on the upstream shared pathway. This results in an upregulation of the total flux through the phenylpropanoid metabolic pathway, leading to increased synthesis of phenylalanine and coumaroyl-CoA in *OsCCR14*-overexpression transgenic plant. We therefore hypothesize that *OsCCR14* overexpression lines may enhance the level of lignin and flavonoids, reducing heat impair and thereby improving thermotolerance. In future studies, integrating transcriptomic and metabolomic analyses to identify key metabolites and elucidating the substrate specificity of OsCCR14 will deepen our understanding of the role of CCR-like proteins in heat stress responses.

4. Materials and Methods

4.1. Plant Materials

To generate *OsCCR14* overexpression plants, the coding sequence of *OsCCR14* (Os09g31514, LOC_09g0491852) was amplified from rice (*Oryza sativa* subsp. *japonica* 'Nipponbare'). The *Japonica* cultivar Zhonghua11 (ZH11) was used as the non-transgenic control. The amplified *OsCCR14* coding sequence was cloned into rice transformation vector pGreenII carrying 35S promoter, and introduced into *Agrobacterium tumefaciens* strains GV3101 (pSoup) and then transformed into ZH11 or Arabidopsis Columbia ecotypes to generate transgenetic plants.

The gene editing sites and T-DNA insertion positions were finally determined by amplifying the target fragments, conducting sequencing, and performing sequence alignment. For the generation of knock-out (*osccr14-2*) mutants, we used the web-based tool CRISPR RGEN Tools (http://www.rgenome.net/, accessed on 8 June 2023). The method was performed as described previously [48]. *osccr14-1* was obtained from the Rice Mutant Database.

4.2. Stress Treatments

To investigate expression profile responses to abiotic stress and phytohormone treatments, seedlings of ZH11 were exposed to heat (HT; 42 °C) for varying durations (0, 3, 5, 12, 24, 36, and 48 h). For phytohormone treatments, seedlings were subjected to root immersion in solutions of different hormones, including 100 μM ABA, 10 mM SA, 10 mM gibberellic acid (GA_3), and 10 mM ethylene (ETH), respectively. Tissue samples were collected during seedlings and flower stages. Samples collected at 0 h served as the control.

4.3. RNA Extraction, RT-PCR and qRT-PCR Analysis

To investigate the spatial and temporal expression patterns of *OsCCR14*, total RNA was isolated from shoots, roots, stem, leaf, spikers and anthers of 8–10 Nipponbare plants at the seedling and flowering stages. Approximately 100 mg of fresh tissue per sample was used, following the manufacturer's instructions for Trizol reagents (Invitrogen, Wuhan, China).

RT–PCR was performed to determine the transcriptional levels of *OsCCR14* in transgenic lines, compared with wild-type plants, using the following cycling conditions: initial denaturation at 94 °C for 5 min, followed by 35 cycles of denaturation at 94 °C for 30 s, annealing at 58 °C for 30 s, and extension at 72 °C for 45 s, with a final extension at 72 °C for 10 min. qRT-PCR was carried out using a SYBR Premix Ex Taq II kit (Takara, Wuhan, China) on a QuantStudio 6 Flex Real-Time PCR System (Thermo Fisher Scientific, Shanghai, China). The amplification reactions were performed at 95 °C for 10 min, followed by 40 cycles of 95 °C for 5 s, 60 °C for 30 s, and 72 °C for 30 s. *18S rRNA* or *Tubulin* was used as an internal control for normalization as previously reported transcript, and two biological and three technical replicates were analyzed. Primers used are listed in Supplementary Table S1.

4.4. Subcellular Localization

The coding sequence of *OsCCR14* was fused to the *GFP* reporter gene under the control of the *35S* promoter, creating a *35S:OsCCR14-GFP* fusion construct using the vector pGreenII-GFP. Rice protoplasts were prepared and transfected following a modified protocol from Yoo et al. [49]. Following transfection, protoplasts were incubated in darkness at 25 °C for 16 h to allow for protein expression and subsequent subcellular targeting. To determine the subcellular localization of OsCCR14, an in vitro approach using transient expression in rice protoplasts was initially employed. GFP fluorescence was then visualized using a Leica TCSSP8 laser confocal microscope (Leica, Wetzlar, Germany), utilizing an excitation wavelength of 488 nm and an emission wavelength of 507 nm. Complementing the in vitro analysis, the 35S:OsCCR14-GFP construct was transformed into the Arabidopsis Columbia to assess OsCCR14 localization in vivo. Transgenic positive T_2 plants were selected and confirmed to carry the transgene. GFP fluorescence was then observed in the root and leaf tissues of these transgenic plants to determine the protein's cellular distribution.

4.5. Seedling Survival Rate Assessment

We evaluated the heat tolerance of transgenic, knockout, and non-transgenic plants at the vegetative stage, with experiments conducted in a greenhouse or controlled environment as follows. For *Arabidopsis thaliana*: wild-type (WT) Columbia ecotype and overexpression lines (OE1, OE2, OE3) were transplanted into soil pots (20 × 10 × 6 cm, 32 plants per pot) and grown for 2 weeks under greenhouse conditions (22 °C, 16 h light/8 h dark cycle). Seedlings were then exposed to heat stress at 42 °C for 48 h, followed by a 7-day recovery under normal growth conditions. For transgenic rice: Seeds of mutants, the wild-type ZH11, and T_3 homozygous overexpression (OE) lines were first soaked in prochloraz for 2 h, rinsed thoroughly with distilled water, and soaked overnight. The next day, seeds were germinated in a moist environment at 30 °C. Germinated seeds were transferred to bottomless 96-well plates and hydroponically cultured in 1× nutrient solution in a light incubator (26 °C, 16:8 light–dark photoperiod). At the two-leaf-one-heart stage, seedlings were subjected to 45 °C high-temperature stress in a growth chamber until obvious wilting occurred, followed by a 3-day recovery at 26 °C under normal conditions.

4.6. Ratio of Leaf White Area Assessment

Leaf bleaching was quantified by measuring the ratio of leaf white area (%) in *Arabidopsis thaliana* (Columbia, OE1, OE2, OE3). Digital photographs of leaves under heat stress and control were captured (Leica, Wetzlar, Germany) and analyzed using ImageJ software (version 1.53j). For each leaf, the total leaf area and bleached area were manually delineated using the freehand selection tool in ImageJ. The bleaching rate was calculated as the ratio of bleached area to total leaf area. Experiments included at least three biological replicates, each consisting of 10 plants.

4.7. Determination of Lignin and Flavonoid Level

Lignin content was determined via the thioglycolic acid method, as previously described by Ding et al. [48], using 0.1 g of plant (whole seedlings) powder per sample and a standard curve (0–0.5 mg/mL, AccuStandard, CAS#9005-53-2) for quantification. The cell wall fraction of the samples was prepared by removing interfering compounds through sequential extractions with phosphate buffer, Triton X-100, and NaCl in acetone. For observing lignin autofluorescence in root cross-sections, young root tissues from three-week-old rice plants were fixed in FAA for 48 h. Fixed tissues were gradually dehydrated using a graded ethanol series to remove water, then treated with a clearing agent to eliminate ethanol and render the tissues permeable to paraffin. After paraffin infiltration and embedding, 10 µm-thick sections were cut using an RM2235 microtome (Leica, Wetzlar, Germany). Lignin autofluorescence in these sections was observed under 405 nm UV light using a fluorescence microscope (Leica TCS SP5, Wetzlar, Germany).

The total flavonoid content was determined via the aluminum chloride colorimetric assay, as previously described by Cai et al. [31], using 0.1 g of plant powder per sample and a rutin standard solution (0–1.0 mg/mL, Solarbio, SR8250) for quantification. To assess flavonol accumulation in roots, in situ staining with diphenylboric acid 2-aminoethyl ester (DPBA) was conducted. Roots of 10-day-old seedlings were stained with an ethanol solution containing 0.25% (*w*/*v*) DPBA and 0.01% (*v*/*v*) Triton X-100 until a saturated signal was observed (for at least 1.5 h). The resulting fluorescence was visualized using a laser confocal microscope (Leica, Wetzlar, Germany) with an excitation wavelength of 458 nm.

4.8. Statistical Analysis

Using R (version 4.5), one-way ANOVA was performed to determine mean differences between groups. Statistically significant differences ($p \leq 0.05$) were then evaluated with Duncan's multiple comparison test, implemented through the duncan.test function in the agricolae package (version 1.3-7).

5. Conclusions

In summary, we identified that *OsCCR14*, a CCR-like gene, plays a key role in regulating lignin and flavonoid biosynthesis, thereby enhancing plant heat tolerance. While *OsCCR14* expression correlated with lignin content, its downregulation only slightly reduced lignin levels. In contrast, *OsCCR14* overexpression significantly increased lignin and flavonoid accumulation, particularly under heat stress. Thus, we infer that *OsCCR14* improves rice heat tolerance by promoting root lignification and flavonoid accumulation, which enhance structural stability and antioxidant defense capacity, respectively. Our findings provide new insights into the role of CCR-like genes in rice heat tolerance.

Supplementary Materials: The following supporting information can be downloaded at: https://www.mdpi.com/article/10.3390/plants14233626/s1, Table S1: Primers used in this study.

Author Contributions: Conceptualization, S.D. and H.W.; investigation, W.T., Y.T., Y.X. and S.D.; data curation, S.D., S.Q., J.L. and H.W.; writing—original draft preparation, S.D.; writing—review and editing, H.W. and S.D.; funding acquisition, S.D. All authors have read and agreed to the published version of the manuscript.

Funding: This work was Funded by Hubei Key Laboratory of Food Crop Germplasm and Genetic Improvement (2025lzjj06) and the Engineering Research Center of Ecology and Agricultural Use of Wetland, Ministry of Education (KFT202012).

Institutional Review Board Statement: Not applicable.

Informed Consent Statement: Not applicable.

Data Availability Statement: All relevant data can be found and available within the manuscript.

Conflicts of Interest: The authors declare no conflict of interest.

References

1. Shi, W.; Yin, X.; Struik, P.C.; Solis, C.; Xie, F.; Schmidt, R.C.; Huang, M.; Zou, Y.; Ye, C.; Jagadish, S.V.K. High day- and night-time temperatures affect grain growth dynamics in contrasting rice genotypes. *J. Exp. Bot.* **2017**, *68*, 5233–5245. [CrossRef]
2. Jagadish, S.V.; Murty, M.V.; Quick, W.P. Rice responses to rising temperatures--challenges, perspectives and future directions. *Plant Cell Environ.* **2015**, *38*, 1686–1698. [CrossRef]
3. Mahmood, A.; Wang, W.; Ali, I.; Zhen, F.; Osman, R.; Liu, B.; Liu, L.; Zhu, Y.; Cao, W.; Tang, L. Individual and combined effects of booting and flowering high-temperature stress on rice biomass accumulation. *Plants* **2021**, *10*, 1021. [CrossRef] [PubMed]
4. Zhao, C.; Liu, B.; Piao, S.; Wang, X.; Lobell, D.B.; Huang, Y.; Huang, M.; Yao, Y.; Bassu, S.; Ciais, P.; et al. Temperature increase reduces global yields of major crops in four independent estimates. *Proc. Natl. Acad. Sci. USA* **2017**, *114*, 9326–9331. [CrossRef] [PubMed]
5. Samat, A.T.; Soltabayeva, A.; Bekturova, A.; Zhanassova, K.; Auganova, D.; Masalimov, Z.; Srivastava, S.; Satkanov, M.; Kurmanbayeva, A. Plant responses to heat stress and advances in mitigation strategies. *Front. Plant Sci.* **2025**, *16*, 1638213. [CrossRef] [PubMed]
6. Ren, H.; Bao, J.; Gao, Z.; Sun, D.; Zheng, S.; Bai, J. How rice adapts to high temperatures. *Front. Plant Sci.* **2023**, *14*, 1137923. [CrossRef]
7. Wang, P.; Liu, W.C.; Han, C.; Wang, S.; Bai, M.Y.; Song, C.P. Reactive oxygen species: Multidimensional regulators of plant adaptation to abiotic stress and development. *J. Integr. Plant Biol.* **2024**, *66*, 330–367. [CrossRef]

8. Das, K.; Roychoudhury, A. Reactive oxygen species (ROS) and response of antioxidants as ROS-scavengers during environmental stress in plants. *Front. Environ. Sci.* **2014**, *2*, 53. [CrossRef]
9. Yin, Q.; Qin, W.; Zhou, Z.; Wu, A.M.; Deng, W.; Li, Z.; Shan, W.; Chen, J.Y.; Kuang, J.F.; Lu, W.J. Banana MaNAC1 activates secondary cell wall cellulose biosynthesis to enhance chilling resistance in fruit. *Plant Biotechnol. J.* **2024**, *22*, 413–426. [CrossRef]
10. Li, Z.; Li, Z.; Ji, Y.; Wang, C.; Wang, S.; Shi, Y.; Le, J.; Zhang, M. The heat shock factor 20-HSF4-cellulose synthase A2 module regulates heat stress tolerance in maize. *Plant Cell* **2024**, *36*, 2652–2667. [CrossRef]
11. Lu, C.; Li, W.; Feng, X.; Chen, J.; Hu, S.; Tan, Y.; Wu, L. The dynamic remodeling of plant cell wall in response to heat stress. *Genes* **2025**, *16*, 628. [CrossRef]
12. Dixon, R.A.; Achnine, L.; Kota, P.; Liu, C.J.; Reddy, M.S.; Wang, L. The phenylpropanoid pathway and plant defence-a genomics perspective. *Mol. Plant Pathol.* **2002**, *3*, 371–390. [CrossRef] [PubMed]
13. Yoon, J.; Choi, H.; An, G. Roles of lignin biosynthesis and regulatory genes in plant development. *J. Integr. Plant Biol.* **2015**, *57*, 902–912. [CrossRef] [PubMed]
14. Ninkuu, V.; Aluko, O.O.; Yan, J.; Zeng, H.; Liu, G.; Zhao, J.; Li, H.; Chen, S.; Dakora, F.D. Phenylpropanoids metabolism: Recent insight into stress tolerance and plant development cues. *Front. Plant Sci.* **2025**, *16*, 1571825. [CrossRef] [PubMed]
15. Hendrix, S.; Dard, A.; Meyer, A.J.; Reichheld, J.P. Redox-mediated responses to high temperature in plants. *J. Exp. Bot.* **2023**, *74*, 2489–2507. [CrossRef]
16. Daryanavard, H.; Postiglione, A.E.; Mühlemann, J.K.; Muday, G.K. Flavonols modulate plant development, signaling, and stress responses. *Curr. Opin. Plant Biol.* **2023**, *72*, 102350. [CrossRef]
17. Zhao, S.; Ma, Y.; Ding, Y.; Dong, G.; Liu, C.; Ma, X.; Hou, B. Rice glycosyltransferase UGT706F1 functions in heat tolerance through glycosylating flavonoids under the regulation of transcription factor MYB61. *Plant J.* **2025**, *121*, e17252. [CrossRef]
18. Yang, W.; Yao, D.; Duan, H.; Zhang, J.; Cai, Y.; Lan, C.; Zhao, B.; Mei, Y.; Zheng, Y.; Yang, E.; et al. VAMP726 from maize and Arabidopsis confers pollen resistance to heat and UV radiation by influencing lignin content of sporopollenin. *Plant Commun.* **2023**, *4*, 100682. [CrossRef]
19. Han, J.; Wang, Z.; Wu, X.; Xia, J.; Wang, L.; Wang, Z.; Zhang, Y. Deciphering high-temperature-induced lignin biosynthesis in wheat through comprehensive transcriptome analysis. *Plants* **2024**, *13*, 1832. [CrossRef]
20. Giordano, A.; Liu, Z.; Panter, S.N.; Dimech, A.M.; Shang, Y.; Wijesinghe, H.; Fulgueras, K.; Ran, Y.; Mouradov, A.; Rochfort, S.; et al. Reduced lignin content and altered lignin composition in the warm season forage grass *Paspalum dilatatum* by down-regulation of a Cinnamoyl CoA reductase gene. *Transgenic Res.* **2014**, *23*, 503–517. [CrossRef]
21. Pan, H.; Zhou, R.; Louie, G.V.; Mühlemann, J.K.; Bomati, E.K.; Bowman, M.E.; Dudareva, N.; Dixon, R.A.; Noel, J.P.; Wang, X. Structural studies of cinnamoyl-CoA reductase and cinnamyl-alcohol dehydrogenase, key enzymes of monolignol biosynthesis. *Plant Cell* **2014**, *26*, 3709–3727. [CrossRef]
22. Costa, M.A.; Collins, R.E.; Anterola, A.M.; Cochrane, F.C.; Davin, L.B.; Lewis, N.G. An in silico assessment of gene function and organization of the phenylpropanoid pathway metabolic networks in *Arabidopsis thaliana* and limitations thereof. *Phytochemistry* **2003**, *64*, 1097–1112. [CrossRef]
23. Shi, R.; Sun, Y.H.; Li, Q.; Heber, S.; Sederoff, R.; Chiang, V.L. Towards a systems approach for lignin biosynthesis in *Populus trichocarpa*: Transcript abundance and specificity of the monolignol biosynthetic genes. *Plant Cell Physiol.* **2010**, *51*, 144–163. [CrossRef] [PubMed]
24. Carocha, V.; Soler, M.; Hefer, C.; Cassan-Wang, H.; Fevereiro, P.; Myburg, A.A.; Paiva, J.A.; Grima-Pettenati, J. Genome-wide analysis of the lignin toolbox of Eucalyptus grandis. *New Phytol.* **2015**, *206*, 1297–1313. [CrossRef] [PubMed]
25. Park, H.L.; Bhoo, S.H.; Kwon, M.; Lee, S.W.; Cho, M.H. Biochemical and expression analyses of the rice cinnamoyl-CoA reductase gene family. *Front. Plant Sci.* **2017**, *8*, 2099. [CrossRef] [PubMed]
26. Zhan, W.; Cui, L.; Song, N.; Liu, X.; Guo, G.; Zhang, Y. Comprehensive analysis of cinnamoyl-CoA reductase (CCR) gene family in wheat: Implications for lignin biosynthesis and stress responses. *BMC Plant Biol.* **2025**, *25*, 567. [CrossRef]
27. Song, X.; Liu, D.; Yao, Y.; Tang, L.; Cheng, L.; Yang, L.; Jiang, Z.; Kang, Q.; Chen, S.; Ru, J.; et al. Genome-wide identification and expression pattern analysis of the cinnamoyl-CoA reductase gene family in flax (*Linum usitatissimum* L.). *BMC Genom.* **2025**, *26*, 315. [CrossRef]
28. Kawasaki, T.; Koita, H.; Nakatsubo, T.; Hasegawa, K.; Wakabayashi, K.; Takahashi, H.; Umemura, K.; Umezawa, T.; Shimamoto, K. Cinnamoyl-CoA reductase, a key enzyme in lignin biosynthesis, is an effector of small GTPase Rac in defense signaling in rice. *Proc. Natl. Acad. Sci. USA* **2006**, *103*, 230–235. [CrossRef]
29. Bang, S.W.; Choi, S.; Jin, X.; Jung, S.E.; Choi, J.W.; Seo, J.S.; Kim, J.K. Transcriptional activation of rice CINNAMOYL-CoA REDUCTASE 10 by OsNAC5, contributes to drought tolerance by modulating lignin accumulation in roots. *Plant Biotechnol. J.* **2022**, *20*, 736–747. [CrossRef]
30. Chao, N.; Jiang, W.T.; Wang, X.C.; Jiang, X.N.; Gai, Y. Novel motif is capable of determining CCR and CCR-like proteins based on the divergence of CCRs in plants. *Tree Physiol.* **2019**, *39*, 2019–2026. [CrossRef]

31. Cai, Z.; He, F.; Feng, X.; Liang, T.; Wang, H.; Ding, S.; Tian, X. Transcriptomic analysis reveals important roles of lignin and flavonoid biosynthetic pathways in rice thermotolerance during reproductive stage. *Front. Genet.* **2020**, *11*, 562937. [CrossRef] [PubMed]
32. Borah, P.; Khurana, J.P. The OsFBK1 E3 ligase subunit affects anther and root secondary cell wall thickenings by mediating turnover of a cinnamoyl-CoA reductase. *Plant Physiol.* **2018**, *176*, 2148–2165. [CrossRef] [PubMed]
33. Wang, J.; Zhang, C.; Li, Y. Genome-wide identification and expression profiles of 13 key structural gene families involved in the biosynthesis of rice flavonoid scaffolds. *Genes* **2022**, *13*, 410. [CrossRef] [PubMed]
34. Oh, J.; Kim, E.; Kwag, Y.; An, H.; Kim, H.S.; Shah, S.; Lee, J.H.; Ha, E. Heat wave exposure and increased heat-related hospitalizations in young children in South Korea: A time-series study. *Environ. Res.* **2024**, *241*, 117561. [CrossRef]
35. Pascual, L.S.; Segarra-Medina, C.; Gómez-Cadenas, A.; López-Climent, M.F.; Vives-Peris, V.; Zandalinas, S.I. Climate change-associated multifactorial stress combination: A present challenge for our ecosystems. *J. Plant Physiol.* **2022**, *276*, 153764. [CrossRef]
36. Wagner, A.; Tobimatsu, Y.; Goeminne, G.; Phillips, L.; Flint, H.; Steward, D.; Torr, K.; Donaldson, L.; Boerjan, W.; Ralph, J. Suppression of CCR impacts metabolite profile and cell wall composition in *Pinus radiata* tracheary elements. *Plant Mol. Biol.* **2013**, *81*, 105–117. [CrossRef]
37. Tamasloukht, B.; Wong, Q.; Lam, M.S.; Martinez, Y.; Tozo, K.; Barbier, O.; Jourda, C.; Jauneau, A.; Borderies, G.; Balzergue, S.; et al. Characterization of a cinnamoyl-CoA reductase 1 (CCR1) mutant in maize: Effects on lignification, fibre development, and global gene expression. *J. Exp. Bot.* **2011**, *62*, 3837–3848. [CrossRef]
38. Wang, Z.; Guo, J.; Luo, W.; Niu, S.; Qu, L.; Li, J.; Chen, Y.; Li, G.; Yang, H.; Lu, D. Salicylic acid cooperates with lignin and sucrose signals to alleviate waxy maize leaf senescence under heat stress. *Plant Cell Environ.* **2025**, *48*, 4341–4355. [CrossRef]
39. Gallego-Giraldo, L.; Escamilla-Trevino, L.; Jackson, L.A.; Dixon, R.A. Salicylic acid mediates the reduced growth of lignin down-regulated plants. *Proc. Natl. Acad. Sci. USA* **2011**, *108*, 20814–20819. [CrossRef]
40. Chao, N.; Li, S.; Li, N.; Qi, Q.; Jiang, W.T.; Jiang, X.N.; Gai, Y. Two distinct cinnamoyl-CoA reductases in *Selaginella moellendorffii* offer insight into the divergence of CCRs in plants. *Planta* **2017**, *246*, 33–43. [CrossRef]
41. Bart, R.S.; Chern, M.; Vega-Sánchez, M.E.; Canlas, P.; Ronald, P.C. Rice *Snl6*, a cinnamoyl-CoA reductase-like gene family member, is required for NH1-mediated immunity to *Xanthomonas oryzae* pv. *oryzae*. *PLoS Genet.* **2010**, *6*, e1001123. [CrossRef] [PubMed]
42. Zhang, L.; Zheng, L.; Wu, J.; Liu, Y.; Liu, W.; He, G.; Wang, N. OsCCRL1 is essential for phenylpropanoid metabolism in rice anthers. *Rice* **2023**, *16*, 10. [CrossRef]
43. Dias, M.C.; Pinto, D.C.G.A.; Silva, A.M.S. Plant flavonoids: Chemical characteristics and biological activity. *Molecules* **2021**, *26*, 5377. [CrossRef] [PubMed]
44. Kamble, N.U. Decoding the role of flavonoids in ROS management during heat stress in tomato pollen. *Plant Cell* **2024**, *36*, 4285–4286. [CrossRef] [PubMed]
45. Shen, N.; Wang, T.; Gan, Q.; Liu, S.; Wang, L.; Jin, B. Plant flavonoids: Classification, distribution, biosynthesis, and antioxidant activity. *Food Chem.* **2022**, *383*, 132531. [CrossRef]
46. Liu, W.; Feng, Y.; Yu, S.; Fan, Z.; Li, X.; Li, J.; Yin, H. The flavonoid biosynthesis network in plants. *Int. J. Mol. Sci.* **2021**, *22*, 12824. [CrossRef]
47. Barakat, A.; Yassin, N.B.; Park, J.S.; Choi, A.; Herr, J.; Carlson, J.E. Comparative and phylogenomic analyses of cinnamoyl-CoA reductase and cinnamoyl-CoA-reductase-like gene family in land plants. *Plant Sci.* **2011**, *181*, 249–257. [CrossRef]
48. Ding, S.; Gao, Z.; Chen, Y.; Liu, J.; Xue, Y.; Teng, Y.; Qin, S.; Tian, X.; Wang, H. *OsLAC12*, a laccase-encoding gene, contributes to rice heat tolerance through regulation of antioxidant defense and secondary metabolism. *Plants* **2025**, *14*, 2846. [CrossRef]
49. Yoo, S.D.; Cho, Y.H.; Sheen, J. Arabidopsis mesophyll protoplasts: A versatile cell system for transient gene expression analysis. *Nat. Protoc.* **2007**, *2*, 1565–1572. [CrossRef]

plants

MDPI

Article

Identification of the *Populus euphratica XTHs* Gene Family and the Response of *PeXTH7* to Abiotic Stress

Jing Li [1,†], Hongyan Jin [1,†], Tongrui Song [1], Donghui Miao [1], Qi Ning [1], Jianhao Sun [1], Zhijun Li [1], Peipei Jiao [1,*] and Zhihua Wu [2,*]

1 Xinjiang Production and Construction Corps Key Laboratory of Protection and Utilization of Biological Resources in Tarim Basin, College of Life Science & Technology, Tarim University, Alar 843300, China; jing926819729@163.com (J.L.); j2601803054@163.com (H.J.); songtr20001209@163.com (T.S.); mdh1216g@163.com (D.M.); ningqi20000423@163.com (Q.N.); sunjianhaotea@163.com (J.S.); lizhijun0202@126.com (Z.L.)

2 College of Life Sciences, Zhejiang Normal University, Jinhua 321004, China

* Correspondence: 120070040@taru.edu.cn (P.J.); zhuawu@zjnu.edu.cn (Z.W.)

† These authors contributed equally to this work.

Abstract

Populus euphratica Oliv. serves as a keystone species in desert riparian ecosystems. Owing to its pronounced tolerance to drought and salinity, as well as its robust reproductive capacity, it has become a pioneer species in desert oases. The xyloglucan endotransglucosylase (XET)/hydrolase (*XTH*) gene family plays a critical role in the remodeling of plant cell walls; however, its potential biological functions in poplar remain poorly understood. In this study, we identified the *XTH* gene family in *P. euphratica* and conducted a preliminary functional analysis. A total of 33 *PeXTH* genes were identified, which were unevenly distributed across the chromosomes, with the highest density observed on chromosome 6. Conserved domain analysis indicated that most members contain the typical GH16 domain associated with xyloglucan endotransglucosylase activity. Phylogenetic analysis classified them into four distinct subgroups, exhibiting evolutionary conservation with the model dicot plant of *Arabidopsis thaliana*. Notably, the promoter analysis revealed an abundance of ABA-responsive and stress-related cis-elements, suggesting their potential involvement in response to multiple stresses. Under drought stress, *PeXTH7* (*PeuTF07G00088.1*) exhibited a distinct expression pattern, with transcript levels significantly increasing with persistent treatment. RT-qPCR results confirmed that *PeXTH7* is highly expressed in both roots and leaves. Furthermore, subcellular localization assays demonstrated that the PeXTH7 protein localizes to the secretory pathway and the cell wall, implying a role in cell wall dynamic remodeling through the regulation of xyloglucan metabolism. The *PeXTH7*-overexpressing transgenic lines exhibited a significant increase in root length compared to the wild-type controls. As the first systematic analysis of the *XTH* gene family in *P. euphratica*, this study fills an important knowledge gap and provides new insights into the adaptive mechanisms of desert tree species.

Keywords: *Populus euphratica* Oliv.; xyloglucan endotransglucosylase/hydrolase; gene family identification; stress response; growth and development

1. Introduction

Under the pressure of intensifying global climate change and worsening water scarcity, drought stress has emerged as a major environmental constraint on agricultural productivity [1]. In response, plants have evolved sophisticated adaptive mechanisms that allow them to perceive external stress signals and mount appropriate physiological and molecular responses [2]. Studies have shown that the cell wall acts as the primary interface for plant–environment interactions, detecting external changes and relaying stress signals to the plasma membrane. This signaling cascade leads to changes in cytosolic Ca^{2+}, reactive oxygen species (ROS), and hormone levels—including abscisic acid (ABA)—and further modulates the synthesis and remodeling of cell wall components through regulation of related gene expression, thereby enhancing plant adaptation to abiotic stress [3]. As the first barrier between plant cells and the external environment, the cell wall plays essential roles not only in sensing and transducing abiotic stress signals and maintaining structural integrity, but also in supporting plant growth and development [4]. The primary cell wall consists mainly of cellulose microfibrils, hemicellulose, pectin, and glycoproteins, which form a complex hydrated network. Water deficit can alter the composition of the cell wall, affecting turgor pressure regulation and growth dynamics, thereby modulating plant growth. Under stress conditions, plants upregulate genes involved in cell wall polysaccharide synthesis and modification through mechanisms such as stress-induced Ca^{2+} signaling oscillations and ABA-dependent pathways, thus preserving cell wall integrity and improving stress adaptability [5].

Xyloglucan (XG) serves as the predominant hemicellulose in the primary cell walls of dicotyledons and non-graminaceous monocotyledons. Its backbone consists of a β-(1 $\rightarrow$ 4)-linked D-glucan chain, which interacts with cellulose microfibrils via hydrogen bonds to form a complex network crucial for maintaining cell wall mechanical strength. Research has shown that under various abiotic stress conditions, transcription of genes involved in the lignin biosynthesis pathway is significantly upregulated. This transcriptional activation is accompanied by enhanced lignin deposition, which reinforces cell wall rigidity—an adaptive response that contributes to improved plant stress tolerance [6].

Xyloglucan endotransglycosylase/hydrolase (XTH), a key enzyme belonging to the glycoside hydrolase family 16 (GH16), catalyzes the cleavage and re-ligation of xyloglucan chains within the cellulose framework and is widely recognized as a pivotal regulator of cell wall remodeling. Through its modulation of xyloglucan metabolism, XTH dynamically regulates cell wall loosening and reinforcement, thereby directly influencing cellular elongation and expansion. Substantial evidence has demonstrated that plant roots exhibit greater sensitivity to abiotic stresses compared to other organs [7]. Functioning as the primary site for environmental stress perception, root growth displays a strong correlation with XTH-mediated xyloglucan metabolic dynamics, which may account for their enhanced stress responsiveness. Experimental findings indicate that when xyloglucan content in the Golgi apparatus substantially exceeds XTH levels, XTH incorporates into the cell wall primarily as xyloglucan-XTH complexes, resulting in enhanced wall rigidity and suppressed cell elongation. Conversely, under conditions where XTH abundance surpasses xyloglucan availability, XTH enters the cell wall as free enzymes that facilitate xyloglucan hydrolysis and transglycosylation, leading to wall loosening and promoted cell elongation [8].

Populus euphratica is a forest-forming tree species native to arid desert regions, primarily distributed across saline-alkali wastelands, desert areas, and river alluvial plains in northwestern China. As a pioneer species in desert riparian ecosystems, its remarkable adaptability to arid and saline-alkali environments establishes *P. euphratica* as a key woody model for studying plant responses to abiotic stresses such as drought and salinity [9,10]. Studies have demonstrated that under drought stress, *P. euphratica* seedlings exhibit thickening of the taproot, increased lateral root density, and significant modifications in root system architecture; these morphological adjustments contribute to enhanced water uptake and reduced water loss [11,12]. Furthermore, the dynamic remodeling of root system architecture (RSA) enables plants to directionally grow toward favorable soil conditions while avoiding adverse environments, thereby significantly improving their survival capacity [13].

Preliminary studies suggest that XTH primarily localizes to and functions in the primary cell wall, where it participates in regulating root cap development and the expression of genes related to root cap formation. By modulating cell wall elasticity and extensibility, XTH plays a significant role in plant adaptation to adverse environmental conditions [14]. Currently, 33, 29, 41, and 71 members of the *XTH* gene family members have been identified in *A.thaliana* [15], *Oryza sativa* [16], *Populus tomentosa* [17], and *wheat* [18], respectively. Increased expression of XTH strengthens the connection between primary and secondary walls in mesophyll cells by reducing the length of xyloglucan, which helps plants adapt to high temperatures [19]. In addition, *XTH* genes are also involved in regulating plant growth and development. Five *XTH* genes (*OfXTH24, 27, 32, 35,* and *36*) in *Osmanthus fragrans* respond to changes in environmental temperature and participate in the regulation of flowering [20]. Transcriptomic analysis of *AcXTHs* in *Ananas comosus* indicates that it involved in the regulation of fruit ripening and Crassulacean acid metabolism (CAM) and exhibit tissue specificity. Real-time fluorescent quantitative PCR analysis shows that *AcXTH18* is involved in root growth of *A. comosus* [21]. Research has found that the *SsXTH* gene in *Schima superba* participates in glycoside metabolism through the transfer and hydrolysis of xyloglucan in the cell wall, thereby regulating the elongation of plant fibers [22]. During tomato fruit development, a novel *XTH* gene (*SlXTH5*) was identified with high expression in mature fruits, correlating with xyloglucan depolymerization [23]. Heterologous overexpression of the pepper *CaXTH3* in *A. thaliana* induced severe leaf curling, implying its role in cell wall remodeling and reinforcement, which may help protect mesophyll cells under drought stress [24]. In sugar beet, XTH has been shown to undergo changes during abiotic stress responses and is involved in stomatal development and function by modulating the flexibility and integrity of guard cell walls [25]. These findings collectively provide a foundation for understanding plant stress regulation mechanisms mediated by cell wall signaling. In recent years, with the release of high-quality genomes and multi-omics data from *Populus* plants, it has become easier to identify and analyze the biological function of *XTH* gene family [26–28].

In summary, *XTH* genes are closely associated with multiple aspects of plant growth and development, including root elongation, leaf venation patterning, fruit maturation, and stomatal development. Accumulating evidence demonstrates that XTH plays a crucial role in plant responses to abiotic stress, offering important insights into cell wall signal-mediated stress regulatory pathways. However, the potential function of the XTH family in the desert tree species *P. euphratica* remains elusive. In order to fill this knowledge gap, we identified, analyzed and predicted the *PeXTH* family, and functionally characterized the drought-responsive member *PeXTH7* to elucidate its role in root development and stress adaptation. Therefore, this study aims to identify and characterize the *XTH* gene family in poplar through genomic and bioinformatic analysis, with the objective of elucidating its potential functions in drought resistance. These results provide new insights for further understanding the role of *PeXTH* in response to abiotic stress, and also have important significance for the development and utilization of desert plant germplasm resources.

2. Results

2.1. Identification and Prediction of Physical and Chemical Properties of PeXTHs Family Members

A total of 33 *PeXTH* family members were identified in *P. euphratica*. The physicochemical properties of the encoded PeXTH proteins—including molecular weight (MW), amino acid number, instability index, theoretical isoelectric point (pI), grand average of hydropathicity (GRAVY), and subcellular localization—were analyzed using TBtools (version 2.119) and Expasy (https://web.expasy.org/protparam/; accessed on 24 November 2024). The results revealed that the molecular weights of PeXTH proteins range from 23.83 kDa to 39.49 kDa, with corresponding amino acid lengths varying between 210 and 338 residues. All members exhibit instability indices ranging from 55.81 to 71.41, indicating that they were generally unstable under in vitro conditions. The theoretical pI values vary widely from 4.48 to 9.58, likely reflecting differences in amino acid composition and polarity. Moreover, all GRAVY values were negative, confirming the overall hydrophilic nature of these proteins. Subcellular localization predictions suggest that 21 PeXTH proteins were localized to the cell wall, whereas the remaining 12 were predicted in both the cell wall and cytoplasm, implying potential roles in coordinating cell wall dynamics with intracellular signaling. Detailed physicochemical characteristics are provided in Table 1.

Table 1. Physical and chemical properties of the *PeXTH* gene family.

Gene ID	Molecular Weight (kDa)	Number of Amino Acid (aa)	Aliphatic Index	Theoretical pI	Grand Average of Hydropathicity	Predicted Location (s)
PeuTF09G00081.1	33.05	294	68.03	6.15	−0.486	Cell wall
PeuTF18G00763.1	33.19	291	62.71	8.79	−0.411	Cell wall
PeuTF07G00088.1	24.7	213	60.05	5.95	−0.546	Cell wall
PeuTF09G01635.1	32.28	289	62.77	5.03	−0.362	Cell wall
PeuTF06G01644.1	32.66	287	69.62	6.59	−0.307	Cell wall, Cytoplasm
PeuTF06G01643.1	32.76	287	68.95	7.04	−0.356	Cell wall, Cytoplasm
PeuTF09G00856.1	34.1	294	65.31	8.9	−0.494	Cell wall
PeuTF08G01082.1	38	336	71.37	5.88	−0.346	Cell wall
PeuTF01G00675.1	33.55	289	59.38	6.22	−0.512	Cell wall, Cytoplasm
PeuTF13G00056.1	23.83	210	55.81	5.71	−0.561	Cell wall, Cytoplasm
PeuTF11G00200.1	34.94	299	56.45	5.09	−0.585	Cell wall
PeuTF04G00162.1	34.1	293	56.62	4.76	−0.561	Cell wall
PeuTF02G00374.1	25.69	224	83.57	8.51	−0.242	Cell wall
PeuTF05G01755.1	33.24	295	70.75	8.84	−0.322	Cell wall, Cytoplasm
PeuTF10G01452.1	38.21	335	71.28	6.07	−0.417	Cell wall
PeuTF01G01381.1	39.49	338	57.43	9.1	−0.559	Cell wall
PeuTF11G00627.1	33.37	292	71.44	8.99	−0.399	Cell wall
PeuTF11G00628.1	31.54	274	75.44	9.67	−0.419	Cell wall
PeuTF03G01575.1	35	300	65.33	8.4	−0.387	Cell wall, Cytoplasm
PeuTF14G00471.1	32.25	282	66.77	9.07	−0.33	Cell wall, Cytoplasm
PeuTF14G00528.1	30.82	264	67.95	8.87	−0.539	Cell wall, Cytoplasm
PeuTF03G00903.1	36.3	311	62.77	9	−0.603	Cell wall
PeuTF06G01225.1	33.88	294	60.75	9.58	−0.443	Cell wall
PeuTF02G02211.1	33.03	288	66.04	9.24	−0.34	Cell wall, Cytoplasm
PeuTF02G02300.1	33.55	293	58.6	5.64	−0.606	Cell wall
PeuTF06G00688.1	31.88	285	62.67	7.58	−0.322	Cell wall, Cytoplasm
PeuTF14G00818.1	34.88	311	70.84	6.49	−0.251	Cell wall
PeuTF19G00986.1	33.78	296	69.8	5.89	−0.295	Cell wall
PeuTF13G01387.1	28.12	246	58.66	5.8	−0.498	Cell wall
PeuTF16G00881.1	34.03	294	61.7	9.39	−0.426	Cell wall
PeuTF18G00888.1	33.23	292	71.16	8.95	−0.383	Cell wall, Cytoplasm
PeuTF05G00063.1	33.01	291	71.41	4.48	−0.261	Cell wall, Cytoplasm
PeuTF19G01001.1	33.78	296	69.8	5.89	−0.295	Cell wall

2.2. Analysis of Gene and Protein Structure of PeXTHs

The gene distribution of conserved domains in *PeXTHs* and gene exon and intron group map were completed using the GSDS (Gene Structure Display Server) online platform. The prediction and visualization of conserved protein motifs were achieved with the help of the MEME (Multiple Em for Motif Elicitation) website. Bioinformatics analysis showed that the amino acid sequences of the *P. euphratica XTH* gene family contained 10 different characteristic motifs, among which motifs 1–7 and motif 10 were highly conserved in most members, while motif 9 was only present in three genes and showed low conservation (Figure 1A). Conservative domain analysis further confirmed that this gene family contains three different characteristic functional domains, among which GH16-XET (Glycoside Hydrolase Family 16 Xyloglucan Endotransglucosylase) is the main pfam model.

This result is consistent with the reported catalytic domain characteristics of xyloglucan endotransglucosylase/hydrogenase (XTH) (Figure 1B). Gene structure analysis revealed that the number of exons in each member does not exceed four (Figure 1C); this minimalistic exon-intron arrangement pattern may be related to their functional conservation. The identification of conserved protein domains of *PeXTHs* revealed the presence of ten conserved motifs (motifs 1–10; Figure 1D). These findings suggest that the core functional regions are evolutionarily conserved.

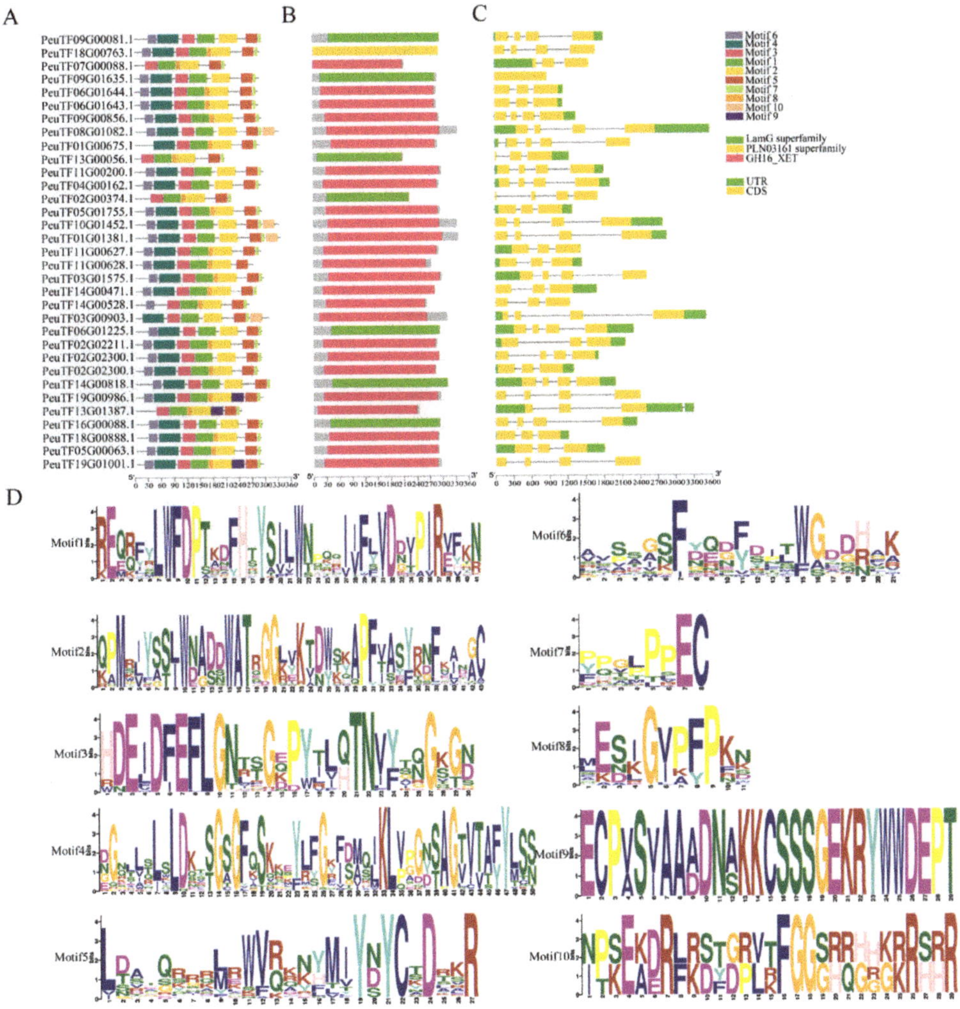

Figure 1. Analysis of gene structure and conserved structural domains of *PeXTHs*. (**A**) Distribution of conserved domains in *PeXTH*. (**B**) Pefam model of *PeXTH*; green indicates LamG superfamily, yellow indicates PLN03161 superfamily, and red indicates GH16_XET. (**C**) Gene structure of *PeXTH*. Yellow boxes indicate exons, black lines indicate introns, and green indicates non-coding regions. (**D**) Ten conserved motifs of PeXTHs, with single-letter abbreviations representing amino acids.

2.3. *Prediction of Cis-Acting Elements in PeXTHs Promoter Region*

The 2 kb promoter sequences upstream of the transcription start site (TSS) of all candidate genes were extracted, and cis-acting elements within these promoter regions were identified using online prediction tools. Systematic analysis revealed a total of 22 types of cis-regulatory elements, which were functionally categorized into four groups: abiotic/biotic stress-related, hormone-responsive, light-responsive, and growth/development-related elements. Among these, hormone-responsive elements were the most abundant, with nine types identified, including recognition sites for abscisic acid (ABA), methyl jasmonate (MeJA), gibberellin (GA), and salicylic acid (SA). ABA-responsive elements constituted the largest proportion within this category. Among the stress-related elements, hypoxia-responsive elements were predominant (Figure 2). These findings suggest that *PeXTH* genes may play diverse and important roles in regulating various physiological processes in *P. euphratica*, particularly under abiotic and biotic stress conditions. The presence of these cis-elements implies that the *PeXTH* gene family may be involved in modulating gene expression and multiple signaling pathways to enhance environmental stress adaptation.

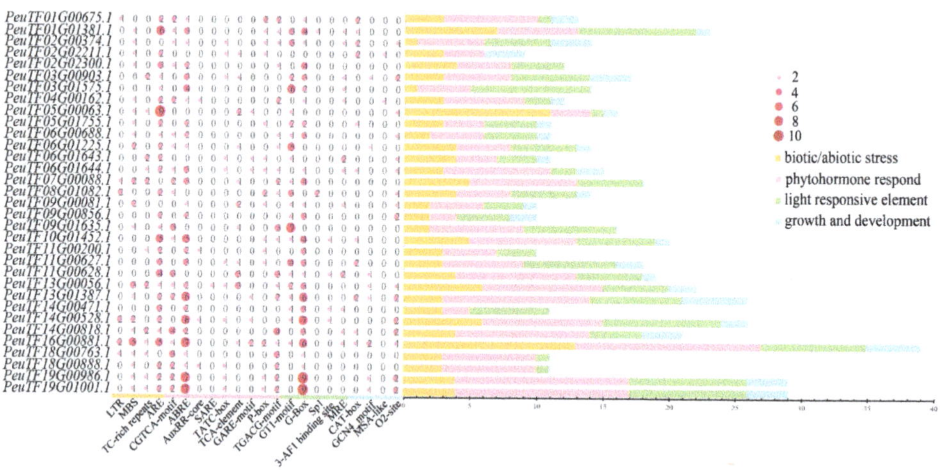

Figure 2. Statistics of cis-elements in the *PeXTHs* promoter region. The number in the figure represents the number of corresponding cis-acting elements predicted in the gene. The color depth and size of the circle are proportional to the number of components. The uncolored circle (number 0) indicates that the component is not detected. The color bars indicate the classification of cis-elements: yellow represents cis-elements related to biotic/abiotic stress, red represents cis-elements related to hormones, green represents cis-elements related to light response, and blue represents cis-elements related to growth and development.

2.4. Collinearity Analysis of XTHs Within P. euphratica and Among Multiple Species

By mapping the collinearity relationships among the 33 member genes within the *P. euphratica* gene family, 13 pairs of collinear *PeXTHs* genes were identified, including *PeuTF05G01755.1*, *PeuTF02G00374.1*, and *PeuTF11G00627.1*, each having two pairs of homologous genes, indicating that during the evolutionary process of *P. euphratica*, *XTHs* underwent multiple segmental duplications or repetitions, which facilitated the evolution and expansion of the gene family (Figure 3). Based on the collinearity relationships, 13 gene pairs were selected for selective pressure analysis, as shown in Table 2, with Ka/Ks values all less than 1, indicating that this gene family has been subjected to strong purifying selection pressure during evolution.

We performed interspecific collinearity analysis between *P. euphratica* and *A. thaliana*, as well as with *Populus deltoides* (a drought-tolerant poplar species of Salicaceae), *Populus pruinosa* (a drought-sensitive poplar species), and *Salix sinopurpurea* (a drought-sensitive willow species of Salicaceae). The analysis revealed 18 collinear gene pairs between *P. euphratica* and *A. thaliana* (Figure 4), whereas the numbers of collinear pairs between *P. euphratica* and *P. deltoides*, *P. pruinosa*, and *S. sinopurpurea* were 30, 28, and 29, respectively. These results indicate that the *XTH* gene family exhibits significantly higher evolutionary conservation within the Salicaceae family than across families (as represented by *A. thaliana*), suggesting that this gene family may have maintained greater structural stability through gene duplication and functional differentiation during adaptive evolution in woody plants.

Table 2. Evolutionary selection pressures of XTHs in *P. euphratica*.

Gene_1	Gene_2	Ka	Ks	Ka/Ks
PeuTF01G01381.1	PeuTF03G00903.1	0.088029963	0.336819885	0.261356192
PeuTF01G00675.1	PeuTF03G01575.1	0.136741168	0.46409146	0.294642715
PeuTF08G01082.1	PeuTF10G01452.1	0.052476004	0.350912053	0.149541755
PeuTF06G01643.1	PeuTF18G00888.1	0.240293693	2.165085817	0.110985759
PeuTF06G01225.1	PeuTF09G00081.1	0.197261823	1.201835773	0.164133759
PeuTF06G01225.1	PeuTF16G00881.1	0.038836235	0.416044352	0.093346383
PeuTF19G01001.1	PeuTF13G01387.1	0.060220985	0.446202234	0.134963432
PeuTF14G00528.1	PeuTF02G02300.1	0.427698912	NaN	NaN
PeuTF05G01755.1	PeuTF11G00627.1	0.211307847	NaN	NaN
PeuTF05G01755.1	PeuTF02G00374.1	0.109048588	0.500680619	0.217800697
PeuTF05G00063.1	PeuTF13G00056.1	0.126769113	0.411406195	0.30813613
PeuTF02G00374.1	PeuTF11G00627.1	0.233638517	NaN	NaN
PeuTF04G00162.1	PeuTF11G00200.1	0.038976214	0.365085239	0.106759218

Non-synonymous is abbreviated as Ka and synonymous is abbreviated as Ks.

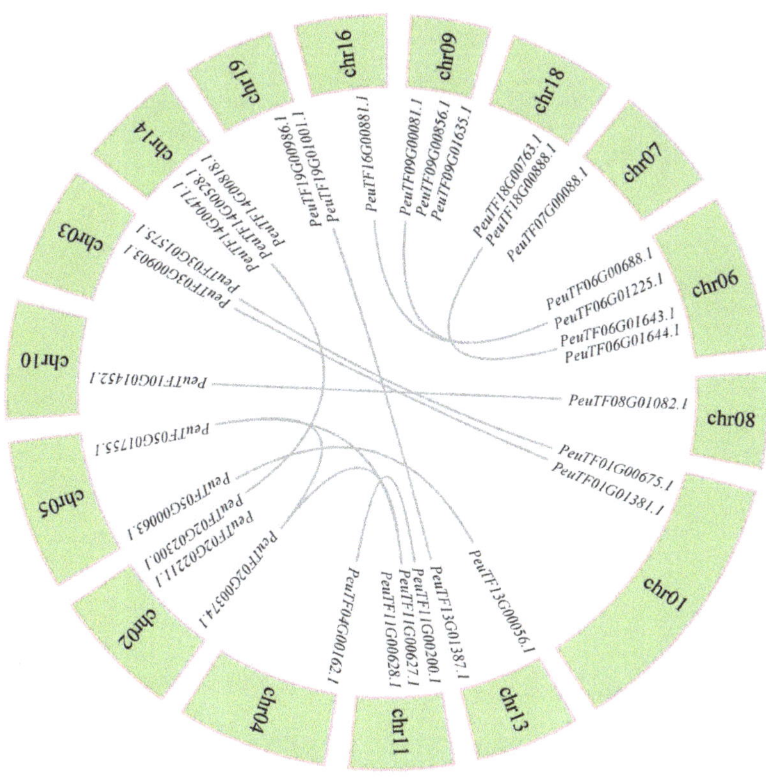

Figure 3. Intraspecific synteny analysis of 33 *PeXTHs* in *P. euphratica*. Green lines represent *P. euphratica* chromosomes, and gray lines indicate syntenic pairs of *PeXTH* fragment repeat genes.

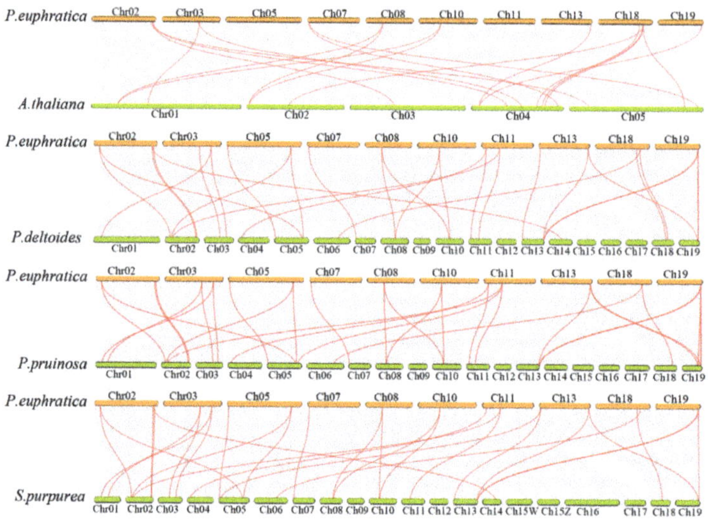

Figure 4. *XTH* colinear analysis among multiple species. Colinear analysis of *P. euphratica* and four other species, *A. thaliana*, *P. deltoides*, *P. pruinosa*, and *S. sinopurpurea*. Gray lines indicate colinear segments between the *P. euphratica* genome and those of other species, while red lines indicate *XTH* colinear pairs among species.

2.5. Chromosomal Localization Analysis of PeXTHs

Chromosomal localization analysis revealed that the 33 identified *PeXTH* genes are unevenly distributed across 16 chromosomes in *P. euphratica* (Figure 5). No family members were detected on chromosomes 12, 15, and 17. Gene distribution density varied among chromosomes, with chromosome 6 harboring the highest number (four *PeXTH* genes). Notably, two genes each were densely clustered on chromosomes 6, 11, and 19, suggesting potential hotspots for *XTH* gene localization in the *P. euphratica* genome. This genomic arrangement suggests that the proteins encoded by these genes may share functional similarities, reflecting potential evolutionary conservation within this gene family.

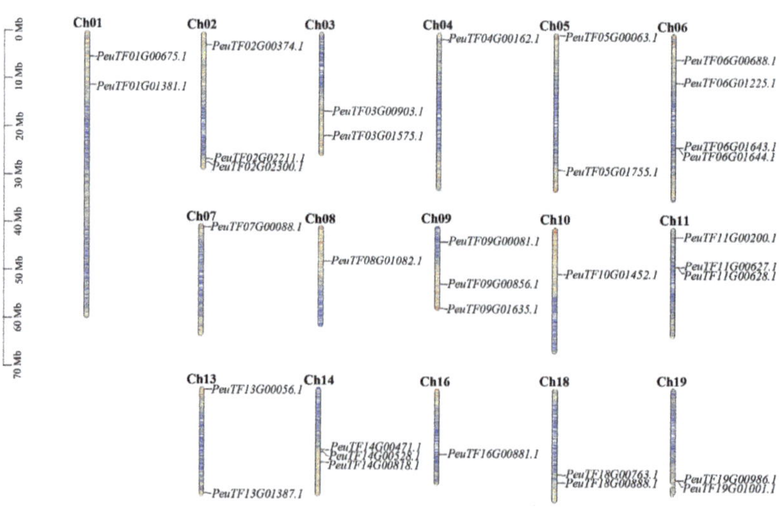

Figure 5. Chromosome location. Blue indicates low gene density on the chromosome, while red indicates high gene density on the chromosome.

2.6. Phylogenetic Tree of PeXTHs

A phylogenetic tree was constructed using the ML (Maximum Likelihood) method based on protein sequences of 33 *XTH* genes from *P. euphratica* and homologous sequences from other species (Figure 6). Following the classification scheme established for *AtXTHs*, the 159 *XTHs* members from five species were categorized into four distinct groups: Group I/II, Group IIIA, Group IIIB, and the Ancestral Group. Members of Groups I, II, and IIIB exhibit xyloglucan endotransglucosylase (XET) activity, enabling cleavage and religation of xyloglucan chains, whereas Group IIIA members display xyloglucan endohydrolase (XEH) activity, specifically hydrolyzing β-1,4-glycosidic bonds within xyloglucan polymers. Among the *PeXTHs*, 23, 5, 4, and 1 members were assigned to these subgroups, respectively. Given that AtXTHs within the same subgroup often share functional similarities, it is plausible that PeXTHs exhibit comparable functional conservation. Notably, XTHs from *P. euphratica* and *P. deltoides* formed highly supported clades, reflecting pronounced sequence homology and evolutionary conservation of function between these two species.

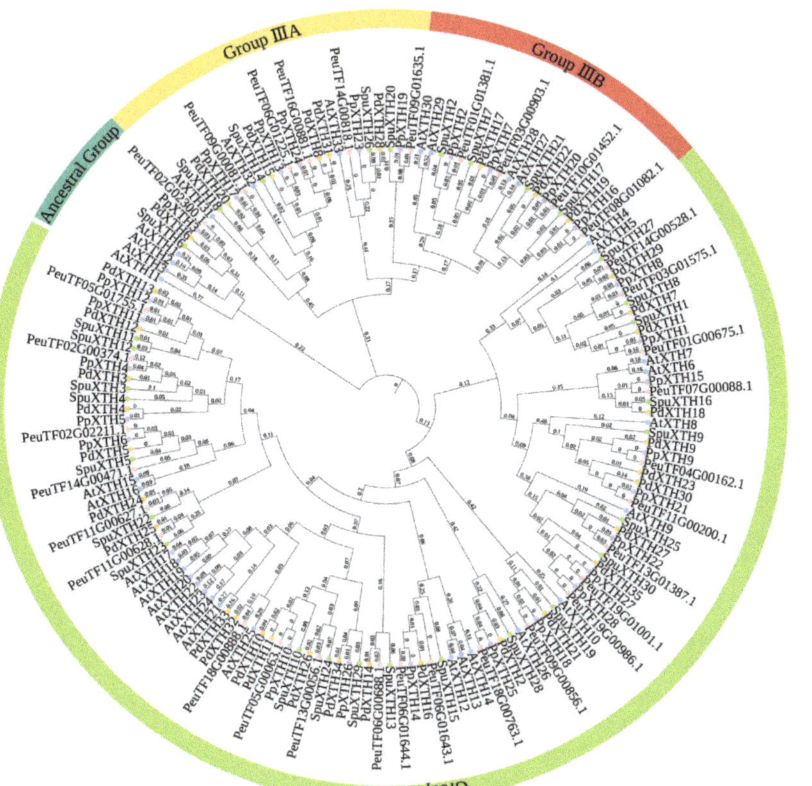

Figure 6. Phylogenetic tree of five species of *XTH* gene family members constructed using the neighbor-joining method. Phylogenetic analysis of XTH family protein sequences from five species (*P. euphratica*, *P. deltoides*, *P. pruinosa*, *S. sinopurpurea*, and *A. thaliana*) was performed using the ML (Maximum Likelihood) method by MEGA11 software (version 11.0.13). The value at the branch node represents the bootstrap value (calculated from 1000 repetitions) and shows only values greater than or equal to 50%. The number on the branch represents the branch confidence. Different colored circles represent different species. Species are color-coded as *P. deltoides* (orange), *A. thaliana* (blue), *S. sinopurpurea* (green), *P. pruinosa* (purple), and *P. euphratica* (pink); based on the grouping of *A. thaliana*, XTH was divided into four categories. Each group is distinguished by a different color.

2.7. *PeXTH* Transcriptome Sequencing and Data Analysis

To investigate the response of *PeXTH* genes to drought stress, we analyzed their transcript levels in *P. euphratica* seedlings under the drought treatment at 0, 4, and 12 h (Figure 7A). Visualization of the expression data using TBtools revealed that *PeuTF07G00088.1* and *PeuTF02G02211.1* exhibited the highest transcript abundance among all *PeXTH* genes. Notably, the expression level of *PeuTF07G00088.1* increased in roots but decreased in leaves under drought stress. In contrast, *PeuTF02G02211.1* showed declining expression trends in both roots and leaves. These results suggest that *PeuTF07G00088.1* may contribute to drought adaptation by enhancing its transcriptional activity in root tissues upon perception of drought signals.

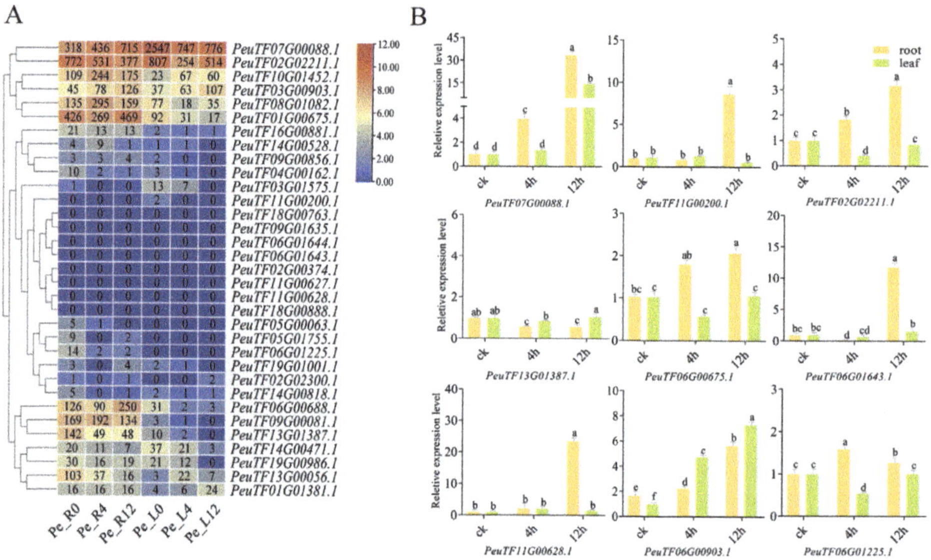

Figure 7. The expression pattern of *PeXTHs* under drought stress treatment. (**A**) Heat map of *PeXTHs* gene family member expression levels in *P. euphratica* seedlings under drought stress. Colors represent relative gene expression levels; red indicates high expression and blue indicates low expression. The number in the heat map is the original FPKM value. Pe-R0 represents gene expression levels in roots without drought treatment, and Pe-L0 represents expression levels in leaves without drought treatment. Pe-R4 indicates gene expression levels in roots after 4 h of drought treatment, while Pe-L4 indicates expression levels in leaves under the same conditions. Pe-R12 indicates gene expression levels in roots after 12 h of drought treatment, and Pe-L12 indicates expression levels in leaves under the same conditions. The color scale directly corresponds to the $\log_2(x + 1)$ converted value of FPKM. (**B**) Validation of the transcriptomic data by RT-qPCR is presented. Yellow and green columns represent the gene expression levels in the root and leaf, respectively. Letters (a, b, c, d) mark statistically distinct groups based on Duncan's multiple range test.

To validate the transcriptome data, the expression patterns of nine *PeXTH* genes in *P. euphratica* seedlings under drought stress were analyzed using quantitative real-time PCR (RT-qPCR). The relative expression of 9 genes from different branches of the phylogenetic tree was measured under simulated drought stress of varying durations. After 12 h of treatment, most genes exhibited generally higher expression levels in roots than in leaves. Notably, *PeXTH7* (*PeuTF07G00088.1*) showed significant time-dependent upregulation in both roots and leaves following 4 h and 12 h of treatment with 25% PEG6000, reaching the highest expression level among all examined genes (Figure 7B). This expression trend was consistent with the transcriptome sequencing results. Furthermore, seven genes displayed significantly higher expression in roots compared to leaves at the end of the treatment period. Based on its markedly high expression in both roots and leaves, *PeXTH7* (*PeuTF07G00088.1*) was selected for further functional investigation.

2.8. Subcellular Localization of PeXTH7 in Nicotiana benthamiana

To further investigate the function of *PeXTH7* (*PeuTF07G00088.1*) and determine its subcellular distribution, we performed experimental validation of its localization. Agrobacterium tumefaciens GV3101 harboring the 35S::*PeXTH7*-YFP construct was infiltrated into *N. benthamiana* leaves, and the subcellular localization was examined using confocal laser scanning microscopy (Figure 8). The fluorescence signal of 35S::*PeXTH7*-YFP showed colocalization with both the plasma membrane and endoplasmic reticulum markers. However, following plasmolysis, the 35S::*PeXTH7*-YFP signal dissociated from the plasma membrane marker. These results indicate that *PeXTH7* is synthesized in the endoplasmic reticulum and subsequently functions in the cell wall.

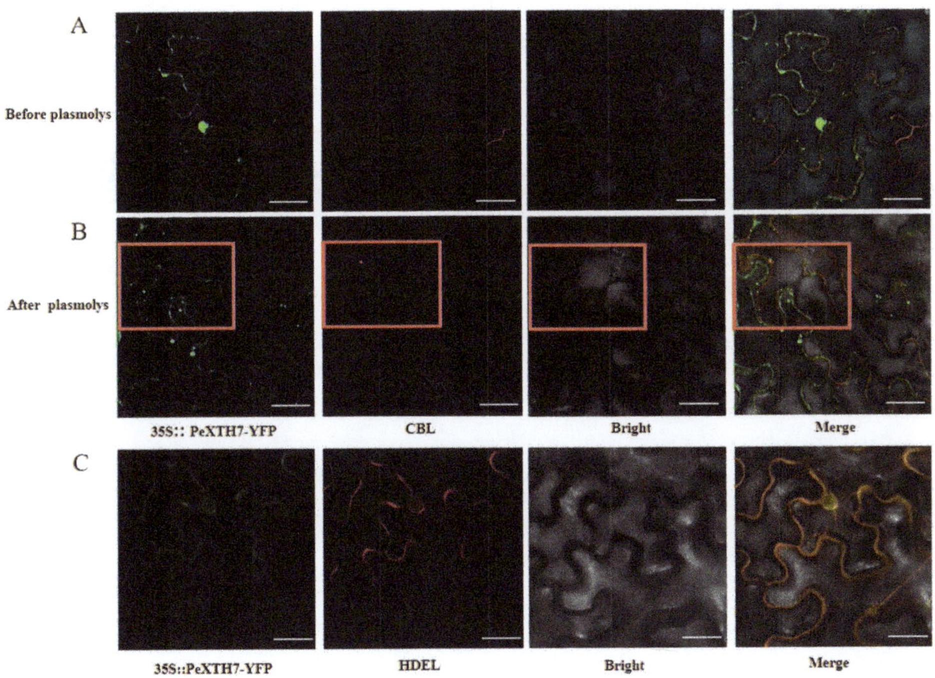

Figure 8. Subcellular localization of the 35S::*PeXTH7*-YFP protein in tobacco leaf epidermal cells. (**A–C**) Observation of membrane localization markers co-expressed with *PeXTH7* before and after plasmodesmal separation, as well as markers for the endoplasmic reticulum, and fusion images. (**A**) Before plasmodesmal separation, the cell membrane marker (CBL) overlaps with the YFP fluorescent label carried by *PeXTH7*. (**B**) After plasma membrane separation, the cell membrane marker (CBL) separates from the YFP fluorescent label carried by *PeXTH7*, The region enclosed by the red box exhibits distinct plasmolysis. (**C**) The endoplasmic reticulum marker (HDEL) fuses with the YFP fluorescent label carried by *PeXTH7* (Scale bar = 50 μm).

2.9. Effect of PeXTH7 Overexpression on the Regulation of Root System Development in Transgenic A. thaliana

To determine the function of *PeXTH7* in root development, we generated transgenic *A. thaliana* (Col-0) plants overexpressing *PeXTH7* driven by the 35S promoter. These plants showed a significant increase in root length. Without sorbitol stress, the overexpression lines exhibited significantly longer primary roots along with an increased density of lateral roots, compared to Col-0 (Figure 9). Under sorbitol-induced osmotic stress, although root growth was suppressed across all lines, the *PeXTH7*-overexpressing lines consistently developed longer roots and sustained better root growth than Col-0. Statistical analysis of three randomly selected independent lines revealed significantly greater root lengths compared to the wild-type. Our results demonstrate that *PeXTH7* promotes root growth and development, which contributes to enhanced drought tolerance.

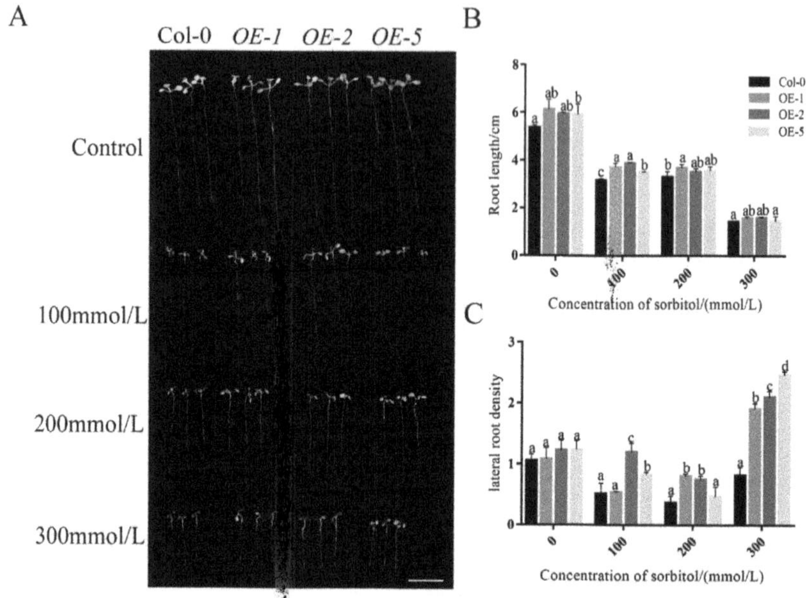

Figure 9. Overexpression of *PeXTH7* enhances root system development under simulated drought stress in *Arabidopsis*. (**A**) Representative root phenotypes of wild-type (Col-0) and three independent *PeXTH7*-overexpressing (OE-1, OE-2 and OE-5) lines under sorbitol-induced osmotic stress. Scale bar = 1 cm. (**B**) Primary root length of the plants shown in (**A**). (**C**) Primary lateral root density of the plants shown in (**A**). Data are presented as mean ± SEM (n ≥ 10). Bars marked with different letters are significantly different ($p < 0.05$, one-way ANOVA).

3. Discussion

3.1. Identification of 33 XTH Family Members in P. euphratica

P. euphratica is predominantly distributed in the Tarim River Basin of Xinjiang, China, where it plays a vital role in maintaining the structure and function of local ecosystems. Through its growth characteristics, it contributes to the regulation of river water resources and serves as an important ecological barrier. Exhibiting remarkable adaptability to the extreme desert conditions of northwestern China, *P. euphratica* has become a key model species for investigating the molecular and physiological mechanisms of abiotic stress tolerance in woody plants. It holds substantial value for research in related scientific fields.

Plant cell walls represent a unique extracellular matrix primarily composed of cellulose, hemicellulose, and pectin, typically supplemented with structural glycoproteins and lignin [29]. Xyloglucan is the main hemicellulose component of primary cell walls in plants [30]. This highly complex organization renders it a critical player in environmental stress perception. Multiple mechanisms regulate plant drought tolerance, with research by Hoson T demonstrating that the cell wall and plasma membrane collaboratively perceive external stress signals. Upon sensing environmental stresses, the mechanical properties of the cell wall undergo substantial modifications to mitigate adverse conditions [31].

Plant cell wall proteins (CWPs) play a major role in cell wall remodeling and signal transduction. Cell wall proteomics and numerous genetic and biochemical studies have revealed the high diversity of CWPs [32]. Enzymatically active protein components, including glycosidases, proteases, and transferases, occupy a central position in dynamic cell wall restructuring and environmental responses [33–35]; these enzymes regulate the assembly and disassembly of cell wall polysaccharide networks through spatiotemporally precise expression patterns [36]. Furthermore, protein–protein and protein–polysaccharide interactions can alter the mechanical properties of the wall and promote its supramolecular assembly [37–39]. These findings reveal how plant cells synthesize cell wall polysaccharides, assemble them into a sturdy fiber network, and regulate cell wall expansion during cell growth [40]. Xyloglucan endotransglucosylase/hydrolase (XTH) mediates the cleavage and re-ligation of xyloglucan, a key network component, thereby regulating cell wall extensibility while maintaining its structural integrity. Comparative genomic analyses suggest a non-plant origin for the XTH family, tracing back to Alphaproteobacteria ExoKs—bacterial enzymes involved in biofilm loosening. The earliest XTHs likely evolved in freshwater algae related to Cosmarium through C-terminal extension of EG16-2 proteins [41]. This ancient evolutionary origin established the foundation for its diverse functional roles in land plants. Previous studies have demonstrated that xyloglucan endotransglucosylase/hydrolase (XTH) contributes to cell wall biosynthesis and is closely associated with plant growth, development, and stress resilience, participating in processes such as root elongation [42], leaf venation patterning [43], fruit ripening [44], and responses to biotic and abiotic stresses [45,46].

Research on PeXTH functions remains relatively limited. In this study, we systematically identified XTH family members in four Salicaceae species: 33 in *P. euphratica*, 28 in *P. deltoides*, 30 in *S. sinopurpurea*, and 35 in *P. pruinosa*. Comparative analysis revealed limited variation in gene copy number among these species. The expansion in *P. pruinosa* may be attributed to gene duplication events during evolution, whereas the contraction in *P. deltoides* and *S. sinopurpurea* likely reflects lineage-specific selection associated with environmental adaptation. Furthermore, bioinformatic analyses of *PeXTHs*, including sequence conservation, gene structure, and potential functions, provide a theoretical foundation for elucidating the functional mechanisms of this gene family.

3.2. Potential Biological Functions of the PeXTH Gene Family

The *XTH* gene family plays crucial roles in plant growth and environmental adaptation. Proteins in this family exhibit three characteristic structural domains: (1) the DEIDFEFLG motif, responsible for transferase activity, in which the first glutamic acid (E) residue serves as a binding site essential for catalysis, while the second E functions as a proton donor [47]; (2) the N-glycosylation motif NXT/S, containing serine (Ser) and threonine (Thr) residues adjacent to the catalytic site, which is critical for enzymatic activity; and (3) a C-terminal cysteine residue capable of forming disulfide bonds that significantly contribute to protein structural stability [48]. Studies have revealed that different XTH members exhibit distinct responses to various hormones. Evidence indicates that gibberellic acid (GA_3) regulates cell wall elongation by modulating the expression and activity of cell wall-modifying enzymes [49,50]. GA3 treatment of *A. thaliana* induced upregulation of *AtXTH21* expression in *A. thaliana* [51]. The expression levels of the rose genes *RbXTH1* and *RbXTH2* were upregulated after ethylene treatment [52]. The expression level of *A. thaliana AtXTH23* was significantly upregulated after treatment with abscisic acid (ABA) [15]. These findings collectively elucidate the molecular mechanisms through which the XTH family participates in plant growth, development, and environmental adaptation via specialized hormone regulatory networks.

Cis-acting elements represent specific DNA sequences situated in the 5′ upstream regulatory regions of structural genes, capable of binding transcription factors to modulate gene expression [53]. Analysis of the promoters of the *PeXTH* gene family revealed that most members contain light-responsive elements (e.g., G-box) and phytohormone-responsive elements (e.g., ABRE and CGTCA-motif), suggesting that their expression is likely regulated by light signaling, abscisic acid (ABA), and methyl jasmonate (MeJA). This inference is supported by previous studies in cherry (*Prunus avium* L.), where exogenous application of ABA, MeJA, and ethylene significantly upregulated the expression of *PavXTH15* [54,55]. Further examination identified that *PeXTH* promoters also harbor multiple cis-elements associated with abiotic stress responses, including drought-inducible elements (MBS), low-temperature response elements (LTR), and defense stress elements (TC-rich repeats), implying their broad involvement in plant adaptation to environmental challenges. Notably, the *PeXTH7* promoter is particularly enriched with stress-responsive elements: the MBS element mediates drought stress responses, while the ABRE element participates in the ABA signaling pathway, which plays a pivotal role in plant adaptation to drought and salt stress [56]. This ABRE-centered regulatory pattern shows remarkable consistency with the *XTH* gene *PagXTH12* in 84K poplar (*Populus alba* × *P. glandulosa*) [21], whose promoter also contains ABRE elements and exhibits significant upregulation under drought and salt stress conditions. Moreover, the presence of numerous growth- and development-related regulatory elements further underscores the functional complexity and diversity of this gene family.

A phylogenetic tree constructed using the neighbor-joining method, integrated with interspecies collinearity analysis, revealed high sequence homology and evolutionary conservation of the *XTH* gene family among closely related species. Phylogenetic classification indicated that *Arabidopsis thaliana XTH* members were distributed across the Group I/II, IIIA, IIIB, and Ancestral branches with 22, 3, 4, and 4 members, respectively, whereas *Populus euphratica* contained 23, 5, 4, and 1 members in the corresponding branches. Although the total number of *XTH* genes was similar between the two species, notable branch-specific distribution differences were observed: *P. euphratica* exhibited expansion in Groups I/II and IIIA but contraction in the Ancestral branch. This branch-specific expansion and contraction may have resulted from gene duplication or loss events during species differentiation [57]. Despite these interspecific distribution variations, analysis of protein characteristics demonstrated that although XTH proteins exhibit considerable polymorphism in molecular weight, isoelectric point, and amino acid sequence length, members within the same evolutionary branch share highly conserved protein motifs and gene structures, such as exon–intron organization, suggesting functional similarities among clade members. Conserved domain analysis further supported this functional differentiation pattern: 80% of Group IIIA members contain the LamG superfamily domain, whereas nearly all members of Groups I/II and IIIB harbor the GH16_XET functional domain. Gene duplication analysis revealed that the expansion of the *PeXTH* gene family primarily arose from segmental duplication events. All duplicated *PeXTH* gene pairs showed Ka/Ks ratios of less than 1.00 (Table 2), indicating that this gene family has undergone strong purifying selection during evolution. This selective pressure maintained relatively slow evolutionary rates and promoted functional conservation. Collectively, these findings demonstrate that the key functional domains and overall architecture of XTH proteins are highly conserved in *P. euphratica*, with purifying selection playing a critical evolutionary role in maintaining this conservation by effectively eliminating random amino acid mutations that could be detrimental to plant adaptability.

Previous studies have collectively demonstrated that *XTH* family members exhibit spatiotemporal expression specificity and participate in multiple plant growth and developmental processes, while being regulated by both hormonal and environmental cues [58]. Overexpression of the *CaXTH3* gene in *Arabidopsis thaliana* and chili peppers significantly enhanced transgenic plant tolerance to water loss, accompanied by increased mesophyll cell density and altered leaf morphology [24]. Earlier research also revealed that heterologous expression of a specific *P. euphratica XTH* gene in tobacco improved drought tolerance by reducing stomatal aperture and consequently decreasing water evaporation [59]. Additionally, studies have shown that *A. thaliana Atxth7* mutants exhibit shorter root length than wild-type plants under salt stress [60]. Based on the phylogenetic analysis and RT-qPCR results presented above, *PeuTF07G00088.1* (named *PeXTH7*) was ultimately selected from *Populus euphratica* for further investigation due to its highest homology with *AtXTH7*. Under drought stress conditions, *PeXTH7* showed sustained upregulation in both roots and leaves, with root tissues displaying significantly higher expression levels than leaf tissues. This tissue-specific expression pattern implies that *PeXTH7* may contribute to drought adaptation in *P. euphratica* primarily through regulating root architecture. Consequently, *PeXTH7* has been identified as a principal target for subsequent functional studies aimed at clarifying its precise role in drought resistance mechanisms of woody plants.

Plant growth and development fundamentally depend on cell wall expansion. Xyloglucan, a major hemicellulosic component of the cell wall, has long been a central focus of research. XTH enzymes play essential roles in xyloglucan cleavage and reconnection during cell wall remodeling. The plant XTH family comprises numerous members that are expressed across various tissues and organs, responding to hormonal and environmental stress signals, thereby underscoring their crucial functions in plant growth and adaptation. Although research on *XTH* genes has achieved considerable progress in recent years, the specific biological functions of this gene family remain incompletely characterized. In this study, we performed a genome-wide analysis of the *XTH* gene family in *P. euphratica*, identified all its members, and conducted preliminary predictions regarding the response of *PeXTH7* to drought stress. Future work should focus on elucidating the molecular mechanisms through which *PeXTH7* contributes to abiotic stress tolerance.

3.3. PeXTH7 Participates in the Regulation of Root Length in Transgenic A. thaliana

To assess whether *PeXTH7* shares a conserved role in root growth regulation, we measured root length and lateral root density in *PeXTH7*-overexpressing *A. thaliana* plants. The results revealed a significant increase in root length compared to wild-type controls. Given the close relationship between root length and water uptake efficiency, we propose that *PeXTH7* may enhance drought tolerance by modulating root elongation through cell wall regulation. However, the precise molecular mechanism underlying this process requires further investigation.

4. Materials and Methods

4.1. Genetic Identification, Physical and Chemical Property Analysis

The XTH protein sequences of *Arabidopsis thaliana* were retrieved from the TAIR database (https://www.arabidopsis.org/; accessed on 19 November 2024) and used to perform a website BLASTp search against the *Populus euphratica* protein database (https://ngdc.cncb.ac.cn/search/specific?db=bioproject&q=PRJCA006811; accessed on 19 November 2024) [13], with an E-value threshold set to 1×10^{-5}. Subsequently, HMM profiles of the Glyco_hydro_16 (PF00722) and XET_C (PF06955) domains were obtained from the Pfam database (https://pfam.xfam.org/; accessed on 22 November 2024) for local HMM-based screening. The results from both approaches were combined, and sequences that did not simultaneously contain the Glyco_hydro_16 and XET_C domains were filtered out using the online bioinformatics tool NCBI CDD Search (https://www.ncbi.nlm.nih.gov/Structure/bwrpsb/bwrpsb.cgi; accessed on 22 November 2024). The remaining sequences were identified as putative *XTH* genes of *P. euphratica*. Based on genomic data available at the National Center for Biotechnology Information (NCBI), *XTH* family members were also identified in three additional poplar species: *P. deltoides* (WV94_445), *P. pruinosa* (PRJNA863418), and *S. sinopurpurea* [61].

The physicochemical properties of the identified PeXTH proteins, including coding sequence length, amino acid number, molecular weight, and theoretical isoelectric point, were analyzed using the Expasy ProtParam tool (https://web.expasy.org/protparam/; accessed on 11 September 2024). Subcellular localization of *P. euphratica* XTH family members was predicted via WOLF PSORT (https://www.genscript.com/; accessed on 11 December 2024). Chromosomal location data were extracted from the genome annotation file, and the physical distribution of *XTH* genes was visualized using TBtools.

4.2. PeXTH Gene Structure and Conserved Motif Analysis

Conserved motifs in the *XTH* gene family proteins were analyzed using the MEME online tool (https://meme-suite.org/meme/tools/meme; accessed on 11 September 2024). The analysis was conducted with the optimum motif width set to 10–150 residues and the maximum number of motifs set to 10, while other parameters remained at their default values. The resulting gene structures and motif distributions were visualized using TBtools software (v2.119) for subsequent interpretation.

4.3. Analysis of Cis-Acting Elements of PeXTH Promoter

The promoter sequences of *PeXTH* genes, defined as the 2000 bp regions upstream of the translation start sites, were extracted using TBtools (v2.119). Putative cis-regulatory elements within these promoter sequences were identified using the PlantCARE database (https://bioinformatics.psb.ugent.be/webtools/plantcare/html; accessed on 18 October 2024). The predicted cis-acting elements were subsequently visualized using TBtools software (v2.119).

4.4. Collinearity and Chromosome Mapping Among Multiple Species

Orthologous gene pairs between *P. euphratica* and the other four species (*P. deltoides*, *P. pruinosa*, *S. sinopurpurea*, and *A. thaliana*) were identified using BLASTP alignment. Subsequently, TBtools (version 2.119) software was used to screen for collinearity blocks between *P. euphratica* and these four species and visualize them.

Chromosome length information (Fasta statistics), *PeXTH* gene IDs and positional data (GFF3 parsing/text block extraction and filtering), and gene density distribution were extracted from the *P. euphratica* genome using TBtools (v2.119). Subsequently, chromosomal localization patterns were visualized with the TBtools Gene Position Visualization function, and regional gene distributions were further analyzed using the same module.

4.5. PeXTH Phylogenetic Tree Analysis

Protein sequence clustering and phylogenetic analysis of *XTH* gene family members from *P. euphratica*, *P. pruinosa*, *P. deltoides*, *S. sinopurpurea*, and *A. thaliana* were performed using MEGA11 software (version 11.0.13). The phylogenetic tree was edited and beautified using the iTOL online tool (https://itol.embl.de/, accessed on 19 January 2025).

4.6. Transcriptome Sequencing and Data Analysis of PeXTHs

Transcriptome sequencing data were derived from four-month-old *P.euphratica* seedlings treated with 25% PEG 6000 and their control conditions. The expression profiles of *PeXTHs* gene family members were extracted from the quantitative expression data of each treatment [62]. *P. euphratica* seedlings were treated with a 25% polyethylene glycol 6000 solution, and leaf and root samples were collected at 0, 4, and 12 h post-treatment. The samples were immediately frozen in liquid nitrogen for storage and sequenced using the Ion Proton sequencing platform at the Shenzhen Huada Gene Research Institute. Gene expression levels were calculated using the FPKM (read per million mapped reads per thousand base pairs) method. The RNA-seq data have been uploaded to the NCBI database (https://www.ncbi.nlm.nih.gov/bioproject/PRJNA580347/, accessed on 11 September 2024). Transcriptome data visualization analysis was performed using TBtools software (version 2.119), and the data were subjected to numerical standardization and log transformation. The log transformation parameters were set as follows: base = 2.0; offset = 1.0.

P. euphratica seedlings were grown under a 16-h photoperiod at 25 ± 1 °C and 75% relative humidity, with bi-weekly irrigation. After three months, seedlings were treated with 25% PEG6000 for 0, 4, and 12 h (n = 3 per group). Post-treatment, roots and leaves were sampled, flash-frozen in liquid nitrogen, and stored at −80 °C. Total RNA was extracted using a SPARKscript II RT Plus Kit (SparkJade, Jinan, China), and cDNA was synthesized from 1 μg RNA. RT-qPCR was carried out in 10 μL reactions with SYBR Green mix and gene-specific primers (Table S1). Amplification was achieved using the following cycle settings: 95 °C for 30 s, then 45 cycles of 95 °C for 10 s, 55 °C for 10 s, and 72 °C for 30 s for plate reading. RT-qPCR yielded products between 80 bp and 150 bp in size. After normalization with *PeActin*, relative expression was assessed. The $2^{-\Delta\Delta Ct}$ method was applied to data evaluation.

4.7. Subcellular Localization of PeXTH7

The constructed 35S::*PeXTH7*-YFP vector was transformed into *Agrobacterium tumefaciens* strain GV3101. Strains carrying the target plasmid (CBL-mCherry) were cultured overnight in LB medium supplemented with appropriate antibiotics. Healthy leaves from 4-week-old tobacco plants were selected for infiltration. Agrobacterium cultures containing *PeXTH7* and the helper strain 19K (used to enhance fluorescence signals) were grown overnight in LB liquid medium. When the OD_{600} of the PeXTH7-containing Agrobacterium culture reached 1.0–1.2, cells were harvested by centrifugation at 10,000 rpm for 1–2 min.

The supernatant was discarded and the bacterial cells were collected. An infiltration buffer containing MES, $MgCl_2 \cdot 6H_2O$, and acetosyringone was prepared and adjusted to pH 6.0 using KOH. This buffer was mixed with the *Agrobacterium* suspension and incubated at 28 °C for 1 h. The bacterial suspension was then infiltrated into the abaxial surface of tobacco leaves, with injection sites clearly marked. Following 24 h of dark treatment, the plants were transferred to a controlled-environment growth chamber for 2–3 days. Subcellular localization was analyzed using a laser scanning confocal microscope (Nikon TS100, Tokyo, Japan). For YFP signal detection, an excitation wavelength of 514 nm was used with emission collection between 524–574 nm. After inducing plasmolysis with 30% sucrose solution, fluorescence distribution was re-evaluated under the same optical settings. For visualization of the endoplasmic reticulum marker, excitation at 587 nm was employed with emission detection in the 597–650 nm range.

4.8. Effect of PeXTH7 Overexpression on Root Length in A. thaliana

The wild-type A. thaliana seeds used in this study was the Columbia-0 (Col-0) ecotype. The *PeXTH7* gene was heterologously expressed in Arabidopsis using the pgreenII 0179 (35S NOS)-YFP overexpression vector in the GV3101 Agrobacterium tumefaciens strain, with *Escherichia coli*—TOP10 used for plasmid propagation. The detailed transgenic methodology followed a previously described protocol [63]. Seeds of the generated lines—Col-0 and three independent *PeXTH7*-overexpressing lines (OE-1, OE-2, OE-5)—were surface-sterilized and sown on 1/2 MS medium. For each line, 20 seeds were plated per biological replicate, with five replicates established per treatment. A 0 mmol/L sorbitol condition served as the control. After stratification at 4 °C for 2–3 days, the plates were transferred to a growth chamber and positioned vertically. Following 4 days of growth under standard conditions, healthy and uniformly sized seedlings were selected and carefully transferred to fresh 1/2 MS media containing either 0, 100, 200, or 300 mmol/L sorbitol. The seedlings were grown for an additional 5 days, after which representative healthy seedlings from each group were photographed. Primary root length was subsequently measured and analyzed from the images using the software ImageJ (version 1.44p).

5. Conclusions

In this study, a total of 33 *PeXTHs* genes were identified in *P. euphratica*, and their gene structures, phylogenetic relationships, collinearity, and cis-acting elements were systematically analyzed. The results indicated that *PeXTHs* gene family members are involved in drought resistance by regulating root growth and development. RNA sequencing and RT-qPCR analyses further confirmed the critical role of *PeXTH7* in the drought stress response. Heterologous overexpression of *PeXTH7* in wild-type Arabidopsis significantly promoted root elongation and enhanced drought tolerance. These findings suggest that *PeXTH7* may improve drought adaptability in *P. euphratica* by facilitating root system development. Therefore, this study not only provides a theoretical basis for understanding the function of *PeXTHs* genes in root growth, but also provides a reference for the subsequent exploration of the molecular response of *P. euphratica* to abiotic stress.

Supplementary Materials: The following supporting information can be downloaded at https://www.mdpi.com/article/10.3390/plants14243847/s1, Table S1: Primer Sequences for RT-qPCR.

Author Contributions: Conceptualization, P.J. and Z.W.; methodology, J.L., H.J. and T.S.; software, J.L., H.J., T.S., D.M., Q.N. and J.S.; formal analysis, J.L.; writing—original draft preparation, J.L., H.J.; writing—review and editing, J.L., P.J., Z.W., J.S. and Z.L. All authors have read and agreed to the published version of the manuscript.

Funding: This research was funded by National Natural Science Foundation of China, grant numbers 32160355 and 32371838, "Tianshan Talents" Program—Top Telent in Youth Science and Technology (2024TSYCCX0114), "Tianshan Talents" Program—Science and Technology Innovation Team (2023TSYCTD0019), "Xinjiang Tianchi Talents Doctoral Project" (BT-2025-TCYC-0011), Natural Science Support Program of XPCC (2024DA034), and Guidance Program for Science & Technology Projects of XPCC (2024ZD087).

Data Availability Statement: No large datasets were created in this study.

Acknowledgments: We thank Zhongshuai Gai and Xiao Zhang for giving valuable advice during manuscript preparation.

Conflicts of Interest: The authors declare no conflicts of interest.

References

1. Fang, Y.; Xiong, L. General mechanisms of drought response and their application in drought resistance improvement in plants. *Cell. Mol. Life Sci.* **2015**, *72*, 673–689.
2. Muhammad Aslam, M.; Waseem, M. Mechanisms of Abscisic Acid-Mediated Drought Stress Responses in Plants. *Int. J. Mol. Sci.* **2022**, *23*, 1084–1105. [CrossRef]
3. Taylor-Teeples, M.; Lin, L.; de Lucas, M.; Turco, G.; Toal, T.W.; Gaudinier, A.; Young, N.F.; Trabucco, G.M.; Veling, M.T.; Lamothe, R. An *Arabidopsis* gene regulatory network for secondary cell wall synthesis. *Nature* **2015**, *517*, 571–575. [CrossRef]
4. Zhao, C.; Zayed, O.; Yu, Z.; Jiang, W.; Zhu, P.; Hsu, C.C.; Zhang, L.; Tao, W.A.; Lozano-Durán, R.; Zhu, J.K. Leucine-rich repeat extensin proteins regulate plant salt tolerance in *Arabidopsis*. *Proc. Natl. Acad. Sci. USA* **2018**, *115*, 13123–13128. [CrossRef]

5. Sivan, P.; Urbancsok, J.; Donev, E.N.; Derba-Maceluch, M.; Barbut, F.R. Modification of xylan in secondary walls alters cell wall biosynthesis and wood formation programs and improves saccharification. *Plant Biotechnol. J.* **2025**, *23*, 174–197. [CrossRef] [PubMed]
6. Feng, W.; Lindner, H. Growing Out of Stress: The Role of Cell- and Organ-Scale Growth Control in Plant Water-Stress Responses. *Plant Cell Physiol.* **2016**, *28*, 1769–1782. [CrossRef] [PubMed]
7. De Caroli, M.; Manno, E.; Piro, G. Ride to cell wall: *Arabidopsis* XTH11, XTH29 and XTH33 exhibit different secretion pathways and responses to heat and drought stress. *Plant J. Cell Mol. Biol.* **2021**, *107*, 448–466. [CrossRef]
8. Nishitani, K.; Tominaga, R. Endo-xyloglucan transferase, a novel class of glycosyltransferase that catalyzes transfer of a segment of xyloglucan molecule to another xyloglucan molecule. *J. Biol. Chem.* **1992**, *267*, 21058–21064. [CrossRef]
9. Zhai, J.; Li, Z.; Si, J.; Zhang, S.; Han, X.; Chen, X. Structural and Functional Responses of the Heteromorphic Leaves of Different Tree Heights on *Populus euphratica* Oliv. to Different Soil Moisture Conditions. *Plants* **2022**, *11*, 2376–2392. [CrossRef] [PubMed]
10. Ge, X.L.; Zhang, L.; Du, J.J.; Wen, S.S.; Qu, G.Z.; Hu, J.J. Transcriptome Analysis of *Populus euphratica* under Salt Treatment and PeERF1 Gene Enhances Salt Tolerance in Transgenic *Populus alba* × *Populus glandulosa*. *Int. J. Mol. Sci.* **2022**, *23*, 3727–3747. [CrossRef]
11. Korver, R.A.; Koevoets, I.T.; Testerink, C. Out of Shape During Stress: A Key Role for Auxin. *Trends Plant Sci.* **2018**, *23*, 783–793. [CrossRef] [PubMed]
12. Cao, D.; Li, J.; Huang, Z.; Baskin, C.C.; Baskin, J.M.; Hao, P.; Zhou, W.; Li, J. Reproductive characteristics of a *Populus euphratica* population and prospects for its restoration in China. *PLoS ONE* **2012**, *7*, 39121–39128.
13. Zhang, Z.; Chen, Y.; Zhang, J.; Ma, X.; Li, Y.; Li, M.; Wang, D.; Kang, M.; Wu, H.; Yang, Y.; et al. Improved genome assembly provides new insights into genome evolution in a desert poplar (*Populus euphratica*). *Plant Physiol.* **2020**, *20*, 781–794.
14. Lin, Z.; Guo, Y.; Zhang, R.; Li, Y.; Wu, Y.; Sheen, J. ABA-activated low-nanomolar Ca^{2+}-CPK signalling controls root cap cycle plasticity and stress adaptation. *Nat. Plants* **2024**, *11*, 90–104. [CrossRef]
15. Yokoyama, R.; Nishitani, K. A comprehensive expression analysis of all members of a gene family encoding cell-wall enzymes allowed us to predict cis-regulatory regions involved in cell-wall construction in specific organs of *Arabidopsis*. *Plant Cell Physiol.* **2001**, *42*, 1025–1033.
16. Yokoyama, R.; Rose, J.K.; Nishitani, K. A Surprising Diversity and Abundance of Xyloglucan Endotransglucosylase/Hydrolases in Rice. Classification and Expression Analysis. *Plant Physiol.* **2004**, *134*, 1088–1099. [CrossRef]
17. Yang, L.; Chen, Y.; Liu, X.; Zhang, S.; Han, Q. Genome-wide identification and expression analysis of xyloglucan endotransglucosylase/hydrolase genes family in *Salicaceae* during grafting. *BMC Genom.* **2023**, *24*, 676–691. [CrossRef]
18. Liu, Y.; Liu, D.; Zhang, H.; Gao, H.; Guo, X.; Wang, D.; Zhang, X.; Zhang, A. The alpha- and beta-expansin and xyloglucan endotransglucosylase/hydrolase gene families of wheat: Molecular cloning, gene expression, and EST data mining. *Genomics* **2007**, *90*, 516–529. [CrossRef]
19. Schultink, A.; Liu, L.; Zhu, L.; Pauly, M. Structural Diversity and Function of Xyloglucan Sidechain Substituents. *Plants* **2014**, *3*, 526–542. [CrossRef] [PubMed]
20. Yang, Y.; Miao, Y.; Zhong, S.; Fang, Q.; Wang, Y.; Dong, B.; Zhao, H. Genome-Wide Identification and Expression Analysis of XTH Gene Family during Flower-Opening Stages in *Osmanthus fragrans*. *Plants* **2022**, *11*, 1015–1032. [CrossRef] [PubMed]
21. Li, Q.; Li, H.; Yin, C.; Wang, X.; Jiang, Q.; Zhang, R.; Ge, F.; Chen, Y.; Yang, L. Genome-Wide Identification and Characterization of Xyloglucan Endotransglycosylase/Hydrolase in *Ananas comosus* during Development. *Genes* **2019**, *10*, 537–549. [CrossRef]
22. Yang, Z.; Zhang, R.; Zhou, Z. The XTH Gene Family in *Schima superba*: Genome-Wide Identification, Expression Profiles, and Functional Interaction Network Analysis. *Front. Plant Sci.* **2022**, *13*, 911761–911776. [CrossRef]
23. Saladié, M.; Rose, J.K.; Cosgrove, D.J.; Catalá, C. Characterization of a new xyloglucan endotransglucosylase/hydrolase (XTH) from ripening tomato fruit and implications for the diverse modes of enzymic action. *Plant J. Cell Mol. Biol.* **2006**, *47*, 282–295. [CrossRef]
24. Choi, J.Y.; Seo, Y.S.; Kim, S.J.; Kim, W.T.; Shin, J.S. Constitutive expression of CaXTH3, a hot pepper xyloglucan endotransglucosylase/hydrolase, enhanced tolerance to salt and drought stresses without phenotypic defects in tomato plants (*Solanum lycopersicum* cv. Dotaerang). *Plant Cell Rep.* **2011**, *30*, 867–877. [CrossRef]
25. Chen, J.; Wan, H.; Zhao, H.; Dai, X.; Wu, W.; Liu, J.; Xu, J.; Yang, R.; Xu, B.; Zeng, C.; et al. Identification and expression analysis of the Xyloglucan transglycosylase/hydrolase (XTH) gene family under abiotic stress in oilseed (*Brassica napus* L.). *BMC Plant Biol.* **2024**, *24*, 400–414. [CrossRef]
26. Zhang, S.; Wu, Z. Chromosome-scale assemblies of the male and female *Populus euphratica* genomes reveal the molecular basis of sex determination and sexual dimorphism. *Commun. Biol.* **2022**, *5*, 1186–1192. [CrossRef]
27. Sun, J.; Xu, J.; Qiu, C.; Zhai, J.; Zhang, S.; Zhang, X.; Wu, Z.; Li, Z. The chromosome-scale genome and population genomics reveal the adaptative evolution of *Populus pruinosa* to desertification environment. *Hortic. Res.* **2024**, *11*, 34–48. [CrossRef]
28. Wu, Z.; Jiang, Z. Multi-omics analysis reveals spatiotemporal regulation and function of heteromorphic leaves in Populus. *Plant Physiol.* **2023**, *192*, 188–204. [CrossRef]

29. Zhang, B.; Gao, Y.; Zhang, L.; Zhou, Y. The plant cell wall: Biosynthesis, construction, and functions. *J. Integr. Plant Biol.* **2021**, *63*, 251–272. [CrossRef] [PubMed]
30. Cosgrove, D.J. Structure and growth of plant cell walls. *Proc. Natl. Acad. Sci. USA* **2024**, *25*, 340–358. [CrossRef] [PubMed]
31. Hoson, T. Apoplast as the site of response to environmental signals. *J. Plant Res.* **1998**, *111*, 167–177. [CrossRef]
32. Barbut, F.R. Plant-Insect Interactions. *Plants* **2022**, *11*, 1140–1144.
33. Zhong, R.; Ye, Z.H. Secondary cell walls: Biosynthesis, patterned deposition and transcriptional regulation. *Plant Cell Physiol.* **2015**, *56*, 195–214. [CrossRef]
34. Jamet, E.; Canut, H.; Boudart, G.; Pont-Lezica, R.F. Cell wall proteins: A new insight through proteomics. *Trends Plant Sci.* **2006**, *11*, 33–39. [CrossRef] [PubMed]
35. Calderan-Rodrigues, M.J.; Guimarães Fonseca, J.; de Moraes, F.E.; Vaz Setem, L.; Carmanhanis Begossi, A.; Labate, C.A. Plant Cell Wall Proteomics: A Focus on Monocot Species, *Brachypodium distachyon*, *Saccharum* spp. and *Oryza sativa*. *Int. J. Mol. Sci.* **2019**, *20*, 1975–1995. [CrossRef]
36. Franková, L.; Fry, S.C. Biochemistry and physiological roles of enzymes that 'cut and paste' plant cell-wall polysaccharides. *J. Exp. Bot.* **2013**, *64*, 3519–3550. [CrossRef] [PubMed]
37. Hijazi, M.; Roujol, D.; Nguyen-Kim, H.; Del Rocio Cisneros Castillo, L.; Saland, E.; Jamet, E.; Albenne, C. Arabinogalactan protein 31 (AGP31), a putative network-forming protein in *Arabidopsis thaliana* cell walls. *Ann. Bot.* **2014**, *114*, 1087–1097. [CrossRef] [PubMed]
38. Tan, L.; Eberhard, S.; Pattathil, S.; Warder, C.; Glushka, J.; Yuan, C.; Hao, Z.; Zhu, X.; Avci, U.; Miller, J.S.; et al. An *Arabidopsis* cell wall proteoglycan consists of pectin and arabinoxylan covalently linked to an arabinogalactan protein. *Plant Cell* **2013**, *25*, 270–287. [CrossRef]
39. Cannon, M.C.; Terneus, K.; Hall, Q.; Tan, L.; Wang, Y.; Wegenhart, B.L.; Chen, L.; Lamport, D.T.; Chen, Y.; Kieliszewski, M.J. Self-assembly of the plant cell wall requires an extensin scaffold. *Proc. Natl. Acad. Sci. USA* **2008**, *105*, 2226–2231. [CrossRef]
40. Cosgrove, D.J. Growth of the plant cell wall. *Nat. Rev. Mol. Cell Biol.* **2005**, *6*, 850–861. [CrossRef]
41. Shinohara, N.; Nishitani, K. Cryogenian Origin and Subsequent Diversification of the Plant Cell-Wall Enzyme XTH Family. *Plant Cell Physiol.* **2021**, *62*, 1874–1889. [CrossRef]
42. Gao, Y.; Wang, L.; Li, D.; Qi, D.; Fang, F.; Luo, Y.; Zhang, H.; Zhang, S. Genome-wide characterization of the xyloglucan endotransglucosylase/hydrolase family genes and their response to plant hormone in sugar beet. *Plant Physiol. Biochem.* **2024**, *206*, 108239–108246. [CrossRef]
43. Sasidharan, R.; Keuskamp, D.H.; Kooke, R.; Voesenek, L.A.; Pierik, R. Interactions between auxin, microtubules and XTHs mediate green shade- induced petiole elongation in *Arabidopsis*. *PLoS ONE* **2014**, *9*, e90587. [CrossRef]
44. Han, Y.; Han, S.; Ban, Q.; He, Y.; Jin, M.; Rao, J. Overexpression of persimmon DkXTH1 enhanced tolerance to abiotic stress and delayed fruit softening in transgenic plants. *Plant Cell Rep.* **2017**, *36*, 583–596. [CrossRef]
45. Li, Q.; Hu, A.; Dou, W.; Qi, J.; Long, Q.; Zou, X.; Lei, T.; Yao, L.; He, Y.; Chen, S. Systematic Analysis and Functional Validation of Citrus XTH Genes Reveal the Role of Csxth04 in Citrus Bacterial Canker Resistance and Tolerance. *Front. Plant Sci.* **2019**, *10*, 1109–1115. [CrossRef]
46. Qiao, T.; Zhang, L.; Yu, Y.; Pang, Y.; Tang, X.; Wang, X.; Li, L.; Li, B.; Sun, Q. Identification and expression analysis of xyloglucan endotransglucosylase/hydrolase (XTH) family in grapevine (*Vitis vinifera* L.). *PeerJ* **2022**, *10*, e13546. [CrossRef] [PubMed]
47. Geisler-Lee, J.; Geisler, M.; Coutinho, P.M.; Segerman, B.; Nishikubo, N.; Takahashi, J.; Aspeborg, H.; Djerbi, S.; Master, E.; Andersson-Gunnerås, S.; et al. Poplar carbohydrate-active enzymes. Gene identification and expression analyses. *Plant Physiol.* **2006**, *140*, 946–962. [CrossRef] [PubMed]
48. Thompson, J.E.; Fry, S.C. Restructuring of wall-bound xyloglucan by transglycosylation in living plant cells. *Plant J. Cell Mol. Biol.* **2001**, *26*, 23–34. [CrossRef]
49. Sauret-Güeto, S.; Calder, G.; Harberd, N.P. Transient gibberellin application promotes *Arabidopsis thaliana* hypocotyl cell elongation without maintaining transverse orientation of microtubules on the outer tangential wall of epidermal cells. *Plant J. Cell Mol. Biol.* **2012**, *69*, 628–639. [CrossRef] [PubMed]
50. Genovesi, V.; Fornalé, S.; Fry, S.C.; Ruel, K.; Ferrer, P.; Encina, A.; Sonbol, F.M.; Bosch, J.; Puigdomènech, P.; Rigau, J.; et al. ZmXTH1, a new xyloglucan endotransglucosylase/hydrolase in maize, affects cell wall structure and composition in *Arabidopsis thaliana*. *J. Exp. Bot.* **2008**, *59*, 875–889. [CrossRef]
51. Liu, Y.B.; Lu, S.M.; Zhang, J.F.; Liu, S.; Lu, Y.T. A xyloglucan endotransglucosylase/hydrolase involves in growth of primary root and alters the deposition of cellulose in *Arabidopsis*. *Planta* **2007**, *226*, 1547–1560. [CrossRef] [PubMed]
52. Singh, A.P.; Tripathi, S.K.; Nath, P.; Sane, A.P. Petal abscission in rose is associated with the differential expression of two ethylene-responsive xyloglucan endotransglucosylase/hydrolase genes, RbXTH1 and RbXTH2. *J. Exp. Bot.* **2011**, *62*, 5091–5103. [CrossRef]
53. Wittkopp, P.J.; Kalay, G. Cis-regulatory elements: Molecular mechanisms and evolutionary processes underlying divergence. *Nat. Rev. Genet.* **2011**, *13*, 59–69. [CrossRef]

54. Thessen, A.E.; Cooper, L.; Swetnam, T.L.; Hegde, H.; Reese, J.; Elser, J.; Jaiswal, P. Using knowledge graphs to infer gene expression in plants. *Front. Artif. Intell.* **2023**, *6*, 1201002. [CrossRef]
55. Zhai, Z.; Feng, C. Genome-Wide Identification of the Xyloglucan endotransglucosylase/Hydrolase (XTH) and Polygalacturonase (PG) Genes and Characterization of Their Role in Fruit Softening of Sweet Cherry. *Int. J. Mol. Sci.* **2021**, *22*, 12331–12342. [CrossRef]
56. Zhu, J.K. Abiotic Stress Signaling and Responses in Plants. *Cell* **2016**, *167*, 313–324. [CrossRef]
57. Zhai, X.; Wang, X.; Yang, X.; Huang, Q.; Wu, D.; Wang, Y.; Kang, H.; Sha, L.; Fan, X.; Zhou, Y.; et al. Genome-wide identification of bHLH transcription factors and expression analysis under drought stress in *Pseudoroegneria libanotica* at germination. *Physiol. Mol. Biol. Plants Int. J. Funct. Plant Biol.* **2024**, *30*, 467–481. [CrossRef]
58. Yokoyama, R.; Nishitani, K. Endoxyloglucan transferase is localized both in the cell plate and in the secretory pathway destined for the apoplast in tobacco cells. *Plant Cell Physiol.* **2001**, *42*, 292–300. [CrossRef] [PubMed]
59. Han, Y.; Wang, W.; Sun, J.; Ding, M.; Zhao, R.; Deng, S.; Wang, F.; Hu, Y.; Wang, Y.; Lu, Y.; et al. *Populus euphratica* XTH overexpression enhances salinity tolerance by the development of leaf succulence in transgenic tobacco plants. *J. Exp. Bot.* **2013**, *64*, 4225–4238. [CrossRef]
60. Cai, H.; Xu, Y.; Yan, K.; Zhang, S.; Yang, G.; Wu, C.; Zheng, C. BREVIPEDICELLUS Positively Regulates Salt-Stress Tolerance in *Arabidopsis thaliana*. *Int. J. Mol. Sci.* **2023**, *24*, 1054–1070. [CrossRef]
61. Zhou, R.; Zhou, R.; Macaya-Sanz, D.; Carlson, C.H.; Schmutz, J. A willow sex chromosome reveals convergent evolution of complex palindromic repeats. *Nat. Plants* **2020**, *21*, 38–57. [CrossRef] [PubMed]
62. Jiao, P.; Wu, Z.; Wang, X.; Jiang, Z.; Wang, Y.; Liu, H.; Qin, R.; Li, Z. Short-term transcriptomic responses of *Populus euphratica* roots and leaves to drought stress. *J. For. Res.* **2021**, *32*, 841–853. [CrossRef]
63. Qu, Y.; Bi, C.; He, B.; Ye, N.; Yin, T.; Xu, L.A. Genome-wide identification and characterization of the MADS-box gene family in Salix suchowensis. *PeerJ* **2019**, *7*, e8019. [CrossRef] [PubMed]

plants

MDPI

Article

Integrated Evaluation of Alkaline Tolerance in Soybean: Linking Germplasm Screening with Physiological, Biochemical, and Molecular Responses

Yongguo Xue [1], Zichun Wei [2], Chengbo Zhang [2], Yudan Wang [2], Dan Cao [1], Xiaofei Tang [1], Yubo Yao [1], Wenjin He [1], Chao Chen [2], Zaib_un Nisa [3,*] and Xinlei Liu [1,*]

[1] Soybean Research Institute, Heilongjiang Academy of Agricultural Sciences, Harbin 150086, China; xyg81@126.com (Y.X.); caodan825@163.com (D.C.); yaoyubo2009@aliyun.com (Y.Y.); nkydds@126.com (W.H.)

[2] Department of Chemistry and Molecular Biology, School of Life Science and Technology, Harbin Normal University, Harbin 150025, China; wzcfyywzy@163.com (Z.W.); zcblss@126.com (C.Z.); wangyudan0662@163.com (Y.W.); chchao@hrbnu.edu.cn (C.C.)

[3] General Botany Laboratory, Institute of Molecular Biology and Biotechnology, The University of Lahore, Defence Road Campus, Lahore 54000, Pakistan

* Correspondence: zaib.nisa@imbb.uol.edu.pk (Z.N.); nkyddslxl@163.com (X.L.); Tel.: +86-0451-86609753 (X.L.)

Abstract

Soybean (*Glycine max* L.) is an essential food and economic crop in China, yet its growth and yield are severely constrained by saline–alkali stress. A saline–alkali soil exacerbates root absorption barriers, leading to 30–50% yield losses. Understanding the mechanisms underlying alkali tolerance is therefore crucial for developing stress-resilient soybean varieties and improving the productivity of saline–alkali land. In our previous study, we evaluated 99 soybean germplasms from Northeast China and obtained the alkali-tolerant varieties HN48 and HN69, along with the alkali-sensitive varieties HNWD4 and HN83. In this study, fifteen-day-old soybean seedlings were subjected to (30 mM $NaHCO_3$) alkali stress for 72 h, and whole plants were sampled to assess their morphology and physiology, while leaf tissues were harvested for biochemical analysis. For transcriptomic analysis, soybean seedlings were exposed to alkali stress (50 mM $NaHCO_3$, pH 9.0) for 6 h, and leaf and root tissues were harvested for RNA sequencing. The results showed that alkali-tolerant varieties mitigated these effects by suppressing excessive ROS generation by 55–63%, decreasing malondialdehyde (MDA) accumulation by 37–39%, and increasing photosynthetic efficiency by 18.3%, as well as accumulating more osmoprotectants and activating antioxidant enzymes such as superoxide dismutase (SOD) and catalase (CAT) under alkaline stress. Transcriptome analysis showed that the alkali-tolerant variety HN69 exhibited cultivar-specific enrichment of metabolism cytochrome P450, estrogen signaling, and GnRH signaling pathways under alkali stress. These results collectively indicate that alkali-tolerant soybean varieties adapt to alkali stress through coordinated multi-pathway responses, with differential pathway enrichment potentially underlying the variation in alkali tolerance between cultivars. Overall, this study elucidates the physiological and molecular mechanisms of alkali tolerance in soybean, providing a theoretical foundation for breeding stress-tolerant germplasms.

Keywords: soybean; alkali tolerance; transcriptome; physiological mechanism

1. Introduction

Soil salinization and alkalization represent major abiotic stress factors that severely restrict global crop productivity and pose an escalating threat to food security. According to the second national soil survey, China possesses approximately 37 million hectares of usable saline–alkali soil resources, with the western Songliao Plain being identified as a major saline–alkali hotspot encompassing about 16.47 million hectares [1]. These soils are characterized by strong alkalinity and a high Na^+ concentration, frequently causing sharp reductions in soybean yield of 30–50%. Soybean (*Glycine max* L.), an essential source of food, vegetable oil, and animal feed, remains in short domestic supply. In 2021, China's soybean production reached only 16.4 million tons, while annual consumption exceeded 100 million tons, revealing a significant production–consumption gap [2]. Addressing this imbalance requires not only agronomic improvements but also genetic innovations that enhance crops' resilience to environmental stresses. Therefore, it is imperative to identify and utilize alkali-tolerant genetic resources within indigenous soybean germplasms and elucidate the underlying molecular and physiological mechanisms of alkali stress adaptation [3]. Such insights will provide a scientific foundation for breeding stress-resilient varieties, expanding soybean cultivation beyond its current geographic limits, and strengthening national self-sufficiency in soybean production [4].

Alkali stress triggers multi-dimensional physiological responses in plants. Morphologically, it inhibits cell division in the root apical meristem [5], restricts primary root elongation (shortening by 20–40%), and significantly reduces lateral root initiation frequency (>30%), accompanied by leaf chlorosis and plant dwarfing [6]. Damage to subcellular organelles is another defining symptom [7], characterized by thylakoid swelling within chloroplasts and the disintegration of mitochondrial cri. These structural alterations severely hinder photophosphorylation and compromise overall energy metabolism. At the physiological level, metabolic imbalances are evident through a sharp rise in reactive oxygen species (ROS), which induces elevated levels of malondialdehyde (MDA), a key indicator of lipid peroxidation, and results in disrupted membrane stability, reflected by a rise in electrolyte leakage [8,9]. As a result, photosynthetic organ function declines, evidenced by accelerated chlorophyll degradation, reduced maximum quantum efficiency of photosystem II (PSII, Fv/Fm), and a significant drop in net CO_2 assimilation rate [10,11].

To cope with alkali stress, soybean has evolved coordinated multi-level defense strategies [12]. First, the antioxidant system is activated, significantly upregulating the activity of key enzymes like superoxide dismutase (SOD) and peroxidase (POD) to effectively scavenge excess ROS [13,14]. Second, osmotic regulators (e.g., proline, soluble sugars) are synthesized abundantly, with accumulation levels increasing 2-3-fold, maintaining cellular hydration status through osmotic potential adjustment. Third, ion homeostasis is primarily maintained via the Salt Overly Sensitive (SOS) signaling pathway. This pathway drives cytoplasmic Na^+ efflux and vacuolar sequestration, thereby ensuring dynamic K^+/Na^+ balance [15]. Notably, alkali-tolerant genotypes typically maintain a higher K^+/Na^+ ratio than sensitive ones [16]. Nevertheless, the identification of alkali-tolerant soybean varieties and the elucidation of their physiological and molecular mechanisms of alkali tolerance are still pressing issues in soybean production.

Considerable genetic variation in alkali tolerance exists among soybean germplasms, as previous screenings of over 140 accessions have identified several tolerant lines. However, most studies focus narrowly on the germination stage and lack standardized evaluation systems. To address these gaps, in our previous study, we evaluated 99 soybean germplasms from Northeast China and obtained the alkali-tolerant varieties HN48 and HN69, along with the alkali-sensitive varieties HNWD4 and HN83. We hypothesized that alkali-tolerant soybean varieties (HN48 and HN69) and alkali-sensitive varieties (HNWD4

https://doi.org/10.3390/plants15020222

and HN83) exhibited differences in physiological, biochemical, and antioxidant enzyme activity profiles, as well as in their regulatory gene expression under alkali stress. Therefore, we elucidated the mechanisms of alkali-tolerant and alkali-sensitive varieties in response to alkaline stress by integrating phenotypic, biochemical, and molecular analyses.

2. Results

2.1. Assessment of Morphological Alterations Under Alkali Stress

To evaluate the morphological responses of soybean seedlings (HN48, HN69, HN83, and HNWD4) to alkaline stress, *Glycine max* accessions were cultivated hydroponically in Hoagland's nutrient solution. The treatment group was exposed to 30 mM $NaHCO_3$. The results showed that alkaline stress markedly inhibited seedling growth compared to the control, as evidenced by pronounced wilting symptoms and a substantial reduction in overall biomass accumulation (Figure 1A,B). Across all tested accessions, exposure to alkaline conditions significantly decreased all key morphological parameters, including plant height, root length, fresh weight, and dry weight (Figure 1E–H). In particular, HNWD4 exhibited a 55.6% reduction in root length and a 66.7% decline in fresh weight, whereas HN83 showed a 44.4% reduction in fresh biomass (Figure 1E,H). In contrast, the more tolerant accessions maintained better morphological stability under stress. HN48 displayed no significant difference in plant height or dry weight relative to the control, while HN69 sustained comparable values for both height and root elongation (Figure 1G,H). Overall, the results demonstrate clear accession-specific variations in seedling morphogenesis under alkaline stress, with a strong positive correlation between morphological performance and genotypic tolerance capacity.

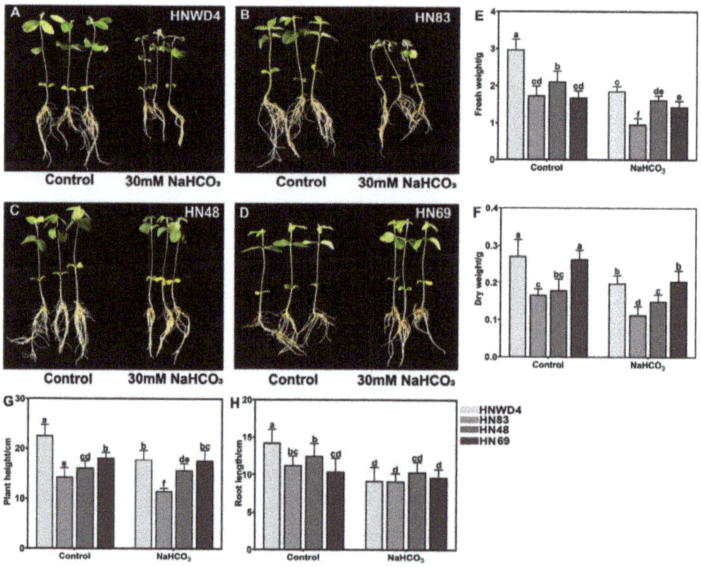

Figure 1. Phenotypic responses of soybean seedlings to alkaline stress. (**A–D**) Phenotypic analysis of soybean seedlings (HN48, HN69, HN83, and HNWD4) under alkaline stress. Seedlings were grown hydroponically in Hoagland's nutrient solution, with the control group or the 30 mM $NaHCO_3$ treatment group. Quantitative analysis of key growth parameters, including plant fresh weight (**E**), dry weight (**F**), plant height (**G**), and root length (**H**), in soybean accessions subjected to 30 mM $NaHCO_3$, compared with control conditions. Different lowercase letters above the bars indicate significant differences. Significant differences between control and treatment groups were determined using one-way ANOVA, followed by Duncan's multiple range test ($p < 0.05$).

https://doi.org/10.3390/plants15020222

2.2. Measurement of Physiological Attributes in Soybean Seedlings Under Alkali Stress

Chlorophyll content analysis provides critical insights into how alkaline stress influences photosynthetic performance in soybean seedlings. As shown in Figure 2A, exposure to alkaline conditions led to a notable decline in chlorophyll content across different soybean accessions, with the extent of reduction varying by genotype. The alkali-sensitive varieties, HNWD4 and HN83, exhibited significant decreases of 38.8% and 16.6%, respectively, compared with the control. In contrast, the alkali-tolerant varieties, HN69 and HN48, experienced only modest reductions of 11.4% and 7.14%, respectively, suggesting a stronger capacity to preserve chlorophyll and sustain photosynthetic activity under stress. Chlorophyll fluorescence measurements further elucidated the physiological effects of alkaline stress on the photosynthetic apparatus. Assessment of the maximum photochemical quantum yield of PSII (Fv/Fm) revealed a marked decline in alkali-sensitive accessions, particularly in HNWD4 and HN83 (Figure 2B). In contrast, the tolerant variety HN69 exhibited only a minor reduction in Fv/Fm, maintaining near-normal photosynthetic efficiency. Overall, these findings indicate that alkali-tolerant soybean genotypes mitigate stress-induced damage by maintaining higher chlorophyll content, thereby enhancing their photosynthetic resilience under alkaline conditions.

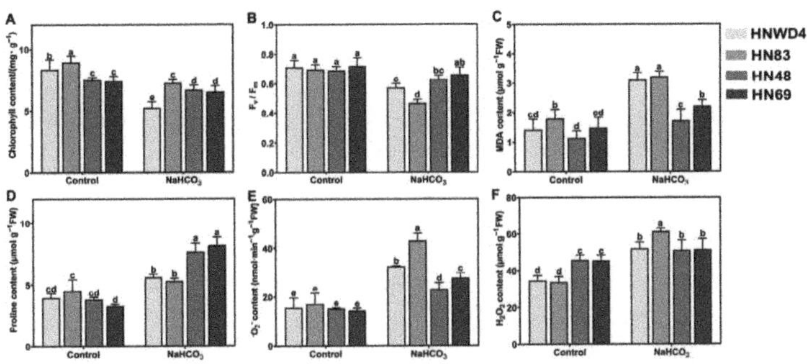

Figure 2. Determination of physiological indices in soybean seedlings under alkaline stress. Panels represent (**A**) chlorophyll content, (**B**) Fv/Fm, (**C**) MDA content, (**D**) proline content, (**E**) superoxide anion (O_2^-) production, and (**F**) hydrogen peroxide (H_2O_2) accumulation in different soybean varieties exposed to 30 mM $NaHCO_3$ compared with controls. Data are expressed as mean ± standard error. Different lowercase letters above the bars indicate significant differences ($p < 0.05$), while identical letters denote no statistically significant differences.

2.3. Measurement of Stress Indicators (MDA, H_2O_2, O_2^-, and Proline Contents)

To elucidate the physiological responses of soybean seedlings to alkaline stress, changes in malondialdehyde (MDA) and hydrogen peroxide (H_2O_2) levels were quantified in leaf tissues. As shown in Figure 2C, the MDA content increased across all tested varieties under alkaline conditions, reflecting enhanced lipid peroxidation. The alkali-sensitive cultivars, HNWD4 and HN83, exhibited pronounced increases of 1.21-fold and 80%, respectively, whereas the alkali-tolerant cultivars, HN48 and HN 69, showed smaller increments of 38.9% and 37.5%, suggesting reduced oxidative damage and better membrane stability in the tolerant genotypes. As presented in Figure 2D, proline accumulation in tolerant varieties (HN48: 150.1%; HN69: 108.3%) was higher than that in sensitive varieties (40.7% and 10.6%). We hypothesized that the elevated proline content may play a role in alleviating alkali stress. Figure 2E illustrates that superoxide anion (O_2^-) production increased by 115.2–146.5% in the sensitive cultivars, compared to a more moderate rise of 52.1–90.6% in the tolerant cultivars. Similarly, Figure 2F shows that H_2O_2 levels rose sharply in sensitive varieties (50.3–81.2%), whereas tolerant varieties maintained rel-

https://doi.org/10.3390/plants15020222

atively stable levels, with only slight increases (12.3–13.5%). Collectively, these results demonstrated that alkali-tolerant soybean varieties possess stronger antioxidant defense mechanisms, characterized by lower accumulation of MDA and reactive oxygen species (ROS). This physiological resilience effectively minimized the membrane lipid peroxidation and contributed to improved stress tolerance under alkaline conditions.

2.4. Determination of Antioxidant Enzyme Activities of Soybean Seedlings (HN48, HN69, HN83, and HNWD4) Under Alkaline Stress

Alkaline stress triggers the activation of the plant antioxidant defense system, in which changes in the activities of key enzymes, POD, CAT, SOD, and APX serve as important physiological markers of stress adaptation. To examine the differential antioxidant responses among soybean genotypes with varying levels of alkali tolerance, the activities of these enzymes were measured in the leaves and roots of four cultivars: HNWD4, HN83, HN48, and HN69. As shown in Figure 3A–H, alkali stress markedly influenced the antioxidant enzyme activities in soybean leaves. The tolerant cultivar HN48 exhibited a significant increase in APX activity (89.2% above the control), while the sensitive cultivar HNWD4 displayed only slight, non-significant increases in CAT and SOD activities (8.1% and 8.69%, respectively). In the roots, the tolerant cultivar HN69 demonstrated substantial enhancements in SOD and APX activities (75% and 50%, respectively), which were significantly higher than those observed in the sensitive cultivars. Overall, these findings suggest that alkali-tolerant soybean varieties alleviate oxidative damage by antioxidant enzymes, particularly POD and APX, in both leaves and roots. This enhanced enzymatic defense likely represents a key physiological mechanism underlying their superior tolerance to alkaline stress.

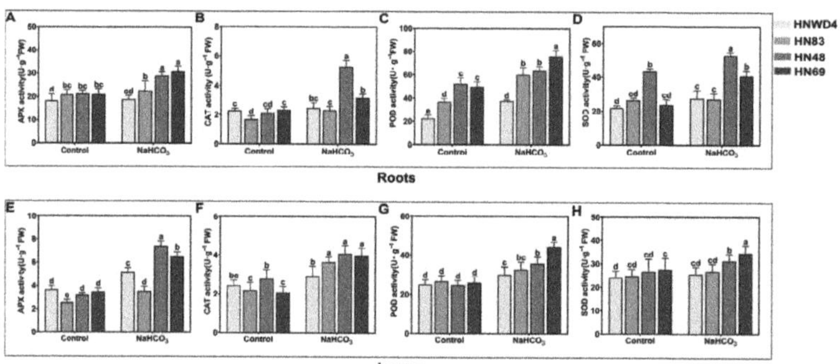

Figure 3. Changes in antioxidant enzyme activities in soybean leaves and roots under alkali stress; the enzyme activities of (**A**) APX, (**B**) SOD, (**C**) POD, and (**D**) CAT were detected in roots. The enzyme activities of (**E**) APX, (**F**) SOD, (**G**) POD, and (**H**) CAT were detected in leaves. Different lowercase letters above the bars indicate significant differences ($p < 0.05$), while the same letters denote no statistically significant differences.

2.5. Differential Transcriptional Analysis of Soybean Seedlings (HN69 and HN83) Under Alkaline Stress

To further explore the difference between alkali-tolerant and alkali-sensitive soybean, we detected the transcriptome data of the alkali-tolerant HN69 and the alkali-sensitive HN83 under alkaline stress. Table S1 provides a summary of transcriptome sequencing data statistics for soybean cultivars under control and alkaline stress conditions. The table presents raw and clean read numbers, quality metrics (Q20 and Q30), clean-to-raw read ratios, and genome alignment rates for alkali-tolerant (HN69) and alkali-sensitive (HN83) soybean cultivars. Each treatment included three biological replicates. High-quality

 https://doi.org/10.3390/plants15020222

sequencing and alignment rates (>90%) confirm the reliability and suitability of the datasets for subsequent transcriptomic analyses.

Differentially expressed genes (DEGs) were identified using a threshold of $|\log_2$ (fold change)$| > 1$. We found that 4134 DEGs were differentially expressed jointly between Alkaline-HN69 vs. Control-HN69 (3247 upregulated and 887 downregulated), 3324 genes were specifically expressed in the Alkaline-HN83 vs. Control-HN83 (3247 upregulated and 887 downregulated) group, and 1353 genes were specifically expressed in the Alkaline-HN69 vs. Alkaline-HN83 group (495 upregulated and 859 downregulated). Heatmap analysis clustered DEGs across groups (Figure 4). Samples or genes with shorter branch distances showed more similar expression patterns. The key findings were clear separation between HN69 and HN83 under both control and alkali stress conditions, indicating alkali-induced transcriptional reprogramming.

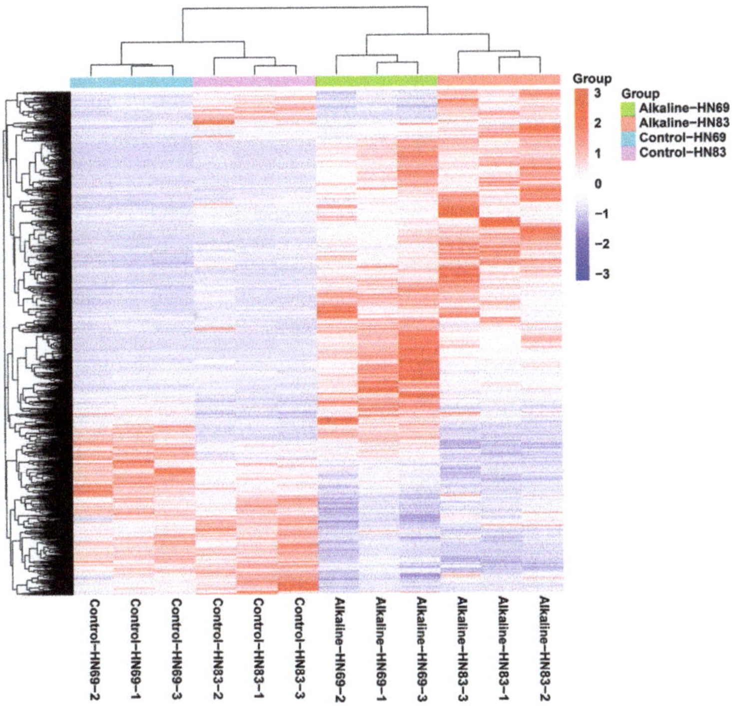

Figure 4. Cluster plot of the differentially expressed genes.

2.6. Functional Annotation and Enrichment Analysis of Differentially Expressed Genes

Hypergeometric testing using R's phyper function (FDR-adjusted $Q \leq 0.05$) identified statistically significant enrichment profiles. Upregulated DEGs in HN 69 were significantly enriched in biosynthesis, metabolism, and organic substance biosynthesis terms (Figure S1), while HN83 showed major enrichment in biosynthesis, metabolism, and biological processes (Figure S2), both indicating critical functional roles in alkali stress response.

KEGG pathway enrichment analysis ($Q \leq 0.05$) further revealed that secondary metabolite biosynthesis and plant hormone signal transduction pathways contained the highest DEG numbers in both cultivars (Figures S3 and S4). Notably, HN 69 uniquely activated the metabolism-cytochrome P450, estrogen signaling, and GnRH signaling pathways. These results collectively indicate that soybean adapts to alkali stress through coordinated

https://doi.org/10.3390/plants15020222

multi-pathway responses, with differential pathway enrichment potentially underlying the variation in alkali tolerance between cultivars.

2.7. Gene Expression Analysis Using qRT-PCR (Quantitative Real-Time PCR)

To validate the reliability of transcriptome data, eight upregulated peroxidase (POD)-associated genes with high fold changes were selected for qRT-PCR verification. Whole-plant samples of HN69 and the alkali-sensitive HN83 were collected after 6 h of alkali treatment and control conditions, followed by RNA extraction and reverse transcription. As shown in Figure 5, the expression patterns of all eight genes were consistent with the transcriptome results. Notably, the peroxidase genes *Glyma.03G038600* and *Glyma.09G277800* exhibited no significant changes between stress and control conditions in HN83 but showed significant upregulation in HN69, suggesting enhanced antioxidant defense capacity in the tolerant cultivar. An integrated schematic representation of morphological, physiological, biochemical, and transcriptomic responses underlying alkali stress tolerance in soybean is presented in Figure 6.

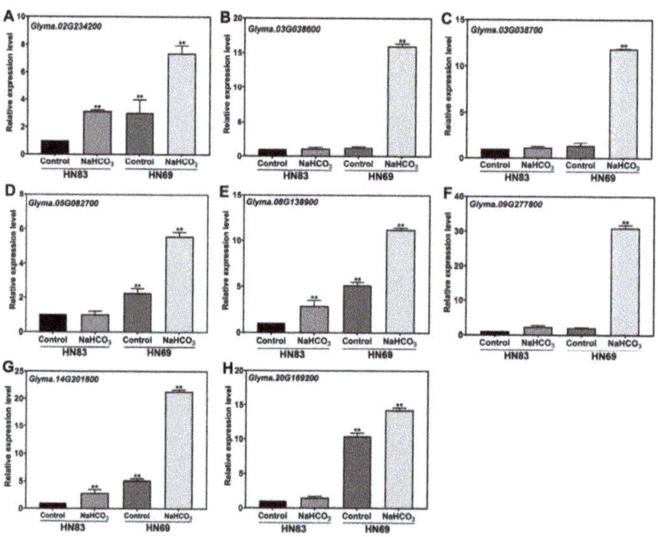

Figure 5. Identification of upregulated differentially expressed genes by qRT-PCR. (**A–H**) Whole-plant samples of HN69 and alkali-sensitive HN83 were collected after 6 h alkali treatment and control conditions. *GADPH* was used as an internal control. Data are the means ± SD. ** $p < 0.01$.

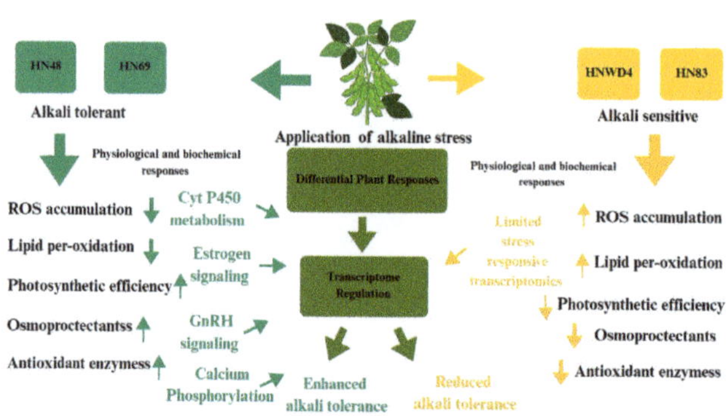

Figure 6. Integrated schematic representation responses underlying alkali stress tolerance in alkali-tolerant and alkali-sensitive varieties in soybean.

https://doi.org/10.3390/plants15020222

These results confirm high concordance between qRT-PCR and transcriptome data. The significantly upregulated expression of *Glyma.03G038600* in alkali-stressed HN69 suggests its potential as a candidate gene for further investigation of alkali tolerance mechanisms.

3. Discussion

Soil alkalization represents a major abiotic stress factor that severely restricts global crop productivity and poses an escalating threat to food security. Soybean is an essential source of food, vegetable oil, and animal feed. Soil alkalization frequently causes sharp reductions in soybean yield. Therefore, the identification of alkali-tolerant soybean varieties and the elucidation of their physiological and molecular mechanisms of alkali tolerance are still pressing issues in soybean production. In our previous study, we evaluated 99 soybean germplasms from Northeast China and obtained the alkali-tolerant varieties HN48 and HN69, along with the alkali-sensitive varieties HNWD4 and HN83. By integrating multi-stage phenotypic, biochemical, and molecular analyses, we aim to uncover the regulatory mechanisms of alkali tolerance and provide a foundation for breeding resilient soybean varieties for saline–alkali soils.

In the present study, the results showed that the alkali-tolerant varieties HN48 and HN69 displayed better growth morphological parameters, including plant height, root length, fresh weight, and dry weight following exposure to alkaline conditions (Figure 1E–H). Physiological regulation constitutes the core mechanism underlying plants' adaptation to environmental changes. The metabolism of chlorophyll is highly susceptible to abiotic stress, such as salt, drought, and alkali stresses [17]. We also found that the alkali-sensitive varieties HNWD4 and HN83 exhibited significant chlorophyll decreases compared with the alkali-tolerant varieties HN48 and HN69 (Figure 2A). As a key indicator of chloroplast function, the chlorophyll content also directly determines the photosynthetic efficiency of plant leaves under both normal and stress conditions [18]. We found that the tolerant variety exhibited only a minor reduction in Fv/Fm, maintaining near-normal photosynthetic efficiency (Figure 2B). These findings indicate that alkali-tolerant soybean genotypes mitigate stress-induced damage by maintaining higher chlorophyll content and preserving PSII functionality, thereby enhancing their photosynthetic resilience under alkaline conditions.

Under stress conditions, the plant cells undergo oxidative damage and lipid peroxidation, leading to the accumulation of MDA [19]. In the present study, we showed that the contents of MDA significant were significantly increased in the alkali-sensitive varieties HNWD4 and HN83 (Figure 2C), indicating increased oxidative damage, superoxide anion (O_2^-) production, and H_2O_2 levels (Figure 2E,F). Therefore, alkali-tolerant varieties mitigate injury through membrane integrity maintenance, osmolyte accumulation, and antioxidant system activation (SOD/CAT/POD) (Figure 3) [20,21].

Transcriptomic analysis revealed divergent gene expression patterns between the alkali-tolerant 'HN 69' and the sensitive HN83 under stress. qRT-PCR validation confirmed substantial upregulation of antioxidant enzyme-related genes in HN69. In alkali-tolerant soybean cultivar HN69, GO/KEGG enrichment analysis revealed activation of three pathways: phenylpropanoid biosynthesis eliminates ROS via phenolic compounds [22], MAPK signaling cascade and phytohormone transduction pathway. Key transcription factors were specifically induced, including 22 *ERF* family members (e.g., *GmERF71* enhancing tolerance through *H^+-ATPase* regulation), 27 NAC members, 31 bHLH members, and 55 MYB members (e.g., GmMYB68 conferring dual salt–alkali resistance). KEGG analysis further identified specific activation of cytochrome P450-mediated drug metabolism and GnRH/estrogen signaling pathways. The enrichment of the estrogen and GnRH signaling pathways in soybeans under alkali stress should be understood in the context of KEGG

https://doi.org/10.3390/plants15020222

annotation principles rather than as proof of animal-specific hormonal signaling, even if these pathways are traditionally described in animal systems. Calcium-dependent kinases, MAPK cascades, phosphatidylinositol signaling elements, and cytochrome P450 enzymes are among the conserved molecular components shared by eukaryotes that form the basis of KEGG pathway mapping. Many of these components are essential for plants' stress perception, signal transduction, and metabolic regulation [23]. Genes associated with the estrogen signaling pathway in plants frequently overlap with redox homeostasis, cytochrome P450-mediated detoxification mechanisms, and secondary metabolite biosynthesis—all of which are known to support abiotic stress tolerance [24]. Therefore, the particular enrichment of these pathways in the alkali-tolerant cultivar HN69 probably indicates the activation of metabolic networks and conserved stress-responsive signaling, which may contribute to cultivar-dependent variations in alkali stress adaption. Similarly, despite variations in upstream ligands, the GnRH signaling pathway contains conserved Ca^{2+}-dependent signaling and phosphorylation modules that are functionally relevant to plant hormone crosstalk and stress-responsive transcriptional regulation. These molecular components form a multi-tiered tolerance strategy wherein phenylpropanoid metabolism enhances ROS clearance [25]. MAPK/hormone networks amplify stress signaling, and P450 enzymes potentiate detoxification. This integrated regulatory framework establishes the molecular basis for alkali-tolerant phenotypes [26]. This study employs phenotypically screened alkali-tolerant and sensitive soybean germplasms to elucidate physiological adaptation mechanisms. Integrated transcriptome analysis identifies pivotal tolerance genes [27], establishing a theoretical foundation for molecular breeding of alkali-resistant soybean cultivars [28].

4. Conclusions

Taken together, we obtained the alkali-tolerant varieties HN48 and HN69, along with the alkali-sensitive varieties HNWD4 and HN83. By integrating multi-stage phenotypic, biochemical, and molecular analyses, we aim to uncover the regulatory mechanisms of alkali tolerance and provide a foundation for breeding resilient soybean varieties for saline–alkali soils. However, the functions of key regulatory genes identified via transcriptome analysis still require further investigation. Further investigations aim to identify candidate alkali-tolerant genes from alkali-tolerant soybean varieties. Secondly, metabolomic profiling of alkali-tolerant and alkali-sensitive soybean varieties under alkali stress could help reveal differences in their molecular tolerance mechanisms.

5. Materials and Methods

5.1. Plant Materials

The soybean accessions used in this study were obtained from the Soybean Research Institute, Heilongjiang Academy of Agricultural Sciences. Among these, the alkaline-tolerant cultivars Heinong48 (HN48) and Heinong69 (HN69), along with the alkaline-sensitive cultivars HeinongWangdou4 (HNWD4) and Heinong83 (HN83), were selected to evaluate morphological responses under alkaline stress and to conduct physiological assessments at the seedling stage. For transcriptome analysis, seedlings of HN 69 and HN 83 were specifically chosen to represent tolerant and sensitive genotypes, respectively.

5.2. Morphological Assessment of Soybean Under Alkaline Stress

Soybean seeds were surface-sterilized, soaked in distilled water, and germinated in Petri dishes for four days. Uniform seedlings were subsequently transferred to Hoagland's nutrient solution, placed on floating supports, and cultivated under controlled environmental conditions (25/20 °C day/night temperature and a 16/8 h light/dark photoperiod).

https://doi.org/10.3390/plants15020222

After fifteen days of germination and growth, the seedlings were divided into two treatment groups: (1) a control group (CK), maintained in standard Hoagland's solution, and (2) an alkali stress group, exposed to 30 mM $NaHCO_3$. The nutrient solutions were replaced every three days to ensure stable nutrient availability and consistent experimental conditions. Seventy-two hours after treatment, twenty seedlings from each group were randomly selected for morphological evaluation. Root length and shoot height were measured immediately, and the plants were subsequently oven-dried at 65 °C for 12 h to determine dry biomass. For biochemical analyses, approximately 0.1 g of fresh leaf tissue was rapidly frozen in liquid nitrogen and stored at −80 °C until further use.

5.3. Physiological Assessment of Soybean Under Alkaline Stress

Chlorophyll content was quantified spectrophotometrically following extraction with acetone by using the method described in [29], while the maximum photochemical efficiency of photosystem II (Fv/Fm) was measured using an Imaging-PAM chlorophyll fluorometer (Walz, Nürnberg, Germany).

5.4. Measurement of MDA, H_2O_2, O_2^-, and Proline Contents

Leaves from both alkali-stressed and control seedlings were collected for biochemical assays following established protocols to assess oxidative stress and osmolyte accumulation. Lipid peroxidation rate was quantified by using the thiobarbituric acid (TBA) reaction method, as described by Heath and Packer [30]. For this purpose, 0.5 g of leaf tissue was homogenized in 5 mL of 10% (w/v) trichloroacetic acid (TCA) and centrifuged at 11,000 $r \cdot min^{-1}$ for 15 min. The resulting supernatant was mixed with 0.5% (w/v) TBA prepared in 10% TCA, and the mixture was heated in boiling water for 30 min. After rapid cooling, absorbance was recorded at 532 nm and corrected for nonspecific turbidity at 600 nm. MDA content was expressed as nmol g^{-1} fresh weight. Free proline content was estimated following the ninhydrin–acetic acid method of Bates [31]. According to this method, fresh leaf samples (0.5 g) were homogenized in 3% sulfosalicylic acid and centrifuged at 3000 $r \cdot min^{-1}$ for 10 min. The supernatant was reacted with acid ninhydrin and glacial acetic acid in equal volumes and incubated at 70 °C for 55 min. After cooling, the reaction mixture was extracted with toluene, and absorbance was measured at 520 nm. Proline concentration was calculated using a standard curve of L-proline. H_2O_2 concentration was determined according to the method of Velikova [32]. Leaf tissue (0.5 g) was homogenized in 0.1% TCA and centrifuged at 12,000 $r \cdot min^{-1}$ for 22 min at 4 °C. The supernatant was mixed with 10 mM potassium phosphate buffer (pH 7.0) and 1 M potassium iodide (KI). The absorbance was recorded at 390 nm, and H_2O_2 concentration was calculated using a standard curve prepared with known concentrations of H_2O_2. Furthermore, O^{2-} generation rate was measured following the hydroxylamine oxidation method described by Elstner and Heupel [33]. In this method, fresh samples were incubated in phosphate buffer containing hydroxylamine hydrochloride at room temperature. After extraction with diethyl ether, sulfanilic acid and α-naphthylamine were added sequentially, and the mixture was left to react at 25 °C for 20 min. The absorbance was measured at 530 nm to estimate O^{2-} levels. All biochemical measurements were performed with twenty seedlings from each group per treatment to ensure data accuracy and reproducibility.

5.5. Determination of Antioxidant Enzyme Activities

Leaves and roots (0.1 g) from both alkali-stressed and control soybean seedlings were collected to assess antioxidant enzyme activities following standard protocols. Peroxidase (POD) activity was determined using the method of Chance and Maehly [34]. Tissues were homogenized in 50 mM phosphate buffer (pH 7.0) under chilled conditions and centrifuged. The reaction mixture contained guaiacol and H_2O_2 in phosphate buffer, and the

https://doi.org/10.3390/plants15020222

increase in absorbance due to tetraguaiacol formation was recorded at 470 nm. Superoxide dismutase (SOD) activity was measured according to Giannopolitis and Ries [35], based on the inhibition of nitroblue tetrazolium (NBT) photoreduction. Extracts prepared in 50 mM phosphate buffer (pH 7.8) were centrifuged, and the reaction mixture contained methionine, NBT, riboflavin, EDTA, and phosphate buffer. Absorbance was recorded at 560 nm after illumination. One unit of SOD activity corresponded to 50% inhibition of NBT reduction. Catalase (CAT) activity was determined following Aebi [36]. Samples were homogenized in phosphate buffer (pH 7.0) and centrifuged at 4 °C. The decomposition of H_2O_2 was monitored by a decline in absorbance at 240 nm, and CAT activity was expressed as the change in absorbance per minute per gram fresh weight. Ascorbate peroxidase (APX) activity was assayed using the method of Nakano and Asada [37]. Tissues were homogenized in phosphate buffer (pH 7.0) containing 1 mM EDTA and 1% polyvinylpyrrolidone (PVP), followed by centrifugation at 4 °C. The reaction mixture contained phosphate buffer, ascorbic acid, and H_2O_2, and the decrease in absorbance at 290 nm due to ascorbate oxidation was recorded.

5.6. Stress Treatments, Cultivation of Soybean, and Total RNA Extraction

Soybean seedlings were cultivated under controlled conditions following the procedure of Xu [38], with slight modifications. Uniform seedlings were exposed to either alkali stress (50 mM $NaHCO_3$, pH 9.0) or control treatment (distilled water) for 6 h. After treatment, leaf and root samples were collected, immediately frozen in liquid nitrogen, and stored at −80 °C until further analysis. Total RNA was isolated using the Plant Total RNA Extraction Kit (Omega Bio-Tek, Bei Jing, China, R6827-01) according to the manufacturer's instructions. RNA purity and concentration were assessed using a NanoDrop 2000 spectrophotometer (Thermo Fisher Scientific, Shang Hai, China) with A260/A280 ratios > 1.8, and integrity was confirmed by 1.5% agarose gel electrophoresis. High-quality RNA samples were sent to AB Life, Inc. (Wuhan, China) for library construction using the NEBNext® Ultra™ RNA Library Prep Kit (New England Biolabs, Ipswich, MA, USA). For transcriptome validation, total RNA was extracted from three biological replicates per treatment using the same protocol.

5.7. Analysis of RNA Sequencing Data

After passing quality control, high-quality RNA samples were enriched and reverse-transcribed to synthesize cDNA for library construction following standard Illumina protocols. The resulting cDNA libraries were subjected to high-throughput sequencing, and the clean reads were aligned to the *Glycine max* reference genome obtained from the Phytozome database (https://phytozome-next.jgi.doe.gov/ (accessed on 14 January 2024)) using default alignment parameters to ensure accurate mapping of genomic regions. Gene expression levels were quantified as read counts and normalized for differential expression analysis using DESeq2 [39]. Genes with a false discovery rate (FDR) < 0.05 and an absolute $\log_2$ fold change ($|\log_2 FC|$) > 1 were considered significantly differentially expressed (upregulated if $\log_2 FC > 1$, downregulated if $\log_2 FC < -1$). Identified DEGs were subjected to Gene Ontology (GO) enrichment analysis using the CARMO web-based platform [40], which categorizes DEGs into molecular function, cellular component, and biological process terms. Enrichment significance was assessed using Fisher's exact test against the background genome. Functional and pathway enrichment analyses were further performed using the Kyoto Encyclopedia of Genes and Genomes (KEGG) database to identify biological pathways and metabolic processes associated with DEGs.

https://doi.org/10.3390/plants15020222

5.8. Differential Gene Expression Analysis

In the differential gene expression analysis, genes with |log2(fold change)| > 1 were identified as DEGs. The DEGs screened from the alkali-tolerant soybean cultivar HN69 and alkali-sensitive cultivar HN83 under both control and 50 mM $NaHCO_3$ conditions were visualized. Complementary DNA (cDNA) synthesis was performed using the PrimeScript™ RT Reagent Kit (Takara, Bei Jing, China, RR047A), and quantitative real-time PCR was carried out on a QuantStudio 5 Real-Time PCR System (Applied Biosystems, Shang Hai, China) with SYBR® Premix Ex Taq™ (Takara). Relative gene expression levels were calculated using the $2^{-\Delta\Delta Ct}$ method of Livak and Schmittgen [41], with *GADPH* (*Glyma.06G172700*) as the internal reference gene.

5.9. Statistical Analysis

For morphological evaluation and biochemical analyses, the data were analyzed using one-way ANOVA, followed by Duncan's multiple range test ($p < 0.05$). For qRT-PCR data analysis, the data were analyzed with at least three independent biological replicates. qRT-PCR results were calculated using the $2^{-\Delta\Delta CT}$ method and Student's *t*-test. All statistical analyses were conducted using SPSS 21.0 software. Graphical data were visualized using GraphPad Prism 8.0.1.

Supplementary Materials: The following supporting information can be downloaded at https://www.mdpi.com/article/10.3390/plants15020222/s1: Figure S1: GO enrichment bubble plot of differentially expressed genes in HN69. Figure S2: GO enrichment bubble plot of differentially expressed genes in HN83. Figure S3: Bubble map of KEGG enrichment of differentially expressed genes in HN69. Figure S4: Bubble map of KEGG enrichment of differentially expressed genes in HN83. Table S1: Summary of sequencing data. Table S2: Gene-specific primers used in this study.

Author Contributions: Conceptualization, Y.X., Z.N., and X.L.; methodology, Z.N. and X.L.; software, C.Z.; validation, Y.W. and Z.W.; formal analysis, Y.W.; investigation, Y.W.; resources, C.C., Z.N. and Y.X.; data curation, D.C.; writing—original draft preparation, Z.W., Z.N., and X.T.; writing—review and editing, Z.N., X.L., Y.Y., and C.C.; visualization, W.H.; supervision, Y.X. and Z.N.; project administration, Y.X. and X.L.; funding acquisition, X.L. All authors have read and agreed to the published version of the manuscript.

Funding: This research was funded by the Agricultural Science and Technology Innovation Leaping Project in Heilongjiang Province (Grant No. CX23ZD04); The Breeding of New Soybean Varieties with Breakthrough in High Oil Content and High Yield (Grant No. 2025JLGJ0009NC); and the Major Agricultural Science and Technology Project of the Ministry of Agriculture and Rural Affairs.

Data Availability Statement: Data are contained within the article are available on request from the corresponding author. The data are not publicly available due to privacy and ethical restrictions.

Acknowledgments: The authors wish to thank Nuo Xu for format editing assistance.

Conflicts of Interest: The authors declare no conflicts of interest.

Abbreviations

The following abbreviations are used in this manuscript:

HN48	Heinong 48
HN69	Heinong 69
HNWD4	Heinong Wangdou 4
HN83	Heinong 83
SOD	Superoxide dismutase
POD	Peroxidase

https://doi.org/10.3390/plants15020222

CAT	Catalase
MDA	Malondialdehyde
ROS	Reactive oxygen species
qRT-PCR	Quantitative real-time PCR
O_2^-	Superoxide anion

References

1. Wang, Z.; Li, Z.; Wang, Z.; Liu, T.; Zhang, P.; Li, S.; Ye, S.; Yang, K.; Gai, Z.; Liu, L. Alkaline stress suppresses soybean waterlogging tolerance by exacerbating energy expenditure and ROS accumulation. *Plant Physiol. Biochem.* **2025**, *229*, 110381. [CrossRef]
2. Xu, X.; Guo, L.; Wang, S.; Ren, M.; Zhao, P.; Huang, Z.; Jia, H.; Wang, J.; Lin, A. Comprehensive evaluation of the risk system for heavy metals in the rehabilitated saline-alkali land. *J. Environ. Manage.* **2023**, *347*, 119117. [CrossRef]
3. Hameed, A.; Ahmed, M.Z.; Hussain, T.; Aziz, I.; Ahmad, N.; Gul, B.; Nielsen, B.L. Effects of Salinity Stress on Chloroplast Structure and Function. *Cells* **2021**, *10*, 2023. [CrossRef]
4. Dai, H.; Fang, Y.; Chen, C.; Liu, K.; Li, X.; Yang, Z.; Fang, N.; Zhang, Y. Geochemical evaluation and driving factor analysis of soil salinization in Northeast China Plain. *Front. Environ. Sci.* **2025**, *13*, 1614178. [CrossRef]
5. Liu, Y.; Mu, C.; Du, D.; Yang, Y.; Li, L.; Xuan, W.; Kircher, S.; Palme, K.; Li, X.; Li, R. Alkaline stress reduces root waving by regulating PIN7 vacuolar transport. *Front. Plant Sci.* **2022**, *13*, 1049144. [CrossRef]
6. Wang, G.; Shen, W.; Zhang, Z.; Guo, S.; Hu, J.; Feng, R.; Zhao, Q.; Du, J.; Du, Y. The Effect of Neutral Salt and Alkaline Stress with the Same Na+ Concentration on Root Growth of Soybean (*Glycine max* (L.) Merr.) Seedlings. *Agronomy* **2022**, *12*, 2708. [CrossRef]
7. Li, J.; Xu, H.H.; Liu, W.C.; Zhang, X.W.; Lu, Y.T. Ethylene Inhibits Root Elongation during Alkaline Stress through AUXIN1 and Associated Changes in Auxin Accumulation. *Plant Physiol.* **2015**, *168*, 1777–1791. [CrossRef]
8. Yang, S.; Xu, Y.; Tang, Z.; Jin, S.; Yang, S. The Impact of Alkaline Stress on Plant Growth and Its Alkaline Resistance Mechanisms. *Int. J. Mol. Sci.* **2024**, *25*, 13719. [CrossRef]
9. Duan, G.; Liang, C.; Su, J.; Liang, Y.; Li, W.; Zhang, J.; Zhang, Y. Integrative transcriptomic and metabolomic analysis elucidates the vital pathways underlying the differences in salt stress responses between two chickpea (*Cicer arietinum* L.) varieties. *BMC Plant Biol.* **2025**, *25*, 903. [CrossRef]
10. Nie, W.; He, Q.; Ma, J.; Guo, H.; Shi, Q. Exogenous 2,4-Epibrassinolide Alleviates Alkaline Stress in Cucumber by Modulating Photosynthetic Performance. *Plants* **2024**, *14*, 54. [CrossRef]
11. Xing, C.J.; Wang, S.W.; Zhang, C.F.; Guo, T.; Hao, H.L.; Zhang, Z.Y.; Wang, S.; Shu, J.M. Effects of leaf scorch on chlorophyll fluorescence characteristics of walnut leaves. *J. Plant Dis. Prot.* **2022**, *130*, 115–124. [CrossRef]
12. Zhang, Y.; Xu, J.; Li, R.; Ge, Y.; Li, Y.; Li, R. Plants' Response to Abiotic Stress: Mechanisms and Strategies. *Int. J. Mol. Sci.* **2023**, *24*, 10915. [CrossRef]
13. Zhu, S.; Shi, F.; Li, H.; Ding, Y.; Chang, W.; Ping, Y.; Song, F. Piriformospora indica alleviates soda saline-alkaline stress in *Glycine max* by modulating plant metabolism. *Front. Plant Sci.* **2024**, *15*, 1406542. [CrossRef]
14. Hameed, A.; Khan, M.A.; Tahir, M.H.N.; Lodhi, M.S.; Muzammil, S.; Shafiq, M.; Gechev, T.; Faisal, M. In Silico identification and characterization of SOS gene family in soybean: Potential of calcium in salinity stress mitigation. *PLoS ONE* **2025**, *20*, e0317612. [CrossRef]
15. Cai, X.; Jia, B.; Sun, M.; Sun, X. Insights into the regulation of wild soybean tolerance to salt-alkaline stress. *Front. Plant Sci.* **2022**, *13*, 1002302. [CrossRef]
16. Wu, J.; Fang, Y.; Xu, L.; Jin, X.; Iqbal, A.; Nisa, Z.U.; Ali, N.; Chen, C.; Shah, A.A.; Gatasheh, M.K. The Glycine soja cytochrome P450 gene *GsCYP82C4* confers alkaline tolerance by promoting reactive oxygen species scavenging. *Physiol. Plant* **2024**, *176*, e14252. [CrossRef]
17. Li, X.; Zhang, W.; Niu, D.; Liu, X. Effects of abiotic stress on chlorophyll metabolism. *Plant Sci.* **2024**, *342*, 112030. [CrossRef]
18. Yu, L.; Luo, X.; Croft, H.; Rogers, C.A.; Chen, J.M. Seasonal variation in the relationship between leaf chlorophyll content and photosynthetic capacity. *Plant Cell Environ.* **2024**, *47*, 3953–3965. [CrossRef]
19. Hnilickova, H.; Kraus, K.; Vachova, P.; Hnilicka, F. Salinity Stress Affects Photosynthesis, Malondialdehyde Formation, and Proline Content in *Portulaca oleracea* L. *Plants* **2021**, *10*, 845. [CrossRef]
20. Molassiotis, A.; Job, D.; Ziogas, V.; Tanou, G. Citrus Plants: A Model System for Unlocking the Secrets of NO and ROS-Inspired Priming Against Salinity and Drought. *Front. Plant Sci.* **2016**, *7*, 229. [CrossRef]
21. El-Esawi, M.A.; Alaraidh, I.A.; Alsahli, A.A.; Alamri, S.A.; Ali, H.M.; Alayafi, A.A. Bacillus firmus (SW5) augments salt tolerance in soybean (*Glycine max* L.) by modulating root system architecture, antioxidant defense systems and stress-responsive genes expression. *Plant Physiol. Biochem.* **2018**, *132*, 375–384. [CrossRef]
22. Inthaisong, S.; Boonchuen, P.; Tharapreuksapong, A.; Tittabutr, P.; Teaumroong, N.; Tantasawat, P.A. Transcriptome Profiling Reveals Mungbean Defense Mechanisms Against Powdery Mildew. *Agronomy* **2025**, *15*, 1871. [CrossRef]

 https://doi.org/10.3390/plants15020222

23. Kanehisa, M.; Sato, Y.; Kawashima, M. KEGG mapping tools for uncovering hidden features in biological data. *Protein Sci.* **2022**, *31*, 47–53. [CrossRef]
24. Jun, X.U.; Xin-Yu, W.; Wang-Zhen, G. The cytochrome P450 superfamily: Key players in plant development and defense. *J. Integr. Agric.* **2015**, *14*, 1673–1686. [CrossRef]
25. Chen, Q.; Yu, L.; Zhang, W.; Cheng, S.; Cong, X.; Xu, F. Molecular and physiological response of chives (*Allium schoenoprasum*) under different concentrations of selenium application by transcriptomic, metabolomic, and physiological approaches. *Plant Physiol. Biochem.* **2025**, *221*, 109633. [CrossRef]
26. Zhang, J.; Huang, A.; Du, H. Transcriptome analysis reveals the effects of hydrogen gas on NaCl stress in purslane roots. *Sci. Hortic.* **2023**, *319*, 112142. [CrossRef]
27. Wu, J.S.; Mu, D.W.; Feng, N.J.; Zheng, D.F.; Sun, Z.Y.; Khan, A.; Zhou, H.; Song, Y.W.; Liu, J.X.; Luo, J.Q. Integrated Analyses Reveal the Physiological and Molecular Mechanisms of Brassinolide in Modulating Salt Tolerance in Rice. *Plants* **2025**, *14*, 1555. [CrossRef]
28. Li-Xin, Z.; Wei, L.; Tsegaw, M.; Xin, X.U.; Yan-Ping, Q.I.; Sapey, E.; Lu-Ping, L.; Ting-Ting, W.U.; Shi, S.; Tian-Fu, H. Principles and practices of the photo-thermal adaptability improvement in soybean. *J. Integr. Agric.* **2020**, *19*, 295–310. [CrossRef]
29. Arnon, D.I. Copper Enzymes in Isolated Chloroplasts. Polyphenoloxidase in *Beta vulgaris*. *Plant Physiol.* **1949**, *24*, 1–15. [CrossRef]
30. Heath, R.L.; Packer, L. Reprint of: Photoperoxidation in Isolated Chloroplasts I. Kinetics and Stoichiometry of Fatty Acid Peroxidation. *Arch. Biochem. Biophys.* **2022**, *726*, 109248. [CrossRef]
31. Bates, L.S.; Waldren, R.P.; Teare, I.D. Rapid Determination of Free Proline for Water-Stress Studies. *Plant Soil* **1973**, *39*, 205–207. [CrossRef]
32. Velikova, V.; Yordanov, I.; Edreva, A. Oxidative stress and some antioxidant systems in acid rain-treated bean plants. Protective role of exogenous polyamines. *Plant Sci. An. Int. J. Exp. Plant Biol.* **2000**, *151*, 59–66. [CrossRef]
33. Elstner, E.F.; Heupel, A. Formation of hydrogen peroxide by isolated cell walls from horseradish (*Armoracia lapathifolia Gilib*). *Planta* **1976**, *130*, 175–180. [CrossRef]
34. Maehly, A.C. *The Assay of Catalases and Peroxidases*; John Wiley & Sons, Ltd.: Hoboken, NJ, USA, 1954. [CrossRef]
35. Giannopolitis, C.N.; Ries, S.K. Superoxide dismutases: I. Occurrence in higher plants. *Plant Physiol.* **1977**, *59*, 309–314. [CrossRef]
36. Tükenmez, H.; Magnussen, H.M.; Rogne, P.; Byström, A.; Wolf-Watz, M. Bridging in Vitro with in Vivo Enzymology. *Biophys. J.* **2016**, *110*, 223a. [CrossRef]
37. Li, C.; Liu, Y.; Liu, X.; Mai, K.K.K.; Li, J.; Guo, X.; Zhang, C.; Li, H.; Kang, B.H.; Hwang, I.; et al. Chloroplast thylakoid ascorbate peroxidase PtotAPX plays a key role in chloroplast development by decreasing hydrogen peroxide in *Populus tomentosa*. *J. Exp. Bot.* **2021**, *72*, 4333–4354. [CrossRef]
38. Chen, S.A.; Hou, J.; Yao, N.; Xie, C.; Li, D. Comparative transcriptome analysis of *Triplophysa yarkandensis* in response to salinity and alkalinity stress. *Comp. Biochem. Physiol. Part. D Genom. Proteom.* **2020**, *33*, 100629. [CrossRef]
39. Love, M.I.; Huber, W.; Anders, S. Moderated estimation of fold change and dispersion for RNA-seq data with DESeq2. *Genome Biol.* **2014**, *15*, 550. [CrossRef]
40. Wang, J.; Qi, M.; Liu, J.; Zhang, Y. CARMO: A comprehensive annotation platform for functional exploration of rice multi-omics data. *Plant J.* **2015**, *83*, 359–374. [CrossRef]
41. Livak, K.J.; Schmittgen, T.D. Analysis of Relative Gene Expression Data Using Real-Time Quantitative PCR and the $2^{-\Delta\Delta CT}$ Method. *Methods* **2001**, *25*, 402–408. [CrossRef]

https://doi.org/10.3390/plants15020222

plants

MDPI

Article

GsEXPA8 Improves Alkaline Tolerance in *Lupinus angustifolius* by Modulating Root Architecture, Stress-Responsive Gene Expression, and Rhizosphere Microbiome

Mengyu Liu [1,†], Yujing Liu [1,†], Hongli Wang [1], Yijia Ruan [1], Xiaoyu Wang [1], Xinlei Du [1], Mengyu Zhou [1], Yishan Fu [1], Jixiang Tang [1], Junfeng Zhang [2,*] and Lei Cao [1,*]

1 College of Horticulture and Landscape Architecture, Northeast Agricultural University, No. 600, Changjiang Road, Xiangfang District, Harbin 150036, China; liumm0730@163.com (M.L.); yujingliu0000@163.com (Y.L.); whl22228888@163.com (H.W.); ryj2504019012@163.com (Y.R.); wxy766696114@163.com (X.W.); duxinlei0101@163.com (X.D.); zhoumengyu2025@163.com (M.Z.); fys01305611@163.com (Y.F.); jxtang1019@163.com (J.T.)

2 School of Geography and Tourism, Harbin University, No. 109, Zhongxing Avenue, Nangang District, Harbin 150030, China

* Correspondence: leicao@neau.edu.cn (L.C.); jfzhang@hrbu.edu.cn (J.Z.)

† These authors contributed equally to this work.

Abstract

Lupinus angustifolius is an important leguminous ornamental species, but its productivity is often compromised by alkaline soil stress. *GsEXPA8*, an expansin gene identified in wild soybean (*Glycine soja*), has been implicated in alkali stress tolerance. In this study, we examined how heterologous expression of *GsEXPA8* in lupinus affects its biochemical, molecular, and rhizospheric responses to alkali stress. Under $NaHCO_3$-induced alkaline conditions, transgenic lines overexpressing *GsEXPA8* displayed improved leaf vigor, greater root biomass and length, elevated activities of antioxidant enzymes (CAT and POD), increased proline accumulation, and reduced malondialdehyde levels compared to the wild type. Expression analysis revealed time-dependent up-regulation of several alkali-responsive genes (*LaSOS1*, *LaNCED3*, *LaMYB39*, *LaNAC56*, *LaNHX6*, and *LaP5CS*). Moreover, the rhizosphere microbial community was significantly restructured, with a marked increase in beneficial microbial taxa such as Pseudomonas and Lysobacter. We also found that the endogenous lupinus homolog *LaEXPA8* is alkali-inducible. Overexpression of *LaEXPA8* similarly enhanced alkaline tolerance, whereas CRISPR/Cas9 knockout lines showed no clear phenotypic alteration, suggesting potential functional redundancy within the expansin family. Notably, *LaEXPA8* and *GsEXPA8* differed in their temporal regulation of downstream genes, indicating both conserved and distinct regulatory roles. Our results demonstrate that *GsEXPA8* improves alkali tolerance in lupinus through integrated mechanisms: promoting root growth, enhancing antioxidant and osmotic adjustment capacity, dynamically modulating stress-related gene expression, and enriching beneficial rhizosphere microbiota. This work provides the critical report of modifying alkali tolerance by manipulating an expansin gene alongside the associated rhizosphere microbiome, offering a combined strategy for breeding stress-resistant ornamentals.

Keywords: alkaline stress; *Lupinus angustifolius*; rhizosphere microbiome; EXPA8; functional characterization

1. Introduction

Soil salinization–alkalization is one of the primary abiotic stresses constraining global agricultural sustainability [1]. Alkaline stress, characterized by high pH, elevated Na^+ levels, and high concentrations of alkaline salts such as $NaHCO_3$, leads to reduced nutrient availability, ionic imbalance, and physiological drought, severely impairing root development and plant growth [2]. Root systems are particularly vulnerable under alkaline conditions, making root-related adaptive mechanisms central to plant tolerance.

Narrow-leaf lupinus (*Lupinus angustifolius*) is a leguminous plant valued as forage, green manure, and an ornamental. Its root system possesses biological nitrogen-fixation capacity, playing a significant role in improving soil structure and reducing dependence on nitrogen fertilizers. For instance, yellow lupinus (*Lupinus luteus* L.), with its high protein content, holds notable economic and ecological value [3]. It fixes atmospheric nitrogen through symbiosis with rhizobia, reducing fertilizer inputs and serving as a protein-rich feed source [3]. However, the growth and yield of lupinus are also significantly inhibited by alkaline soils. In response to alkaline stress, plants adapt not only through their intrinsic physiological and molecular regulatory networks, but also by actively modifying the rhizosphere microenvironment through changes in root morphology and exudate composition [4]. Current research on lupinus has primarily focused on its nutritional properties [5], mechanisms of organic acid secretion [6], drought stress [3], and heat stress [7], while studies on the alkali-tolerance mechanisms of narrow-leaf lupinus remain insufficient.

Expansins constitute a multigene family, encompassing α-expansins (EXPA), β-expansins (EXPB), expansin-like α, and expansin-like β subfamilies, which play crucial roles in plant growth and development, primarily through promoting root system expansion [8]. Recent studies indicate that cell wall remodeling-associated proteins are key players in plant responses to environmental stresses. For instance, overexpression of *ZmEXPA3* in maize enhances growth under both normal and saline conditions via modifications to cell wall architecture [9]. As cell wall-loosening proteins, expansins are instrumental not only in regulating growth and development but also in mediating plant adaptation to diverse environmental challenges. The structural features, regulatory mechanisms, and functional roles of expansins exhibit considerable diversity across plant species. They modulate plant growth, organogenesis, and adaptation to stresses such as drought [10], heat [11], and plant diseases [12]. Wild soybean (*Glycine soja*) represents a valuable genetic reservoir harboring numerous genes conferring tolerance to saline-alkaline stress. *GsEXPA8*, a member of the expansin family in wild soybean belonging to the EXPA subfamily, has been demonstrated to enhance alkaline stress tolerance and participate in regulating root architecture in soybean seedlings [13]. However, the function of *LaEXPA8*, the homologous gene of *GsEXPA8* in narrow-leaf lupinus (*Lupinus angustifolius*), remains unexplored.

Plant adaptation to alkaline stress involves not only intrinsic physiological and molecular regulation but also active modification of the rhizosphere environment through changes in root architecture and exudate composition. The rhizosphere microbial community, often regarded as the plant's 'second genome', plays a crucial role in assisting nutrient acquisition, enhancing disease resistance, and improving abiotic stress tolerance [14].The introduction of exogenous genes into plants can alter soil microbial composition, though the underlying mechanisms remain incompletely understood [15–17]. 16S rRNA gene sequencing, a high-throughput metagenomic approach, enables the analysis of microbial community diversity, taxonomic composition, and ecological interactions in the rhizosphere soil of herbaceous plants [18], and further facilitates the isolation of beneficial strains [19]. Under stress conditions, plants often modulate root exudation to recruit beneficial microorganisms that aid in stress adaptation. For example, lotus expansin EXPA1 is recruited during intracellular and intercellular colonization by rhizobia [20]. Most leguminous plants estab-

https://doi.org/10.3390/plants15050679

lish symbiotic relationships with rhizobia, a group of nitrogen-fixing bacteria. Expansins, which promote cell enlargement by loosening the cell wall, hold considerable breeding potential for improving biological nitrogen fixation in legumes. *GmEXPA11*, regulated by *GmPTF1*, promotes soybean nodule enlargement and nitrogen fixation via interaction with *GmNOD20* [21]. Therefore, elucidating how overexpression of the key gene *GsEXPA8* drives the reassembly of the rhizosphere microbial community through modified root phenotypes and physiological metabolism in other legumes is essential for a comprehensive understanding of the mechanisms underlying plant alkali tolerance, stress-induced changes in root development, and the regulatory influence of transgenic root systems on the rhizosphere microbiome.

Based on previous findings regarding the potential role of *GsEXPA8* in plant alkali tolerance [13], we propose that *GsEXPA8* overexpression enhances adaptation of narrow-leaf lupinus to alkaline stress by promoting root development and physiological resilience, potentially accompanied by changes in rhizosphere microbial community structure. To test this hypothesis, we generated transgenic lupinus lines overexpressing *GsEXPA8* and systematically assessed their phenotypic, biochemical analysis, and molecular responses under alkaline conditions. Furthermore, using 16S rDNA high-throughput sequencing, we compared the compositional changes in the rhizosphere microbial communities among transgenic soybean, transgenic lupinus, and their corresponding wild-type plants. From a novel 'plant–microbe' interaction perspective, this study aims to elucidate the integrated mechanism through which *GsEXPA8* mediates rhizosphere microbial community dynamics to regulate alkali tolerance in lupinus. Additionally, through overexpression, CRISPR/Cas9-based knockout, and hairy root induction assays, we functionally characterized the homologous gene *LaEXPA8* in its native host. These findings are expected to provide a new theoretical foundation and genetic resources for synergistically improving crop stress resistance through genetic modification and rhizosphere microbiome engineering.

2. Results

2.1. *GsEXPA8* Positively Regulates Alkaline Tolerance in Lupinus

Previous studies have shown that overexpression of alkali-tolerance genes from wild soybean, such as *GsHZ4*, in the root system of narrow-leaf lupinus (*Lupinus angustifolius*) can enhance plant alkaline tolerance [4]. Heterologous expression of the wild soybean *GsEXPA8* gene in soybean also improves alkaline stress resistance. To investigate the function of *GsEXPA8* in the ornamental plant lupinus, this study generated transgenic lupinus lines overexpressing *GsEXPA8* in roots. Positive transgenic plants were identified by RT-PCR (Figure S1). RT-PCR analysis demonstrated differential expression levels among the independent OE lines. OE1 showed the lowest expression level and correspondingly exhibited the smallest plant size under control conditions, suggesting a potential positive association between transgene expression and basal growth performance. Both wild-type (WT) and overexpression (OE) lines were treated with 50 mM $NaHCO_3$ for 12 days, and phenotypic changes were analyzed. Under alkaline stress, WT plants exhibited pronounced leaf wilting, whereas OE lines retained visibly greener leaves (Figure 1A). The fresh weight and root length of OE lines were significantly greater than those of WT plants after treatment (Figure 1B,C). Biochemical analysis revealed that CAT activity was significantly higher in OE plants compared to WT following stress treatment (Figure 1D). In contrast, MDA content was markedly lower in OE lines (Figure 1E). POD activity was significantly elevated in OE plants under both control and stress conditions relative to WT (Figure 1F). Pro content was also higher in OE lines after alkali treatment (Figure 1G). These results indicate that *GsEXPA8* overexpression enhances the core antioxidant defense system in roots under stress, effectively protecting cellular membrane integrity and activating osmotic adjustment.

 https://doi.org/10.3390/plants15050679

Phenotypic observation under non-stress conditions indicated that the shoot parts of OE plants were substantially larger than those of WT, suggesting a growth-promoting effect of the transgene. Root architecture was further analyzed using a root scanning system in untreated plants (Figure 2A). OE lines developed more robust and extensive root systems compared to WT. Quantitative analysis showed that transgenic lines exhibited significantly greater root surface area, total root length, nodule number, root tip count, and node number (Figure 2B–F). Overall, transgenic plants displayed taller and more vigorous growth. These findings demonstrate that *GsEXPA8* overexpression promotes root growth, branching, and the formation of root tip tissues in lupinus, which may enhance nutrient and water acquisition. These results are consistent with the effects of *GsEXPA8* overexpression previously reported in soybean.

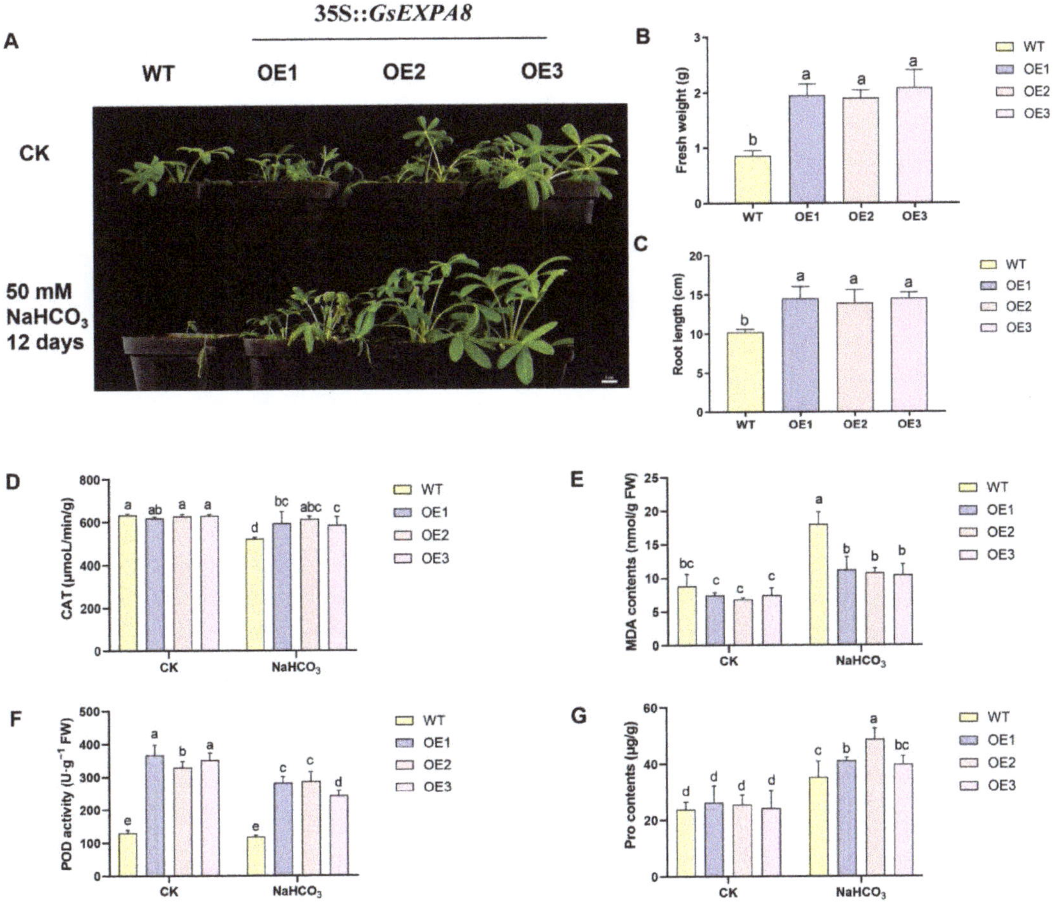

Figure 1. *GsEXPA8* improves alkaline tolerance in lupinus. (**A**) Phenotypic characteristics of OE and WT plants treated with 50 mM $NaHCO_3$. (**B**) Fresh weight of OE and WT plants. (**C**) Root length of OE and WT plants. (**D**) CAT enzyme activity of OE and WT plants. (**E**) MDA content of OE and WT plants. (**F**) POD enzyme activity of OE and WT plants. (**G**) Pro content of OE and WT plants. Data are presented as mean ± SD ($n = 3$). Different letters indicate statistically significant differences as determined by one-way ANOVA followed by post hoc multiple comparison tests ($p < 0.05$).

https://doi.org/10.3390/plants15050679

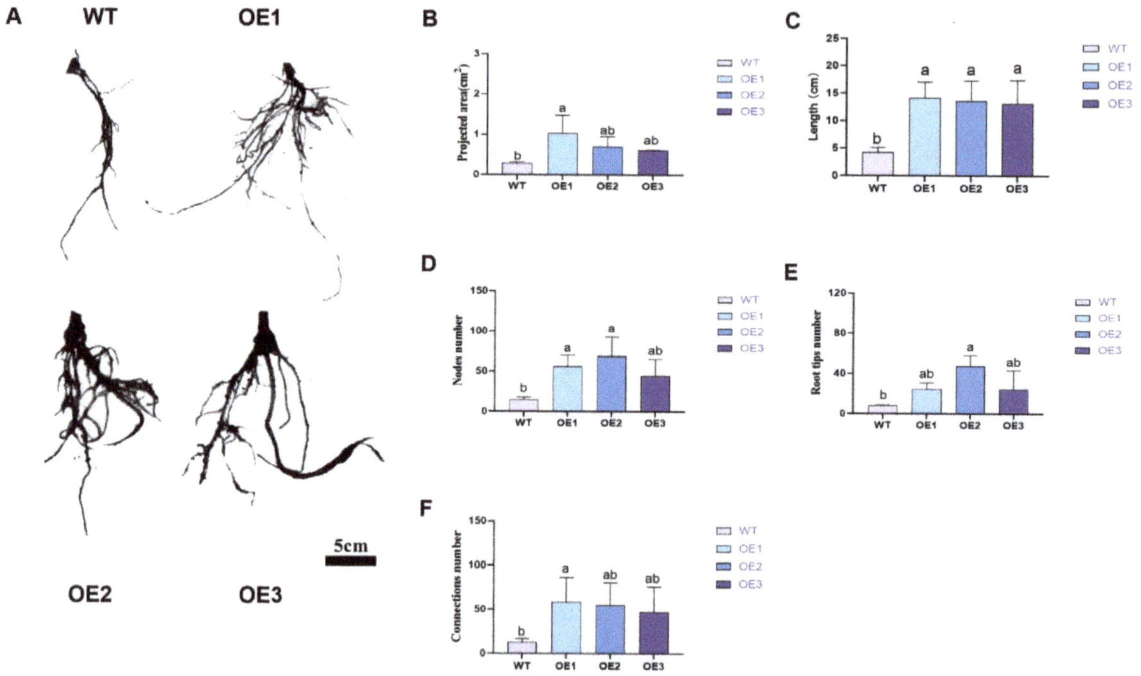

Figure 2. *GsEXPA8* can improve the root system of lupinus. (**A**) Root phenotypes of WT and OE plants. (**B**) Projected area of WT and OE plants. (**C**) Root length of WT and OE plants. (**D**) Number of nodes in WT and OE plants. (**E**) Number of root tips in WT and OE plants. (**F**) Number of root connections in WT and OE plants. Data are shown as mean ± SD ($n = 15$). Different letters indicate significant differences based on one-way ANOVA followed by post hoc tests ($p < 0.05$).

2.2. GsEXPA8 Upregulates the Expression of Alkaline Stress-Related Genes

To investigate the regulatory network through which *GsEXPA8* confers alkali tolerance in lupinus, WT and root-specific *GsEXPA8*-overexpressing (OE) plants were treated with 100 mM $NaHCO_3$. The relative expression levels of key alkali-responsive genes—*LaSOS1*, *LaNAC56*, *LaNHX6*, *LaP5CS*, *LaMYB39*, and *LaNCED3*—were analyzed at 0, 3, 6, and 12 h after treatment (Figure 3A–F). Compared with WT, the expression of *LaSOS1*, *LaNCED3*, *LaNAC56*, and *LaMYB39* was significantly upregulated in OE plants at 12 h post-treatment. *LaNHX6* showed marked induction at earlier time points (3 h and 6 h). In contrast, *LaP5CS* transcript levels were already higher in OE plants before stress and gradually declined with prolonged treatment. These results indicate that *GsEXPA8* enhances alkali tolerance by differentially regulating a suite of stress-related genes at distinct stages of the stress response.

https://doi.org/10.3390/plants15050679

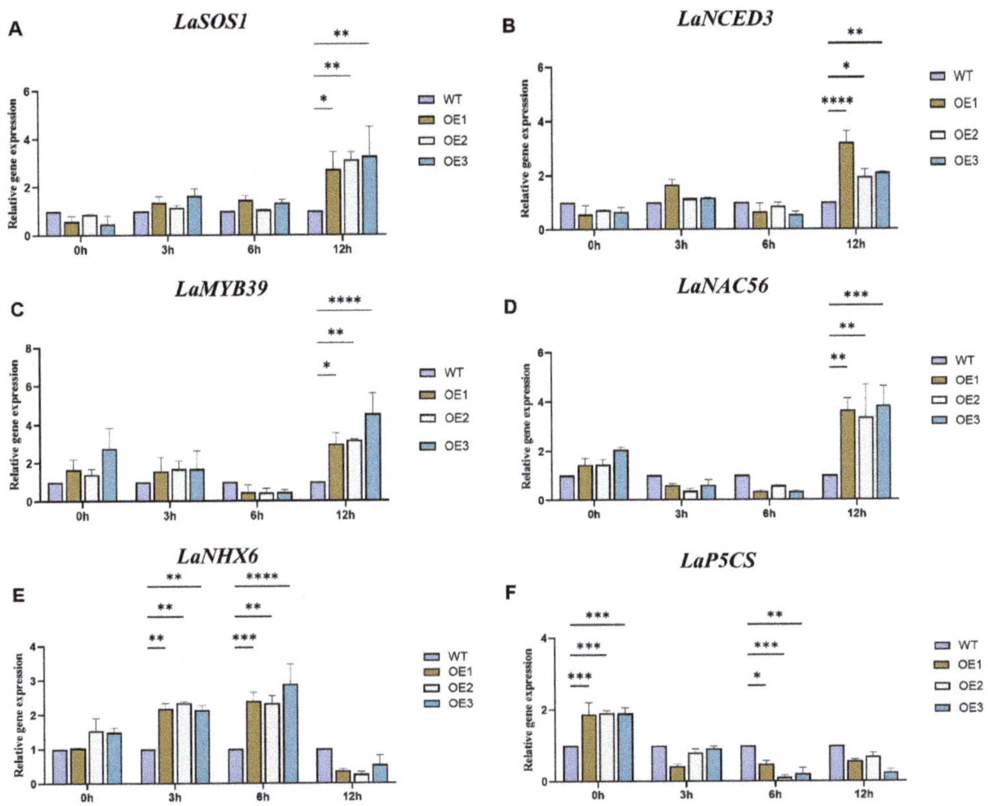

Figure 3. Relative expression levels of marker genes in WT and OE plants under 100 mM $NaHCO_3$ treatment. (**A**) *LaSOS1*. (**B**) *LaNCED3*. (**C**) *LaMYB39*. (**D**) *LaNAC56*. (**E**) *LaNHX6*. (**F**) *LaP5CS*. Data represent mean ± SD (n = 3). Statistical significance was evaluated using two-way ANOVA (genotype × treatment), followed by post hoc multiple comparison tests. Asterisks indicate significant differences (* $p < 0.05$, ** $p < 0.01$, *** $p < 0.001$, **** $p < 0.0001$).

2.3. Heterologous Expression of GsEXPA8 Enriches Distinct Microbial Communities in Different Plant Hosts

Following expression of *GsEXPA8* in soybean roots, the protein was primarily localized in the intercellular space between the cell membrane and wall within the outermost and sub-outermost layers of root tips, suggesting a potential role in modulating root–microbe interactions [13]. To investigate the impact of *GsEXPA8* transformation on the plant-associated microbial environment, we analyzed the rhizosphere soil microbiota of both transgenic soybean and lupinus. In soybean, 1756 Amplicon Sequence Variants (ASVs) were shared between the control and overexpression groups, while 1837 ASVs were unique to the control and 1728 ASVs were unique to the overexpression group (Figure 4A). In lupinus, 1993 ASVs were common to both groups, with 3111 ASVs unique to the wild-type (WT) and 2039 ASVs unique to the transgenic lines (Figure 4B). Venn diagram analysis further revealed that 118 microbial genotypes increased, and 94 decreased, in the rhizosphere of both species. Conversely, 107 genotypes increased in soybean but decreased in lupinus, while 151 increased in lupinus but decreased in soybean (Figure 4C). Heatmap analysis indicated a marked enrichment of beneficial rhizobacteria such as *Pseudomonas* (associated with growth promotion and disease suppression) and *Pseudolabrys* (with potential for environmental remediation). In contrast, taxa including the Allorhizobium-Neorhizobium-Pararhizobium-Rhizobium group (closely related to rhizobia)

https://doi.org/10.3390/plants15050679

and Bdellovibrio (a predator of various pathogens) were reduced (Figure 4D,E). This shift may reflect functional compensation by *GsEXPA8* for certain rhizobial activities. In transgenic lupinus, we observed an increase in genera such as *Lysobacter* (a prolific producer of antimicrobial compounds and a resource for biopesticides), *Phenylobacterium* (capable of degrading recalcitrant aromatic pollutants), and *Ramlibacter* (involved in mineralization of organic phosphorus). Conversely, genera including *Pseudoxanthomonas* (a degrader of persistent pollutants), *Nocardioides* (a decomposer of chitin and organic matter), and the potential phytopathogen *Cellulosimicrobium* showed decreased abundance. Comparative analysis highlighted three genera—*Sphingomonas* (an efficient degrader and potential plant probiotic), *Pseudolabrys* (a common soil inhabitant with unclear plant effects), and *Pseudarthrobacter* (some strains of which promote plant growth or degrade pollutants)—that exhibited consistent directional changes in both soybean and lupinus rhizospheres (marked within red boxes in Figure 4D,E). *Sphingomonas* and *Pseudolabrys* increased in both systems, whereas *Pseudarthrobacter* decreased. Collectively, these results demonstrate that *GsEXPA8* expression reshapes the rhizosphere microbiome, with both common and species-specific effects across different leguminous hosts. The potential benefits and risks associated with such transgenic-driven microbial shifts warrant further evaluation.

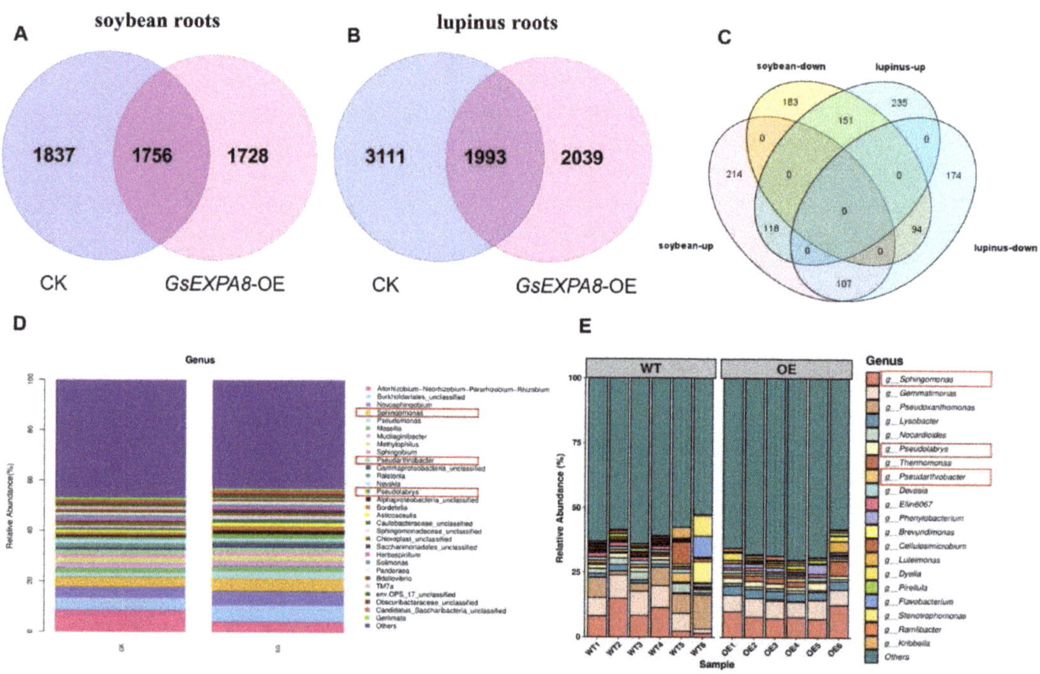

Figure 4. Transformation of exogenous genes alters soil microbial communities around transgenic soybean and lupinus root systems. (**A**) *GsEXPA8* transformation of soybean alters the soil microbial community of soybean root system. (**B**) *GsEXPA8* transformation of lupinus alters the soil microbial community of lupinus root system. (**C**) Venn of soil microbial community increase and decrease in soybean and lupinus. (**D**) Top 30 soil microbial species in soybeans. (**E**) Top 20 soil microbial species in lupinus plants. Red frames indicate the same significantly regulated microbial species around transgenic soybean and lupinus root systems.

2.4. *The Exogenous GsEXPA8 and the Endogenous LaEXPA8 Share High Homology*

While investigating the enhanced alkaline tolerance conferred by the exogenous *GsEXPA8*, we also examined its endogenous homolog in narrow-leaf lupinus, *LaEXPA8*. Sequence analysis revealed that both proteins possess an identical DPBB_1 structural domain

 https://doi.org/10.3390/plants15050679

(Figure S2), suggesting potential functional conservation. Analysis of transcriptomic data under $NaHCO_3$ treatment (pH 8.5) showed that, among the expansin gene family, *LaEXPA8* expression was significantly upregulated at 0, 3, and 6 h post-treatment, with a particularly pronounced increase observed at the 6-h time point (Figure 5A). This induction was confirmed by examining the FPKM values of *LaEXPA8* across different time points. Subsequent qRT-PCR validation using roots treated with 0 or 100 mM $NaHCO_3$ for 6 h demonstrated a significant increase in *LaEXPA8* transcript levels in wild-type plants (Figure 5C). In contrast, the expression of other expansin family members, including *LaEXPA4* (Figure 5D), *LaEXPA13* (Figure 5E), *LaEXPA16* (Figure 5F), and *LaEXPA18* (Figure 5G), was significantly downregulated under the same conditions. *LaEXPB3* expression also showed a decreasing, though not statistically significant, trend (Figure 5H). These results indicate that not all expansin family members are upregulated under alkaline stress. The specific and significant induction of *LaEXPA8* suggests its potential involvement in the alkaline stress response mechanism in lupinus.

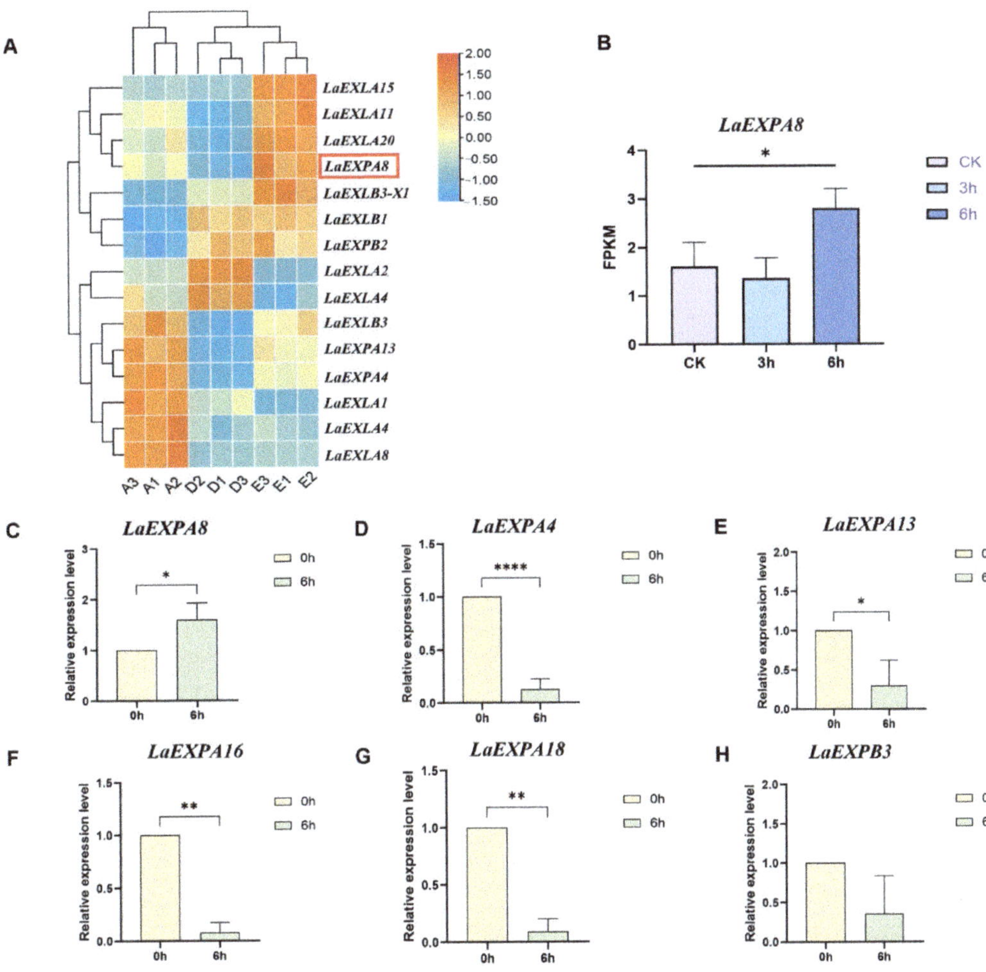

Figure 5. Expression levels and validation of expansin family genes in lupinus under untreated and alkaline treatments. (**A**) The heatmap represents the expression levels of the expansion family of lupinus

 https://doi.org/10.3390/plants15050679

in the transcriptome. Red frame indicates relative expression level of *LaEXPA8* in heatmap. (**B**) The FPKM value of *LaEXPA8* in untreated and $NaHCO_3$ (pH 8.5) treated conditions. (**C–H**) Relative expression levels of expansin family in lupinus treated with 100 mM $NaHCO_3$: (**C**) *LaEXPA8*, (**D**) *LaEXPA4*, (**E**) *LaEXPA13*, (**F**) *LaEXPA16*, (**G**) *LaEXPA18*, (**H**) *LaEXPB3*. Data are presented as mean ± SD ($n = 3$). Statistical significance was determined using Student's t-test. Asterisks indicate significant differences (* $p < 0.05$, ** $p < 0.01$, **** $p < 0.0001$).

2.5. Functional Validation of LaEXPA8 in Positively Regulating Alkaline Tolerance

To determine whether the endogenous *LaEXPA8* gene enhances alkaline tolerance in lupinus, overexpression and CRISPR-Cas9 knockout constructs were generated (Figure S3). Wild-type (WT), overexpression (OE), and CRISPR mutant (CR) plants were treated with 100 mM $NaHCO_3$. Under non-stress conditions, OE plants exhibited larger and taller growth phenotypes compared to WT, while WT and CR lines showed no significant difference. After three days of treatment, CR plants began to show wilting symptoms. By 6 days, WT plants started to wilt, whereas CR plants were completely necrotic (Figure 6A; a top-view comparison is shown in Figure 6B). Biochemical analysis assays under alkaline stress revealed that, relative to WT, OE plants displayed significantly higher activities of POD and CAT, increased proline (Pro) content (Figure 6C–E), and a marked reduction in malondialdehyde (MDA) levels (Figure 6F). In contrast, CR plants showed no significant differences from WT in phenotypic appearance or any measured biochemical parameters under either control or stress conditions. These results indicate that *LaEXPA8* overexpression enhances alkaline tolerance, while its knockout does not confer a negative phenotype, suggesting possible functional redundancy within the expansin gene family in lupinus. Root architecture analysis further demonstrated that OE lines developed more robust root systems (Figure 7A). Quantitative parameters, including total root length (Figure 7B), number of healthy roots (Figure 7C), total surface area (Figure 7D), lateral root count (Figure 7E), surface area (Figure 7F), and total root volume (Figure 7G), were all significantly greater in OE plants than in WT, whereas CR and WT lines did not differ significantly. Histochemical staining with DAB (Figure 8A) and NBT (Figure 8B) on leaves from treated and untreated plants showed that CR leaves consistently accumulated more reactive oxygen species (ROS), indicative of greater oxidative damage, under both conditions. Under non-stress conditions, WT and OE leaves showed minimal staining. After alkaline treatment, OE leaves exhibited only slight staining, whereas WT leaves displayed more severe ROS accumulation. In summary, similar to *GsEXPA8* overexpression, ectopic expression of *LaEXPA8* promotes root growth, branching, and root tissue proliferation, thereby enhancing alkaline tolerance and shoot vitality in lupinus. However, knockout of *LaEXPA8* did not impair root development or alkaline stress tolerance, likely due to genetic redundancy within the expansin family. Since overexpression of both *GsEXPA8* and *LaEXPA8* improved plant biomass, we hypothesize that these genes may also enhance resistance to other abiotic stresses and ornamental traits, although further validation is required.

https://doi.org/10.3390/plants15050679

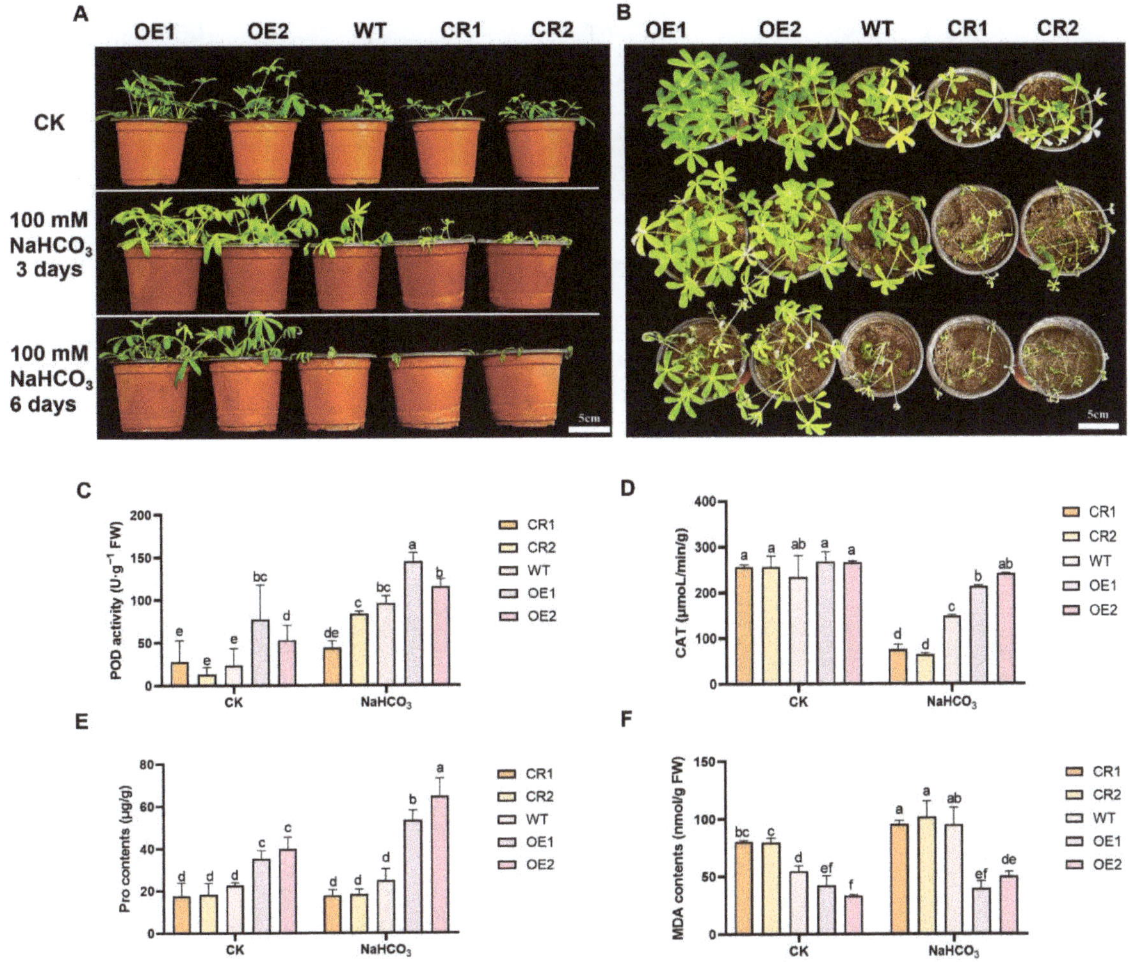

Figure 6. *LaEXPA8* improves alkaline tolerance in Lupinus. (**A**) Positive phenotype of lupinus treated with 100 mM $NaHCO_3$. (**B**) Top view of lupinus treated with 100 mM $NaHCO_3$. (**C**) POD enzyme activity. (**D**) CAT enzyme activity. (**E**) Pro content. (**F**) MDA content. Data are presented as mean ± SD ($n = 3$). Different letters indicate significant differences determined by one-way ANOVA ($p < 0.05$).

2.6. LaEXPA8 Upregulates the Expression of Alkaline Stress-Responsive Genes

To validate the relative expression levels of key alkaline stress-related genes and elucidate the regulatory mechanism of *LaEXPA8*, wild-type (WT) and overexpression (OE) plants were treated with 100 mM $NaHCO_3$. The transcript levels of *LaSOS1* (Figure 8C), *LaP5CS* (Figure 8D), and *LaNHX6* (Figure 8E) were analyzed at 0, 3, 6, and 12 h post-treatment. The results showed that in OE plants, *LaSOS1* was significantly upregulated at 3 h and 6 h, *LaP5CS* was markedly induced at 6 h and 12 h, and *LaNHX6* expression was significantly elevated at 3 h. These findings indicate that, similar to *GsEXPA8*, *LaEXPA8* enhances plant alkaline tolerance by upregulating stress-responsive genes. However, the timing of induction differed between the two genes: *LaSOS1* upregulation occurred earlier, while *LaP5CS* induction was delayed compared to the pattern observed with *GsEXPA8*. This suggests that *LaEXPA8* and *GsEXPA8* may function within distinct regulatory networks despite their conserved role in promoting stress tolerance.

https://doi.org/10.3390/plants15050679

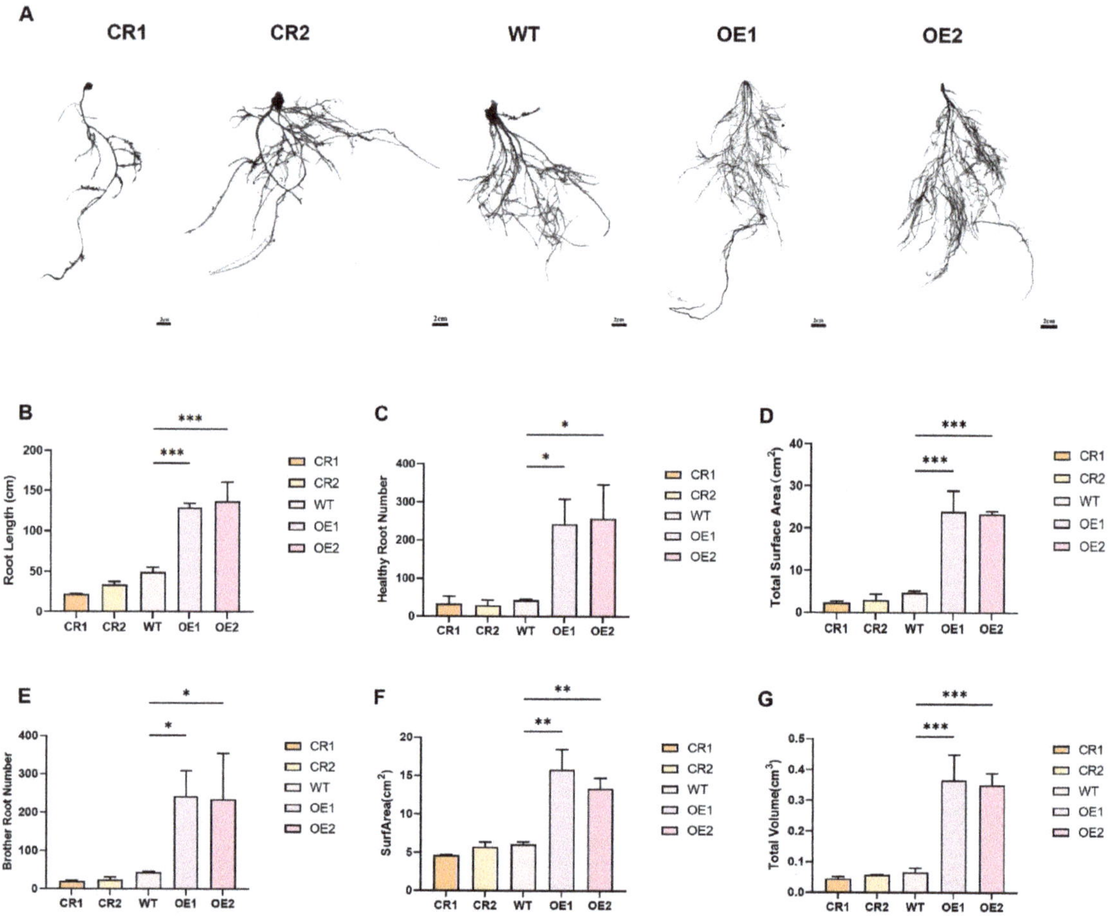

Figure 7. *LaEXPA8* can improve the root system of lupinus. (**A**) Root phenotypes of WT, OE and CR plants. (**B**) Root length of WT, OE and CR plants. (**C**) Number of healthy roots in WT, OE and CR plants. (**D**) Total root surface area of WT, OE and CR plants. (**E**) Brother root number of WT, OE and CR plants. (**F**) Surface area of WT, OE and CR plants. (**G**) Total root volume of WT, OE and CR plants. Values represent mean ± SD (n = 15). Asterisks indicate significant differences (one-way ANOVA, * $p < 0.05$, ** $p < 0.01$, *** $p < 0.001$).

https://doi.org/10.3390/plants15050679

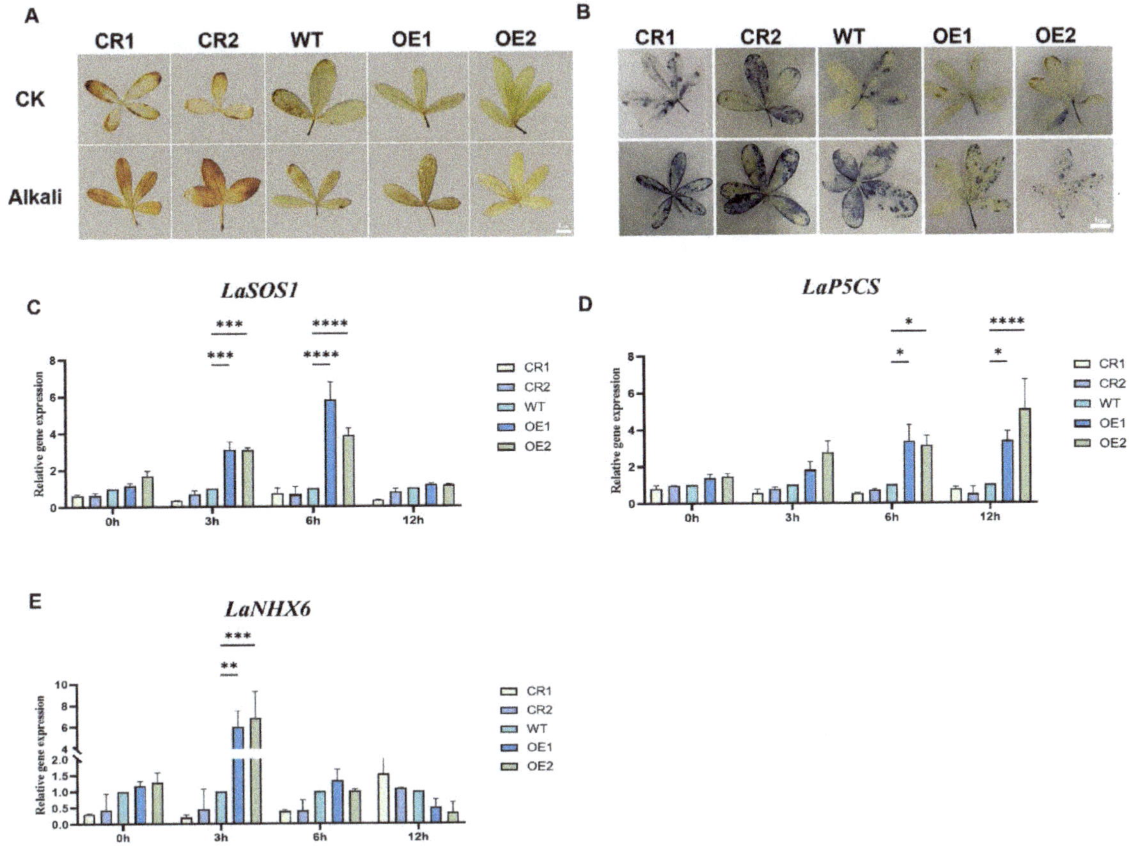

Figure 8. *LaEXPA8* improves the alkaline tolerance of lupinus. (**A**) DAB staining of WT, OE, and CR plants under CK and 100 mM $NaHCO_3$ treatments. (**B**) NBT staining of WT, OE and CR plants under CK and 100 mM $NaHCO_3$ treatments. (**C–E**) Relative expression levels of marker genes (**C**) *LaSOS1*; (**D**) *LaP5CS*; (**E**) *LaNHX6* in WT, OE and CR plants treated with 100 mM $NaHCO_3$. Data are presented as mean ± SD (n = 3). Statistical significance was determined by two-way ANOVA followed by post hoc multiple comparison tests. Asterisks indicate significant differences (* $p < 0.05$, ** $p < 0.01$, *** $p < 0.001$, **** $p < 0.0001$).

3. Discussion

3.1. GsEXPA8 Enhances Alkaline Tolerance in Lupinus Through Coordinated Biochemical and Molecular Mechanisms

As previously established, the root system is a multifunctional organ critical for water and nutrient uptake, metabolite storage, anchorage, mechanical support, and interaction with the soil environment [8]. This study demonstrates that heterologous overexpression of the wild soybean-derived *GsEXPA8* gene in lupinus systemically improves plant tolerance to alkaline stress. The enhanced tolerance was first evident at the phenotypic level. The observed variation in plant size among independent OE lines under control conditions may be associated with differences in transgene expression levels. The positive trend between expression level and plant size suggests a dosage-dependent effect of *GsEXPA8* on plant growth. However, further quantitative correlation analysis would be required to confirm this relationship. Under alkaline stress, transgenic overexpression (OE) plants maintained higher biomass and developed more extensive root architecture compared to wild-type

 https://doi.org/10.3390/plants15050679

(WT) plants. This improvement directly enhances the capacity for water and nutrient acquisition, forming the foundational basis for stress resilience. At the biochemical level, OE plants exhibited a more active antioxidant defense system, characterized by increased catalase (CAT) activity and reduced malondialdehyde (MDA) content, effectively mitigating oxidative damage. Concurrently, elevated proline (Pro) content indicated enhanced osmotic adjustment, alleviating osmotic stress induced by alkalinity. Further analysis showed that the endogenous homolog *LaEXPA8* in *Lupinus angustifolius* is induced under alkaline stress and may participate in the stress response. However, phenotypic comparisons between wild-type and *GsEXPA8*-overexpressing lines under alkaline treatment indicated that the enhanced tolerance observed in the OE plants is primarily associated with ectopic expression of *GsEXPA8*. Meanwhile, the endogenous *LaEXPA8* gene is induced under alkaline stress, and its potential contribution to stress tolerance cannot be completely excluded. Molecular investigations revealed that *GsEXPA8* overexpression positively regulated a suite of key alkaline stress-responsive genes in lupinus. For instance: *LaNHX6*, a homolog of *GmNHX6*, is involved in cellular Na^+/K^+ and pH homeostasis, crucial for salt-alkali tolerance. Heterologous expression of *GmNHX6* in Arabidopsis was shown to enhance alkaline tolerance by maintaining high K^+ levels and a low Na^+/K^+ ratio [22]. *LaP5CS*, homologous to *GmP5CS*, encodes a bifunctional enzyme catalyzing the rate-limiting step in proline biosynthesis, linking its upregulation to the observed osmotic adjustment [23,24]. *LaNCED3* (a homolog of *AtNCED3*) is a key enzyme in abscisic acid (ABA) biosynthesis. Studies indicate that *NCED3* can be strongly induced by NaCl via an ABA-independent pathway, highlighting its role in stress signaling [25,26]. *LaSOS1* encodes a Na^+/H^+ antiporter that contributes to cellular ion homeostasis by facilitating Na^+ efflux, which is vital for salt tolerance [27]. *LaNAC56* and *LaMYB39* are transcription factors implicated in stress response regulation. *NAC56* in rapeseed acts as a transcriptional activator involved in regulating reactive oxygen species (ROS) accumulation and cell death under stress [28,29], while *MYB39* in maize enhances seedling drought tolerance [30]. Their upregulation suggests involvement in transcriptional reprogramming. These genes participate in critical pathways governing ion homeostasis, transcriptional regulation, and hormone synthesis. Therefore, *GsEXPA8* does not function solely through its canonical cell wall-loosening activity. Instead, it acts as a key regulatory node, coordinating a multi-layered defense system. This system integrates the promotion of root growth, enhancement of biochemical resilience, and activation of core stress-signaling networks, collectively conferring strong alkaline tolerance to lupinus.

3.2. Heterologous Expression of GsEXPA8 Specifically Reshapes the Rhizosphere Microbial Community

This study further examined the potential impact of *GsEXPA8* expression on the rhizosphere microbiome of host plants. Our analysis revealed significant, yet host-specific, restructuring of the rhizobacterial community in both transgenic soybean and lupinus hairy root systems. A common trend was the enrichment of certain plant-beneficial genera, such as *Sphingomonas* [31,32] and *Pseudolabrys* [33], in the rhizosphere of both species. However, distinct differences were observed: in soybean, a relative decrease occurred in rhizobial taxa closely associated with symbiotic nitrogen fixation, whereas in lupinus, increases were noted in genera involved in pollutant degradation (e.g., *Phenylobacterium* [34]) and nutrient mobilization (e.g., *Ramlibacter* [35,36]), alongside a reduction in potential pathogenic bacteria. These host-dependent shifts likely stem from inherent differences in root exudate profiles between plant species, compounded by specific alterations in exudate composition induced by the transgene in each host. For instance, *Sphingomonas* species possess diverse functionalities, including environmental remediation and production of high-value phytohormones, and have been implicated in the degradation of organometallic compounds [31]

 https://doi.org/10.3390/plants15050679

and in mitigating pesticide stress by colonizing roots and degrading chlorpyrifos while releasing beneficial metabolites [37]. The enrichment of *Ramlibacter*, which includes alkali-tolerant species such as *R. alkalitolerans* isolated from ginseng soil [38], may be particularly relevant under alkaline conditions. These findings suggest that the enhanced alkaline tolerance mediated by *GsEXPA8* is not solely attributable to intrinsic biochemical adjustments within the plant. It may also involve a 'plant–microbe' interaction mechanism, whereby the gene facilitates the recruitment and enrichment of a beneficial rhizosphere microbiome that indirectly bolsters stress adaptation. Plant growth-promoting rhizobacteria (PGPR) can colonize the rhizosphere and enhance stress tolerance through various mechanisms, such as producing exopolysaccharides (EPS), phytohormones, ACC deaminase, and volatile compounds, inducing the accumulation of osmolytes and antioxidants, modulating stress-responsive gene expression, and altering root morphology [39]. Examples include Bacillus megaterium strains improving drought tolerance in rice [40,41] and various Bacillus species alleviating salt stress damage in maize [42–45]. Similarly, specific Pseudomonas strains have been shown to modulate mercury uptake in lupinus, underscoring the potential of soil–plant–microbe synergy. Collectively, our results indicate that *GsEXPA8*-mediated rhizosphere microbiome remodeling represents a contributory, host-specific layer to the overall alkaline tolerance phenotype, warranting deeper investigation into this synergistic tripartite interaction.

3.3. Structural Homology Between GsEXPA8 and LaEXPA8 Suggests Functional Conservation

Sequence analysis revealed a high degree of homology between the exogenous *GsEXPA8* and the endogenous *LaEXPA8*, with both genes containing the canonical DPBB_1 domain characteristic of functional expansins. This structural conservation provides a critical molecular basis for the successful functionality of *GsEXPA8* in lupinus, indicating that these homologous genes, despite their different species origins, likely encode proteins with similar biochemical activities, namely, cell wall loosening via disruption of hydrogen bonds between wall polysaccharides to promote cell elongation. Both *GsEXPA8* and the endogenous *LaEXPA8* can increase plant root length, which may be related to the enhanced activity of cell wall modifying proteins (such as expansin), and this needs to be directly verified by subsequent cell microscopy studies. Such functional conservation explains how an expansin gene from soybean can be effectively integrated into the growth and stress-response programs of lupinus.

This aligns with reports of expansins playing conserved roles across diverse species: for instance, EXPB2 is associated with cell wall loosening and fruit softening in strawberry [46]; certain expansins positively regulate drought tolerance in wheat [47]; and *OsEXPB2*, a root-predominant gene in rice, plays a key role in root hair formation and holds potential for transgenic root breeding to enhance abiotic stress tolerance [48]. Similarly, transcription factors such as *NAC2* have been shown to improve growth and drought/heat tolerance in transgenic cowpea [49], and have been linked to freezing tolerance in *Aegilops tauschii* [50] and enhanced stress tolerance in *Picea wilsonii* [51]. Furthermore, this homology implies that the core cell wall-modifying function of expansins and their role in stress responses may be evolutionarily conserved across leguminous and broader plant lineages. However, comparative expression analysis revealed distinct kinetic profiles in the regulation of downstream stress-responsive genes. In *LaEXPA8*-overexpressing plants under alkaline stress, *LaSOS1* was significantly upregulated at 3 and 6 h, whereas in *GsEXPA8*-overexpressing plants, its upregulation was pronounced at 12 h. *LaNHX6* showed marked induction at 3 h in *LaEXPA8* OE lines but exhibited a sustained increase from 3 to 6 h in *GsEXPA8* OE lines. For *LaP5CS*, transcript levels peaked at 6 and 12 h in *LaEXPA8* OE plants, while in *GsEXPA8* OE plants, they were already elevated at 0 h and declined signifi-

 https://doi.org/10.3390/plants15050679

cantly by 6 h. These temporally distinct expression patterns suggest that while *GsEXPA8* and *LaEXPA8* share conserved biochemical functions, they may operate within nuanced, gene-specific regulatory networks, fine-tuning the stress response dynamics in lupinus.

3.4. *LaEXPA8* Is a Potential Endogenous Regulator of Alkaline Tolerance

A notable finding of this study is the specific induction of the endogenous *LaEXPA8* gene under alkaline stress, contrasting with the suppressed expression of other members within the same expansin family. This distinct expression pattern strongly suggests that *LaEXPA8* is not merely a conventional expansin but may serve as a key endogenous regulator in the lupinus alkali-stress response. Given that the heterologous expression of *GsEXPA8* significantly enhanced alkaline tolerance, it is plausible to speculate that the upregulation of endogenous *LaEXPA8* represents an adaptive strategy employed by lupinus to counteract alkaline stress. Its function likely parallels that of *GsEXPA8*, potentially enhancing tolerance by promoting adaptive root growth under stress and modulating downstream stress-responsive genes. This insight shifts the perspective from the application of exogenous genes towards understanding intrinsic plant mechanisms. Direct functional validation via CRISPR/Cas9-mediated knockout of *LaEXPA8* resulted in a significant reduction in root size, underscoring its essential role. This finding is crucial for elucidating the natural genetic basis of alkali tolerance in lupinus and may provide a novel target for precision molecular breeding utilizing the plant's own homologous genes. Furthermore, *LaEXPA8* may be functionally linked to critical regulatory pathways, such as ion homeostasis and abscisic acid biosynthesis.

4. Materials and Methods

4.1. Plant Materials and Growth Conditions

The experiment was conducted at the Horticultural Experiment Center of Northeast Agricultural University. *Lupinus angustifolius* seeds used in the research were sourced from the commercial brand 'Hua You Xiu' [4]. The lupinus was cultured in pots containing a 1:1 mixture of nutrient soil and vermiculite. Plants were maintained under controlled conditions at 22–25 °C with a 16-h light/8-h dark photoperiod until the third pair of palmate compound leaves were fully developed. The plant was carefully removed from the pot, and the roots were gently rinsed with water to remove impurities. Subsequently, sterile surgical blades were used to separate the plant into aboveground and belowground parts. The fresh weight was measured using an electronic balance, and the length of the main root was determined by direct measurement. To evaluate the phenotypic differences under alkaline stress, a pot irrigation method was employed, and wild-type and overexpressing plants were treated with 0 mM (control) and 100 mM $NaHCO_3$ solutions, respectively.

Cultivated soybean were planted in the Experimental Practice and Demonstration Center of Northeast Agricultural University, Harbin, China.

4.2. Vector Construction

For generating overexpression lines, the full-length cDNA of *GsEXPA8* was inserted into the *BamH I* and *EcoR I* sites of the pBI121 vector, while the full-length cDNA of *LaEXPA8* was cloned into the *Nco I* site of pCAMBIA1302. For generating knockout lines, a CRISPR/Cas9 system was employed using the pHK2-Cas9-U6 vector (Biorun, Wuhan, China), modified from pCAMBIA1300 for dicot gene editing. A dual-target strategy was used for guide RNA design. The recombinant constructs were amplified in *Escherichia coli* DH5α cells. Purified plasmids were subsequently introduced into *Agrobacterium rhizogenes* strain K599 (Weidi, Shanghai, China) via the freeze–thaw method. Transgenic lupinus plants (overexpression and knockout lines) were generated through root transformation

https://doi.org/10.3390/plants15050679

using A. rhizogenes infection, as previously described. WT soybean seeds were cultured until germination, portions of the hypocotyls were cut off, the seed coats were peeled, and the two cotyledons were separated and clipped with the nodes. Two true leaves were removed, and a knife tip was used to cut to growth point 1–2 to create explants. The recombinant *GsEXP8* plasmid was introduced into *Agrobacterium* K599, followed by propagation. [52,53].

4.3. RNA Extraction and Reverse Transcription

Twenty-five-day-old wild-type (WT), overexpression (OE), and CRISPR knockout (CR) lupinus plants were treated with 100 mM $NaHCO_3$. Root samples were collected at 0, 3, 6, and 12 h post-treatment, with three biological replicates per time point. Total RNA was extracted using the TransZol Up Kit (ET111-01-V2; TransGen Biotech, Beijing, China). First-strand cDNA was synthesized from total RNA using the SPARKscript II RT Plus Kit (with gDNA Eraser) (AG0304-B, Sparkjade, Jinan, China). Gene-specific primers were designed for qRT-PCR analysis, which was performed using SYBR Green dye (AH0104-B, Sparkjade, Jinan, China) on a real-time PCR system. The relative expression levels of target genes (*LaSOS1*, *LaNAC56*, *LaNHX6*, *LaP5CS*, *LaMYB39*, and *LaNCED3*) were calculated using the $2^{-\Delta\Delta CT}$ method [54], with *Ubiquitin* serving as the internal reference gene for normalization. All primer sequences are listed in Supplementary Table S1.

4.4. RNA-Seq Analysis

Lupinus plants were grown in Hoagland solution at pH 5.5 as the control group (Group A/a). For treatment groups, plants were exposed to Hoagland solution supplemented with $NaHCO_3$ (pH 8.5) for 3 h (Group D/d) or 6 h (Group E/e). Root tips were immediately frozen in liquid nitrogen and stored at −80 °C. Each group included three biological replicates, resulting in a total of 15 samples. These samples were subsequently submitted to BGI for transcriptome sequencing. The raw sequencing data have been deposited in the NCBI SRA database under accession number SUB14514279.

4.5. Root Scanning Assay

Wild-type (WT), overexpression (OE), and CRISPR knockout (CR) lupinus plants with uniform growth were selected, with 15 biological replicates per genotype. Root systems were carefully washed clean and scanned using a root scanner (LA-S; Wanshen, Hangzhou, China) according to the manufacturer's instructions. Key morphological parameters were measured from the scanned images [55].

4.6. Microbiome Analysis

Rhizosphere soil samples (5 g each) were collected from 25-day-old untreated transgenic and non-transgenic lupinus and soybean plants, with 6 replicates per group, resulting in a total of 24 samples. Microbial community analysis was performed via 16S rDNA amplicon sequencing to assess composition, diversity, and structure. Sequencing was conducted on the DNBSEQ-G99 platform by LC-Bio Technologies (Hangzhou) Co., Ltd (Hangzhou, China) [56].

4.7. Determination of Biochemical Parameters

Wild-type (WT), overexpression (OE), and CRISPR knockout (CR) lupinus plants were treated with 0, 100 mM $NaHCO_3$ to assess biochemical responses under alkaline stress. Roots were harvested for analysis, with each sample consisting of a pool of five roots and three biological replicates per line. The following commercial assay kits were used according to the manufacturers' protocols: peroxidase (POD) activity kit (No. M0105; Mengxi Biomed, Suzhou, China), malondialdehyde (MDA) content kit (No. G0109W; Geruisi

 https://doi.org/10.3390/plants15050679

Biotechnology, Suzhou, China), catalase (CAT) activity kit (No. G0105W; Geruisi Biotechnology, Suzhou, China), and proline (Pro) content kit (No. G1101W; Geruisi Biotechnology, Suzhou, China).

4.8. Diaminobenzidine (DAB) Staining

DAB staining solution was prepared using DAB powder (Biotopped, Beijing, China). Briefly, 0.1 g of DAB powder was dissolved in a brown bottle containing 45 mL of distilled water. The pH was adjusted to 5.7, and the volume was brought to 50 mL with distilled water. Leaves were vacuum-infiltrated with the DAB solution and incubated for 18 h in the dark. Subsequently, chlorophyll was removed by boiling the samples in absolute ethanol. After complete decolorization, the leaves were mounted on glass slides with glycerol and imaged using a digital camera [57].

4.9. Nitroblue Tetrazolium (NBT) Staining

The staining solution was prepared by dissolving 0.1 g of NBT powder (Biotopped, Beijing, China) in 50 mL of phosphate buffer (pH 7.5). The phosphate buffer was prepared by mixing 1.6 mL of NaH_2PO_4 buffer with 8.4 mL of Na_2HPO_4 buffer and diluting to a final volume of 200 mL with distilled water. Leaves were immersed in the NBT solution and incubated in the dark for 18 h. Decolorization, mounting, and imaging were performed following the same procedures described for DAB staining [57].

4.10. Statistical Analysis

Data are presented as the mean $\pm$ standard deviation (SD). Differences between two groups were evaluated using a two-tailed Student's *t*-test. For multiple comparisons, statistical significance was assessed by either one-way or two-way analysis of variance (ANOVA) using GraphPad Prism 9.5.1 software.

5. Conclusions

This study demonstrates that heterologous overexpression of the *GsEXPA8* gene from wild soybean significantly enhances alkaline tolerance in narrow-leaf lupinus (*Lupinus angustifolius*). *GsEXPA8* promotes root elongation and branching under stress, establishing a more developed root architecture that may improve water and nutrient uptake. Concurrently, it mitigates oxidative damage and osmotic stress induced by alkalinity by elevating antioxidant enzyme (e.g., CAT) activity, reducing membrane lipid peroxidation (MDA), and increasing the accumulation of osmoregulatory compounds such as proline. Functioning as a regulatory node, *GsEXPA8* is associated with the upregulation of multiple key stress-responsive genes involved in ion transport (*LaSOS1*), transcriptional regulation (*LaNAC56*, *LaMYB39*), and hormone synthesis (*LaNCED3*), thereby systemically reinforcing the plant's stress-adaptation network. Notably, this study reveals for the first time that *GsEXPA8* overexpression specifically reshapes the rhizosphere microbial community, enriching beneficial bacterial genera with plant growth-promoting, antimicrobial, and nutrient-mobilizing functions in lupinus. This indicates that the enhanced alkaline tolerance is partly derived from the optimization of beneficial plant-microbe interactions. Furthermore, the specific induction of the endogenous homologous gene *LaEXPA8* under alkaline stress suggests a conserved functional role for this expansin family in alkali tolerance. In summary, *GsEXPA8* enhances alkaline tolerance in lupinus through an integrated strategy involving the promotion of root development, enhancement of biochemical resilience, activation of molecular networks, and recruitment of a beneficial rhizosphere microbiome. These findings provide new insights into utilizing expansin genes for the genetic improvement of crop stress resistance and enhancing crop adaptability through rhizosphere microbiome engineering.

https://doi.org/10.3390/plants15050679

Supplementary Materials: The following supporting information can be downloaded at: https://www.mdpi.com/article/10.3390/plants15050679/s1, Figure S1: RT-PCR analysis of *GsEXPA8* and *Ubiquitin* expression in WT and *GsEXPA8*-OE plants; Figure S2: GsEXPA8 has high homology with LaEXPA8; Figure S3: Identification of overexpression strains and CRISPR/Cas9 induced silencing strains; Table S1: qRT-PCR primers for each gene; Table S2: RT-PCR primers for each gene.

Author Contributions: Conceptualization, L.C.; Formal Analysis, M.L.; Investigation, M.L., Y.L., J.T., X.W., X.D., H.W., Y.R., M.Z. and Y.F.; Methodology, M.L.; Resources, L.C. and J.Z.; Supervision, L.C.; Writing—Original Draft, M.L.; Writing—Review and Editing, L.C. All authors have read and agreed to the published version of the manuscript.

Funding: This work was supported by the Natural Science Foundation of Heilongjiang Province (LH2023C006) to L.C. National Natural Science Foundation of China (32001505) to L.C.

Data Availability Statement: Raw sequencing reads have been deposited in the NCBI Sequence Read Archive under the accession number SUB14514279, PRJNA1402660 and PRJNA1402017.

Acknowledgments: We are grateful to the College of Horticulture and Landscape Architecture at Northeast Agricultural University for providing the experimental platform.

Conflicts of Interest: The authors declare no competing interests.

References

1. Tarolli, P.; Luo, J.; Park, E.; Barcaccia, G.; Masin, R. Soil salinization in agriculture: Mitigation and adaptation strategies combining nature-based solutions and bioengineering. *iScience* **2024**, *27*, 108830. [CrossRef]
2. Yang, S.; Xu, Y.; Tang, Z.; Jin, S.; Yang, S. The Impact of Alkaline Stress on Plant Growth and Its Alkaline Resistance Mechanisms. *Int. J. Mol. Sci.* **2024**, *25*, 13719. [CrossRef]
3. Burchardt, S.; Czernicka, M.; Kućko, A.; Pokora, W.; Kapusta, M.; Domagalski, K.; Jasieniecka-Gazarkiewicz, K.; Karwaszewski, J.; Wilmowicz, E. Exploring the response of yellow lupine (*Lupinus luteus* L.) root to drought mediated by pathways related to phytohormones, lipid, and redox homeostasis. *BMC Plant Biol.* **2024**, *24*, 1049. [CrossRef]
4. Tang, J.; Liu, M.; Liu, Y.; Wang, X.; Du, X.; Liu, X.; Zhou, M.; Mao, H.; Feng, Y.; Gao, Q.; et al. Leveraging Transcriptome Insights and *GsHZ4* Gene Expression to Improve Alkaline Tolerance in *Lupinus angustifolius*. *Plants* **2025**, *14*, 3408. [CrossRef]
5. Frick, K.M.; Foley, R.C.; Kamphuis, L.G.; Siddique, K.H.M.; Garg, G.; Singh, K.B. Characterization of the genetic factors affecting quinolizidine alkaloid biosynthesis and its response to abiotic stress in narrow-leafed lupin (*Lupinus angustifolius* L.). *Plant Cell Environ.* **2018**, *41*, 2155–2168. [CrossRef] [PubMed]
6. Adeleke, R.; Nwangburuka, C.; Oboirien, B. Origins, roles and fate of organic acids in soils: A review. *S. Afr. J. Bot.* **2017**, *108*, 393–406. [CrossRef]
7. Priya, M.; Farooq, M.; Siddique, K.H.M. Enhancing Tolerance to Combined Heat and Drought Stress in Cool-Season Grain Legumes: Mechanisms, Genetic Insights, and Future Directions. *Plant Cell Environ.* **2025**. [CrossRef]
8. Maqbool, S.; Hassan, M.A.; Xia, X.; York, L.M.; Rasheed, A.; He, Z. Root system architecture in cereals: Progress, challenges and perspective. *Plant J.* **2022**, *110*, 23–42. [CrossRef] [PubMed]
9. Leng, B.; Liu, X.; Sun, Y.; Yin, H.; Mu, C.; Ma, S.; Liu, Q.; Hou, J.; Yan, Z.; Yao, G. A Novel Breeding Target for Salt-Tolerant Maize: *ZmEXPA3* Overexpression Enhances Growth of Maize Under Both Non-Stressed and Salt Stress Conditions Through Cell-Wall Architecture Alteration. *Plants* **2025**, *14*, 3697. [CrossRef]
10. Omar, S.A.; Fetyan, N.A.H.; Eldenary, M.E.; Abdelfattah, M.H.; Abd-Elhalim, H.M.; Wróbel, J.; Kalaji, H.M. Alteration in expression level of some growth and stress-related genes after rhizobacteria inoculation to alleviate drought tolerance in sensitive rice genotype. *Chem. Biol. Technol. Agric.* **2021**, *8*, 41. [CrossRef]
11. Chen, Y.; Zhang, B.; Li, C.; Lei, C.; Kong, C.; Yang, Y.; Gong, M. A comprehensive expression analysis of the expansin gene family in potato (*Solanum tuberosum*) discloses stress-responsive expansin-like B genes for drought and heat tolerances. *PLoS ONE* **2019**, *14*, e0219837. [CrossRef]
12. de Souza-Neto, R.R.; Cavalcante, L.N.; Carvalho, I.G.B.; Curtolo, M.; Benedetti, C.E.; Takita, M.A.; Wang, N.; de Souza, A.A. CRISPR/Cas9-Mediated Disruption of *CsLIEXP1* Reveals Expansin as a Key Susceptibility Factor for Citrus Canker Disease. *Mol. Plant-Microbe Interact.* **2025**, *38*, 589–598. [CrossRef] [PubMed]
13. Liu, M.; Tang, J.; Ma, S.; Liu, Y.; Wang, X.; Du, X.; Sun, X.; Zeng, Y.; Zeng, Y.; Ding, X.; et al. *GsEXPA8* Enhances Soybean Tolerance of $NaHCO_3$ Stress by Regulating Root Morphology. *Agronomy* **2025**, *15*, 16. [CrossRef]
14. Qetrani, S.; Bouray, M.; Errahali, S.; Chtouki, M.; Barakat, A.; Oukarroum, A. Phosphorus source addition modulates P-related biochemical processes in the rhizosphere of Lupinus albus grown in alkaline soils. *Rhizosphere* **2025**, *35*, 101173. [CrossRef]

 https://doi.org/10.3390/plants15050679

15. Dunfield, K.E.; Germida, J.J. Impact of genetically modified crops on soil- and plant-associated microbial communities. *J. Environ. Qual.* **2004**, *33*, 806–815. [CrossRef]
16. Pepoyan, A.Z.; Chikindas, M.L. Plant-associated and soil microbiota composition as a novel criterion for the environmental risk assessment of genetically modified plants. *GM Crops Food* **2020**, *11*, 47–53. [CrossRef]
17. Turrini, A.; Sbrana, C.; Giovannetti, M. Belowground environmental effects of transgenic crops: A soil microbial perspective. *Res. Microbiol.* **2015**, *166*, 121–131. [CrossRef]
18. Guo, Y.; Zhang, D.; Qi, W. Bacterial diversity of herbal rhizospheric soils in Ordos desert steppes under different degradation gradients. *PeerJ* **2023**, *11*, e16289. [CrossRef]
19. Sagar, A.; Rai, S.; Sharma, S.; Perveen, K.; Bukhari, N.A.; Sayyed, R.Z.; Mastinu, A. Molecular Characterization Reveals Biodiversity and Biopotential of Rhizobacterial Isolates of *Bacillus* spp. *Microb. Ecol.* **2024**, *87*, 83. [CrossRef]
20. Montiel, J.; García-Soto, I.; Monroy-Morales, E.; Lace, B.; Robledo-Gamboa, M.; Vestergaard, M.; Sandal, N.; Ott, T.; Stougaard, J. The *Lotus japonicus* alpha-expansin EXPA1 is recruited during intracellular and intercellular rhizobial colonization. *Plant J.* **2025**, *124*, e70639. [CrossRef]
21. Xing, X.; Du, H.; Yang, Z.; Zhang, H.; Li, N.; Shao, Z.; Li, W.; Kong, Y.; Li, X.; Zhang, C. *GmEXPA11* facilitates nodule enlargement and nitrogen fixation via interaction with GmNOD20 under regulation of GmPTF1 in soybean. *Plant Sci.* **2025**, *355*, 112469. [CrossRef] [PubMed]
22. Jin, T.; An, J.; Xu, H.; Chen, J.; Pan, L.; Zhao, R.; Wang, N.; Gai, J.; Li, Y. A soybean sodium/hydrogen exchanger *GmNHX6* confers plant alkaline salt tolerance by regulating Na(+)/K(+) homeostasis. *Front. Plant Sci.* **2022**, *13*, 938635. [CrossRef]
23. Rai, A.N.; Penna, S. Molecular evolution of plant *P5CS* gene involved in proline biosynthesis. *Mol. Biol. Rep.* **2013**, *40*, 6429–6435. [CrossRef]
24. Wang, B.; Fang, R.; Zhang, G.; Liu, N.; Feng, Z.; Bu, Y.; Gong, Y. A CCT protein GmCIC5 activates *GmP5CS* to regulate proline accumulation and cadmium tolerance in vegetable soybean. *J. Hazard Mater.* **2025**, *493*, 138319. [CrossRef]
25. Barrero, J.M.; Rodríguez, P.L.; Quesada, V.; Piqueras, P.; Ponce, M.R.; Micol, J.L. Both abscisic acid (ABA)-dependent and ABA-independent pathways govern the induction of *NCED3*, *AAO3* and *ABA1* in response to salt stress. *Plant Cell Environ.* **2006**, *29*, 2000–2008. [CrossRef] [PubMed]
26. Wang, S.; He, J.; Hu, B.; Deng, M.; Li, W.; Guo, J.; Song, Y.; Zheng, Q.; Song, X.; Ma, F.; et al. An integrative multi-omics analysis of histone modifications and DNA methylation reveals the epigenomic landscape in apple under drought stress. *Plant Biotechnol. J.* **2025**, *23*, 4440–4460. [CrossRef]
27. Lu, K.K.; Yang, H.; Liao, C.Y.; Song, R.F.; Hu, X.Y.; Ren, F.; Liu, W.C. A transcriptional recognition site within *SOS1* coding region controls salt tolerance in Arabidopsis. *Dev. Cell* **2025**, *60*, 2626–2642.e2625. [CrossRef] [PubMed]
28. Chen, Q.; Niu, F.; Yan, J.; Chen, B.; Wu, F.; Guo, X.; Yang, B.; Jiang, Y.Q. Oilseed rape *NAC56* transcription factor modulates reactive oxygen species accumulation and hypersensitive response-like cell death. *Physiol. Plant* **2017**, *160*, 209–221. [CrossRef]
29. Zhang, X.; Yang, Q.; Li, H.; Zhang, L.; Sun, C.; Zhang, H.; Li, H.; Li, J.; Zhu, Y.; Du, C.; et al. Cold resistance identification and transcriptomic responses of rapeseed (*Brassica napus* L.) seedlings under cold stress. *Plant Physiol. Biochem.* **2025**, *229*, 110778. [CrossRef]
30. Ren, Z.; Zhang, P.; Su, H.; Xie, X.; Shao, J.; Ku, L.; Tian, Z.; Deng, D.; Wei, L. Regulatory mechanisms used by *ZmMYB39* to enhance drought tolerance in maize (*Zea mays*) seedlings. *Plant Physiol. Biochem.* **2024**, *211*, 108696. [CrossRef]
31. Asaf, S.; Numan, M.; Khan, A.L.; Al-Harrasi, A. *Sphingomonas*: From diversity and genomics to functional role in environmental remediation and plant growth. *Crit. Rev. Biotechnol.* **2020**, *40*, 138–152. [CrossRef]
32. Wang, N.; Ping, L.; Mei, X.; Zhang, Y.; Zhang, Y.; Yang, X.; Guo, Y.; Gao, Y.; Xu, Y.; Shen, Q.; et al. Succinic acid reduces tomato bacterial wilt disease by recruiting *Sphingomonas* sp. *Environ. Microbiome* **2025**, *20*, 85. [CrossRef]
33. Yin, W.; Zhang, B.; Zhang, H.; Zhang, D.; Leiviskä, T. Vertically co-distributed vanadium and microplastics drive distinct microbial community composition and assembly in soil. *J. Hazard. Mater.* **2022**, *440*, 129700. [CrossRef]
34. Li, X.; Yu, Y.; Choi, L.; Song, Y.; Wu, M.; Wang, G.; Li, M. *Phenylobacterium soli* sp. nov., isolated from arsenic and cadmium contaminated farmland soil. *Int. J. Syst. Evol. Microbiol.* **2019**, *69*, 1398–1403. [CrossRef]
35. De Luca, G.; Barakat, M.; Verméglio, A.; Achouak, W.; Heulin, T. The Bacterial Genus *Ramlibacter*: Betaproteobacteria Capable of Surviving in Oligotrophic Environments Thanks to Several Shared Genetic Adaptation Traits. *Environ. Microbiol.* **2025**, *27*, e70059. [CrossRef]
36. Mehmood, M.A.; Wang, J.; Cheng, J.; Xie, J.; Jiang, D.; Fu, Y. Sclerotia-Mediated Soil Microbiome Modulation in Rice-Rapeseed Cropping Systems. *J. Fungi* **2025**, *11*, 755. [CrossRef]
37. Li, Y.; Feng, F.; Mu, Q.; Li, M.; Ma, L.; Wan, Q.; Jousset, A.; Liu, C.; Yu, X. Foliar Spraying of Chlorpyrifos Triggers Plant Production of Linolenic Acid Recruiting Rhizosphere Bacterial *Sphingomonas* sp. *Environ. Sci. Technol.* **2023**, *57*, 17312–17323. [CrossRef] [PubMed]
38. Lee, D.H.; Cha, C.J. *Ramlibacter alkalitolerans* sp. nov., alkali-tolerant bacterium isolated from soil of ginseng. *Int. J. Syst. Evol. Microbiol.* **2017**, *67*, 4619–4623. [CrossRef] [PubMed]

 https://doi.org/10.3390/plants15050679

39. Vurukonda, S.S.; Vardharajula, S.; Shrivastava, M.; Sk, Z.A. Enhancement of drought stress tolerance in crops by plant growth promoting rhizobacteria. *Microbiol. Res.* **2016**, *184*, 13–24. [CrossRef]
40. Devarajan, A.K.; Muthukrishanan, G.; Truu, J.; Truu, M.; Ostonen, I.; Kizhaeral, S.S.; Panneerselvam, P.; Kuttalingam Gopalasubramanian, S. The Foliar Application of Rice Phyllosphere Bacteria induces Drought-Stress Tolerance in *Oryza sativa* (L.). *Plants* **2021**, *10*, 387. [CrossRef] [PubMed]
41. Lee, S.; Kim, J.A.; Song, J.; Choe, S.; Jang, G.; Kim, Y. Plant growth-promoting rhizobacterium *Bacillus megaterium* modulates the expression of antioxidant-related and drought-responsive genes to protect rice (*Oryza sativa* L.) from drought. *Front. Microbiol.* **2024**, *15*, 1430546. [CrossRef]
42. Ali, B.; Wang, X.; Saleem, M.H.; Sumaira; Hafeez, A.; Afridi, M.S.; Khan, S.; Ullah, I.; Amaral Júnior, A.T.D.; Alatawi, A.; et al. PGPR-Mediated Salt Tolerance in Maize by Modulating Plant Physiology, Antioxidant Defense, Compatible Solutes Accumulation and Bio-Surfactant Producing Genes. *Plants* **2022**, *11*, 345. [CrossRef]
43. Ali, B.; Wang, X.; Saleem, M.H.; Azeem, M.A.; Afridi, M.S.; Nadeem, M.; Ghazal, M.; Batool, T.; Qayyum, A.; Alatawi, A.; et al. *Bacillus mycoides* PM35 Reinforces Photosynthetic Efficiency, Antioxidant Defense, Expression of Stress-Responsive Genes, and Ameliorates the Effects of Salinity Stress in Maize. *Life* **2022**, *12*, 219. [CrossRef]
44. Azeem, M.A.; Shah, F.H.; Ullah, A.; Ali, K.; Jones, D.A.; Khan, M.E.H.; Ashraf, A. Biochemical Characterization of Halotolerant *Bacillus safensis* PM22 and Its Potential to Enhance Growth of Maize under Salinity Stress. *Plants* **2022**, *11*, 1721. [CrossRef] [PubMed]
45. Ali, B.; Hafeez, A.; Ahmad, S.; Javed, M.A.; Sumaira; Afridi, M.S.; Dawoud, T.M.; Almaary, K.S.; Muresan, C.C.; Marc, R.A.; et al. *Bacillus thuringiensis* PM25 ameliorates oxidative damage of salinity stress in maize via regulating growth, leaf pigments, antioxidant defense system, and stress responsive gene expression. *Front. Plant Sci.* **2022**, *13*, 921668. [CrossRef]
46. Nardi, C.F.; Villarreal, N.M.; Dotto, M.C.; Ariza, M.T.; Vallarino, J.G.; Martínez, G.A.; Valpuesta, V.; Civello, P.M. Influence of plant growth regulators on Expansin2 expression in strawberry fruit. Cloning and functional analysis of FaEXP2 promoter region. *Postharvest Biol. Technol.* **2016**, *114*, 17–28. [CrossRef]
47. Yang, J.; Zhang, G.; An, J.; Li, Q.; Chen, Y.; Zhao, X.; Wu, J.; Wang, Y.; Hao, Q.; Wang, W.; et al. Expansin gene *TaEXPA2* positively regulates drought tolerance in transgenic wheat (*Triticum aestivum* L.). *Plant Sci.* **2020**, *298*, 110596. [CrossRef]
48. Zou, H.; Wenwen, Y.; Zang, G.; Kang, Z.; Zhang, Z.; Huang, J.; Wang, G. *OsEXPB2*, a β-expansin gene, is involved in rice root system architecture. *Mol. Breed.* **2015**, *35*, 41. [CrossRef]
49. Srivastava, R.; Kobayashi, Y.; Koyama, H.; Sahoo, L. Cowpea *NAC1*/*NAC2* transcription factors improve growth and tolerance to drought and heat in transgenic cowpea through combined activation of photosynthetic and antioxidant mechanisms. *J. Integr. Plant Biol.* **2023**, *65*, 25–44. [CrossRef]
50. Masoomi-Aladizgeh, F.; Aalami, A.; Esfahani, M.; Aghaei, M.J.; Mozaffari, K. Identification of *CBF14* and *NAC2* Genes in *Aegilops tauschii* Associated with Resistance to Freezing Stress. *Appl. Biochem. Biotechnol.* **2015**, *176*, 1059–1070. [CrossRef]
51. Zhang, H.; Cui, X.; Guo, Y.; Luo, C.; Zhang, L. Picea wilsonii transcription factor *NAC2* enhanced plant tolerance to abiotic stress and participated in RFCP1-regulated flowering time. *Plant Mol. Biol.* **2018**, *98*, 471–493. [CrossRef] [PubMed]
52. Xiang, T.H.; Wang, L.L.; Pang, J.L.; Chen, M.; Xu, C. Hairy root induced by wild-type agrobacterium rhizogenes K599 in soybean, cucumber and garden balsam in vivo. *Yi Chuan* **2005**, *27*, 783–786.
53. Chen, L.; Cai, Y.; Liu, X.; Guo, C.; Sun, S.; Wu, C.; Jiang, B.; Han, T.; Hou, W. Soybean hairy roots produced in vitro by *Agrobacterium rhizogenes*-mediated transformation. *Crop J.* **2018**, *6*, 162–171. [CrossRef]
54. Livak, K.J.; Schmittgen, T.D. Analysis of relative gene expression data using real-time quantitative PCR and the 2(-Delta Delta C(T)) Method. *Methods* **2001**, *25*, 402–408. [CrossRef]
55. Liang, F.; Wei, S.; Xue, L.; Yan, S. Adopting green absorbent for CO_2 capture and agricultural utilization: Biogas slurry and biomass ash case. *Green Carbon* **2024**, *2*, 252–261. [CrossRef]
56. Molano, L.A.; Vega-Abellaneda, S.; Manichanh, C. GSR-DB: A manually curated and optimized taxonomical database for 16S rRNA amplicon analysis. *mSystems* **2024**, *9*, e0095023. [CrossRef]
57. Yu, Z.; Chen, X.; Chen, Z.; Wang, H.; Shah, S.H.A.; Bai, A.; Liu, T.; Xiao, D.; Hou, X.; Li, Y. BcSRC2 interacts with BcAPX4 to increase ascorbic acid content for responding ABA signaling and drought stress in pak choi. *Hortic. Res.* **2024**, *11*, uhae165. [CrossRef] [PubMed]

https://doi.org/10.3390/plants15050679

MDPI AG
Grosspeteranlage 5
4052 Basel
Switzerland
Tel.: +41 61 683 77 34

Plants Editorial Office
E-mail: plants@mdpi.com
www.mdpi.com/journal/plants

www.ingramcontent.com/pod-product-compliance
Lightning Source LLC
Chambersburg PA
CBHW061451210926
51850CB00008B/1686
9783725880218